Study Guide and Student Solutions Manual to Accompany

Physics

FOR SCIENTISTS AND ENGINEERS

Volume 1

by Serway

Fourth Edition

Steven Van Wyk
Chapman College

Ralph McGrew
Broome Community College

John R. Gordon
James Madison University

Raymond Serway
James Madison University

Saunders Golden Sunburst Series

SAUNDERS COLLEGE PUBLISHING
Harcourt Brace College Publishers

Fort Worth   Philadelphia   San Diego   New York
Orlando   Austin   San Antonio   Toronto
Montreal   London   Sydney   Tokyo

Printed in the United States of America.

Serway: Study Guide and Student Solutions Manual to
accompany Physics for Scientists and Engineers, 4E.

ISBN 0-03-015664-5

7 021 98765432

Preface

This <u>Student Solution Manual and Study Guide</u> has been written to accompany the textbook **Physics for Scientists and Engineers**, Fourth Edition, by Raymond A. Serway. The purpose of this Student Solution Manual and Study Guide is to provide the students with a convenient review of the basic concepts and applications presented in the textbook, together with solutions to selected end-of-chapter problems from the textbook. This is not an attempt to rewrite the textbook in a condensed fashion. Rather, emphasis is placed upon clarifying typical troublesome points, and providing further drill for methods of problem solving.

Each chapter is divided into several parts, and every textbook chapter has a matching chapter in this Solution Manual. Very often, reference is made to specific equations or figures in the textbook. Every feature of this Solution Manual has been included to insure that it serves as a useful supplement to the textbook. Most chapters contain the following components:

- **Notes From Selected Chapter Sections:** This is a summary of important concepts, newly defined physical quantities, and rules governing their behavior.

- **Equations and Concepts:** This represents a review of the chapter, with emphasis on highlighting important concepts, and describing important equations and formalisms.

- **Suggestions, Skills, and Strategies:** This offers hints and strategies for solving typical problems that the student will often encounter in the course. In some sections, suggestions are made concerning mathematical skills that are necessary in the analysis of problems.

- **Review Checklist:** This is a list of topics and techniques the student should master after reading the chapter and working the assigned problems.

- **Solutions to Selected End-of-Chapter Problems:** Solutions are shown for approximately half of the odd-numbered problems from the text which illustrate the important concepts of the chapter.

An important note concerning significant figures: The answers to all end-of-chapter problems are stated to three significant figures. For uniformity of style, this practice has been followed even in those cases where the data in the corresponding problem statement is given to only two significant figures. In the latter case, your instructor may request that you "round off" your calculations to two significant figures.

We sincerely hope that this Student Solution Manual and Study Guide will be useful to you in reviewing the material presented in the text, and in improving your ability to solve problems and score well on exams. We welcome any comments or suggestions which could help improve the content of this book in future editions; and we wish you success in your study.

John R. Gordon
Raymond A. Serway
James Madison University
Harrisonburg, VA 22807

Ralph McGrew
Broome Community College
Binghamton, NY 13902-1017

Steve Van Wyk
Olympic College
Bremerton, WA 98310

Acknowledgments

It is a pleasure to acknowledge the excellent work of Mrs. Linda Miller who typed the manuscript for this Fourth Edition. Her attention to detail has made our work easier. We thank Michael Rudmin for revising and creating original art work for this fourth edition. His graphics skills and technical expertise have combined to produce illustrations which complement the text. The overall appearance of the Solution Manual reflects his work on the final camera-ready copy.

We also thank the professional staff at Saunders, especially Laura Maier and Travis Westfall for managing all phases of the project. Finally, we express our appreciation to our families for their inspiration, patience, and encouragement.

Suggestions for Study

Very often we are asked "How should I study this subject, and prepare for examinations?" There is no simple answer to this question, however, we would like to offer some suggestions which may be useful to you.

1. It is essential that you understand the basic concepts and principles before attempting to solve assigned problems. This is best accomplished through a careful reading of the textbook before attending your lecture on that material, jotting down certain points which are not clear to you, taking careful notes in class, and asking questions. You should reduce memorization of material to a minimum. Memorizing sections of a text, equations, and derivations does not necessarily mean you understand the material. Perhaps the best test of your understanding of the material will be your ability to solve the problems in the text, or those given on exams.

2. Try to solve as many problems at the end of the chapter as possible. You will be able to check the accuracy of your calculations to the odd-numbered problems, since the answers to these are given at the back of the text. Furthermore, detailed solutions to approximately half of the odd-numbered problems are provided in this Solution Manual. Many of the worked examples in the text will serve as a basis for your study.

3. The method of solving problems should be carefully planned. First, read the problem several times until you are confident you understand what is being asked. Look for key words which will help simplify the problem, and perhaps allow you to make certain assumptions. You should also pay special attention to the information provided in the problem. In many cases a simple diagram is a good starting point; and it is always a good idea to write down the given information before proceeding with a solution. After you have decided on the method you feel is appropriate for the problem, proceed with your solution. If you are having difficulty in working problems, we suggest that you again read the text and your lecture notes. It may take several readings before you are ready to solve certain problems, though the solved problems in this Solution Manual should be of value to you in this regard. However, your solution to a problem does not have to look just like the one presented here. A problem can sometimes be solved in different ways, starting from different principles. If you wonder about the validity of an alternative approach, ask your instructor.

4. After reading a chapter, you should be able to define any new quantities that were introduced, and discuss the first principles that were used to derive fundamental formulas. A review is provided in each chapter of the Solution Manual for this purpose, and the marginal notes in the textbook (or the index) will help you locate these topics. You should be able to correctly associate with each physical quantity the symbol used to represent that quantity (including vector notation if appropriate) and the SI unit in which the quantity is specified. Furthermore, you should be able to express each important formula or equation in a concise and accurate prose statement.

5. We suggest that you use this Solution Manual to review the material covered in the text, and as a guide in preparing for exams. You should also use the Chapter Review, Notes From Selected Chapter Sections, and Equations and Concepts to focus in on any points which require further study. Remember that the main purpose of this Solution Manual is to improve upon the efficiency and effectiveness of your study hours and your overall understanding of physical concepts. However, it should not be regarded as a substitute for your textbook or individual study and practice in problem solving.

Table of Contents

Chapter 1

Physics and Measurement

PHYSICS AND MEASUREMENT

INTRODUCTION

The goal of physics is to provide an understanding of nature by developing theories based on experiments. The theories are usually expressed in mathematical form. Fortunately, it is possible to explain the behavior of a variety of physical systems with a limited number of fundamental laws.

Since following chapters will be concerned with the laws of physics, we must begin by clearly defining the basic quantities involved in these laws. For example, such physical quantities as force, velocity, volume, and acceleration can be described in terms of more fundamental quantities. In the next several chapters we shall encounter three basic quantities: length (L), mass (M), and time (T). In later chapters we will need to add two other standard units to our list, for temperature (the Kelvin) and for electric current (the ampere). In our study of mechanics, however, we shall be concerned only with the units of length, mass and time.

NOTES FROM SELECTED CHAPTER SECTIONS

1.1   Standards of Length, Mass, and Time

Mechanical quantities can be expressed in terms of three fundamental quantities, *mass*, *length*, and *time*, which in the SI system have the units *kilograms* (kg), *meters* (m), and *seconds* (s), respectively. It is often useful to use the *method of dimensional analysis* to check equations and to assist in deriving expressions.

1.3   Density and Atomic Mass

The density of a substance is defined as its *mass per unit volume*. Different substances have different densities mainly because of differences in their atomic masses and atomic arrangements.

The masses of atomic and nuclear particles are expressed in terms of *atomic mass units* (u). The mass of ^{12}C is defined to be exactly 12 u.

$$1 \text{ u} = 1.660 \times 10^{-27} \text{ kg}$$

$$\text{mass of proton} = 1.0073 \text{ u}$$

$$\text{mass of neutron} = 1.0087 \text{ u}$$

The number of molecules in one mole of any element or compound, called Avogadro's number (N_A), is 6.02×10^{23}.

One mole (mol) of an element (or compound) is that quantity of the material which contains Avogadro's number of atoms (or molecules).

1.4 Dimensional Analysis

Dimensional analysis makes use of the fact that *dimensions can be treated as algebraic quantities*. That is, quantities can be added or subtracted only if they have the same dimensions. Furthermore, the terms on both sides of an equation must have the same dimensions. By following these simple rules, you can use dimensional analysis to help determine whether or not an expression has the correct form, because the relationship can be correct only if the dimensions on the two sides of the equation are the same.

1.7 Significant Figures

When one performs measurements on certain quantities, the accuracy of the measured values can vary; that is, the true values are known only to be within the limits of the experimental uncertainty. The value of the uncertainty can depend on various factors such as the quality of the apparatus, the skill of the experimenter, and the number of measurements performed.

When multiplying several quantities, the number of significant figures in the final answer is the same as the number of significant figures in the *least* accurate of the quantities being multiplied, where "least accurate" means "having the lowest number of significant figures." The same rule applies to division.

When numbers are added (or subtracted), the number of decimal places in the result should equal the smallest number of decimal places of any term in the sum.

1.8 Mathematical Notation

It is often convenient to use symbols to represent mathematical operations and/or relationships between or among variables or values of physical quantities. Some symbols which will be used throughout your textbook and in the Solutions Manual are shown below.

Symbol	means
=	equality of two quantities
$\approx$	approximately equal to
$\equiv$	defined as
$\propto$	proportional to
<	less than
>	greater than
Δ	change in a quantity
$\lvert x \rvert$	magnitude of the quantity x
$\sum x_i$	sum of the set x

EQUATIONS AND CONCEPTS

The density of any substance is defined as the ratio of mass to volume.

$$\rho \equiv \frac{m}{V} \tag{1.1}$$

The mass per atom for a given element is found by using Avogadro's number.

$$m_{\text{atom}} = \frac{\text{atomic mass of element}}{N_A} \tag{1.2}$$

SUGGESTIONS, SKILLS, AND STRATEGIES

General Problem-Solving Strategy

Six basic steps are commonly used to develop a problem-solving strategy:

- Read the problem carefully at least twice. Be sure you understand the nature of the problem before proceeding further.

- Draw a suitable diagram with appropriate labels and coordinate axes if needed.

- As you examine what is being asked in the problem, identify the basic physical principle or principles that are involved, listing the knowns and unknowns.

- Select a basic relationship or derive an equation that can be used to find the unknown, and symbolically solve the equation for the unknown.

- Substitute the given values, along with the appropriate units, into the equation.

- Obtain a numerical value for the unknown. The problem is verified if the following questions can be properly answered: Do the units match? Is the answer reasonable? Is the plus or minus sign proper or meaningful?

You should be familiar with some important mathematical techniques:

- Using powers of ten in expressing such numbers as $0.00058 = 5.8 \times 10^{-4}$. Appendix B.1 gives a brief review of such notation, and the algebraic operations of numbers using powers of ten.

- Basic algebraic operations such as factoring, handling fractions, solving quadratic equations, and solving linear equations. For your convenience, some of these techniques are reviewed in Appendix B.2.

- The fundamentals of plane and solid geometry—including the ability to graph functions, calculate the areas and volumes of standard geometric figures and recognize the equations and graphs of a straight line, a circle, an ellipse, a parabola and a hyperbola.

- The basic ideas of trigonometry—definitions and properties of the sine, cosine, and tangent functions; the Pythagorean Theorem, the law of cosines, the law of sines, and some of the basic trigonometric identities.

  For your convenience, reviews of geometry and trigonometry are given in Appendix D of the text.

REVIEW CHECKLIST

▷ Discuss the units of length, mass and time and the standards for these quantities in SI units; and perform a *dimensional analysis* of an equation containing physical quantities whose individual units are known.

▷ *Convert units* from one system to another.

▷ Carry out *order-of-magnitude calculations* or guesstimates.

▷ Become familiar with the meaning of various mathematical symbols and Greek letters. Identify and properly use mathematical notations such as the following: ∝ (is proportional to), < (is less than), ≈ (is approximately equal to), Δ (change in value), etc.

SOLUTIONS TO SELECTED END-OF-CHAPTER PROBLEMS

1. Calculate the density of a solid cube that measures 5.00 cm on each side and has a mass of 350 g.

Solution   Density, $\rho = \dfrac{m}{V} = \dfrac{350 \text{ g}}{(5.00 \text{ cm})^3} = 2.80 \text{ g/cm}^3$

$\dfrac{350}{(5^3)} = 2.8 \text{ g/cm}^3$

or expressed in SI units,

$$\rho = \left(2.80 \ \frac{\text{g}}{\text{cm}^3}\right)\left(\frac{1 \text{ kg}}{1000 \text{ g}}\right)\left(\frac{10^6 \text{ cm}^3}{\text{m}^3}\right) = 2.80 \times 10^3 \text{ kg/m}^3 \ \Diamond$$

---

5. Calculate the mass of an atom of (a) helium, (b) iron, and (c) lead. Give your answers in atomic mass units and in grams. The atomic weights are 4, 56, and 207, respectively, for the atoms given.

Solution

atomic mass units

To solve this problem, use $m = \dfrac{\text{atomic weight}}{N_A}$, and the conversion $1 \text{ u} = 1.66 \times 10^{-24}$ g.

(a)   For He, $m = \dfrac{4 \text{ g/mol}}{6.02 \times 10^{23} \text{ molecules/mol}} = 6.64 \times 10^{-24} \text{ g} = 4.00 \text{ u} \ \Diamond$

$1.66 \times 10^{-24} \text{ g}$

(b)   For Fe, $m = \dfrac{56 \text{ g/mol}}{6.02 \times 10^{23} \text{ molecules/mol}} = 9.30 \times 10^{-23} \text{ g} = 56.0 \text{ u} \ \Diamond$

$1.66 \times 10^{-24} \text{ g}$

(c)   For Pb, $m = \dfrac{207 \text{ g/mol}}{6.02 \times 10^{23} \text{ molecules/mol}} = 3.44 \times 10^{-22} \text{ g} = 207 \text{ u} \ \Diamond$

---

6. A small cube of iron is observed under a microscope. The edge of the cube is 5.00×10^{-6} cm long. Find (a) the mass of the cube and (b) the number of iron atoms in the cube. The atomic mass of iron is 56 u, and its density is 7.86 g/cm³.

Solution Since $\rho = \dfrac{M}{V}$, and $V = L^3$, we have $M = \rho L^3$

(a) $M = \rho L^3 = (7.86 \text{ g/cm}^3)(5.00 \times 10^{-6} \text{ cm})^3 = 9.83 \times 10^{-16} \text{ g}$ ◊

(b) $N = M\left(\dfrac{N_A}{\text{atomic weight}}\right) = (9.83 \times 10^{-16} \text{ g})\left(\dfrac{6.02 \times 10^{23} \text{ atoms / mol}}{56 \text{ g / mol}}\right) = 1.06 \times 10^7 \text{ atoms}$ ◊

---

10. The displacement of a particle when moving under uniform acceleration is some function of the elapsed time and the acceleration. Suppose we write this displacement $s = ka^m t^n$, where k is a dimensionless constant. Show by dimensional analysis that this expression is satisfied if $m = 1$ and $n = 2$. Can this analysis give the value of k?

Solution $[s] = \text{L}$ $[a] = \text{LT}^{-2}$ $[t] = \text{T}$

$s = ka^m t^n$, where k is a dimensionless constant

$[s] = [a]^m [t]^n = (\text{LT}^{-2})^m \text{T}^n = \text{L}^m \text{T}^{-2m+n}$

$\text{L} = \text{L}^m \text{T}^{(n-2m)}$

Comparing the left and right sides, we have $m=1$ ◊

and $-2m + n = 0$.

Therefore, $n = 2m = 2$ ◊

When $m = 1$ and $n = 2$, $s = kat^2$.

The value of the dimensionless constant k *cannot* be determined by this analysis.   ◊

---

13. Which of the equations below is dimensionally correct?

(a) $v = v_0 + ax$ (b) $y = (2\ \text{m})\cos(kx)$, where $k = 2\ \text{m}^{-1}$

Solution

(a) Write out dimensions for each quantity in the equation $v = v_0 + at$.

v and v_0 are expressed in units of m/s; therefore, $[v] = [v_0] = LT^{-1}$.

a is expressed in units of m/s^2 and $[a] = LT^{-2}$; therefore $[ax] = L^2 T^{-2}$

Consider the right-hand member of equation (a):

$$[\text{right-hand member}] = LT^{-1} + L^2 T^{-2}$$

Quantities in a sum must be dimensionally the same. Therefore, equation (a) *is not* dimensionally correct. ◊

(b) Write out dimensions for each quantity in the equation $y = (2\ \text{m})\cos(kx)$

For y, $[y] = L$; for 2 m, $[2\ \text{m}] = L$; and for (kx), $[kx] = [2\ \text{m}^{-1} x] = L^{-1} L$.

For the left-hand member (LHM) of equation,

$$[\text{LHM}] = [y] = L$$

For the right-hand member (RHM) of equation,

$$[\text{RHM}] = [2\ \text{m}][\cos(kx)] = L$$

Therefore, equation (b) *is* dimensionally correct. ◊

---

16. The consumption of natural gas by a company satisfies the empirical equation $V = 1.5t + 0.0080t^2$, where V is the volume in millions of cubic feet and t the time in months. Express this equation in units of cubic feet and seconds. Put the proper units on the coefficients. Assume a month is 30 days.

Solution   The constants in the equation as stated must have units of:

$$\frac{\text{million ft}^3}{\text{month}} \quad \text{and} \quad \frac{\text{million ft}^3}{(\text{month})^2} \quad \text{thus} \quad V = \left(1.5\times10^6\,\frac{\text{ft}^3}{\text{month}}\right)t + \left(0.008\times10^6\,\frac{\text{ft}^3}{\text{month}^2}\right)t^2$$

To convert, use 1 month = 2.59 × 10^6 s, to obtain $\quad V = \left(0.579\,\frac{\text{ft}^3}{\text{s}}\right)t + \left(1.19\times10^{-9}\,\frac{\text{ft}^3}{\text{s}^2}\right)t^2$

$$V = \left(0.579\,\frac{\text{ft}^3}{\text{s}}\right)t + \left(1.19\times10^{-9}\,\frac{\text{ft}^3}{\text{s}^2}\right)t^2 \quad \lozenge$$

---

17. Newton's law of universal gravitation is $F = G\dfrac{Mm}{r^2}$. Here F is the force of gravity, M and m are masses, and r is a length. Force has the SI units kg·m/s^2. What are the SI units of the constant G ?

Solution   Solve the equation for G : $\qquad G = \dfrac{Fr^2}{Mm}$

Recall that the units of force, F, are kg·m/s^2. Find SI units for G by substituting units for each quantity.

$$[G] = \frac{(\text{kg}\cdot\text{m}/\text{s}^2)(\text{m}^2)}{\text{kg}\cdot\text{kg}} = \frac{\text{m}^3}{\text{kg}\cdot\text{s}^2} \quad \lozenge$$

---

19. A rectangular building lot is 100.0 ft by 150.0 ft. Determine the area of this lot in m^2.

Solution   Use the conversion factor 1 m = 3.281 ft to find the length and width in meters:

$$L = \left(\frac{100.0\text{ ft}}{3.281\text{ ft}/\text{m}}\right) = 30.48\text{ m} \qquad \text{and} \qquad w = \left(\frac{150\text{ ft}}{3.28\text{ ft}/\text{m}}\right) = 45.72\text{ m}$$

$$A = Lw = (30.48\text{ m})(45.72\text{ m}) = 1390\text{ m}^2 \quad \lozenge$$

23. A solid piece of lead has a mass of 23.94 g and a volume of 2.10 cm³. From these data, calculate the density of lead in SI units (kg/m³).

Solution   Use conversions:

$$1 \text{ g} = 10^{-3} \text{ kg and } 1 \text{ cm} = 10^{-2} \text{ m}$$

$$\text{Mass, } M = (23.94 \text{ g})(10^{-3} \text{ kg/g}) = 2.394 \times 10^{-2} \text{ kg}$$

$$\text{Volume, } V = (2.10 \text{ cm}^3)(10^{-2} \text{ m/cm})^3 = 2.10 \times 10^{-6} \text{ m}^3$$

Therefore,
$$\rho = \frac{M}{V} = \frac{2.394 \times 10^{-2} \text{ kg}}{2.10 \times 10^{-6} \text{ m}^3} = 1.14 \times 10^4 \text{ kg/m}^3 \quad \lozenge$$

27.   At the time of this book's printing, the U.S. national debt is about $4 trillion. (a) If payments were made at the rate of $1000 per second, how many years would it take to pay off the debt, assuming no interest were charged? (b) A dollar bill is about 15.5 cm long. If 4 trillion dollar bills were laid end to end around the Earth's equator, how many times would they encircle the Earth? Take the radius of the Earth at the equator to be 6378 km. (*Note:* Before doing any of these calculations, try to guess at the answers. You may be very surprised.)

Solution   (a) Time to repay debt will be calculated by dividing the total debt by the rate at which it is repaid.

$$T = \frac{\$4 \text{ trillion}}{\$1000/\text{s}} = \frac{\$4 \times 10^{12}}{\$1 \times 10^3/\text{s}} = 4.00 \times 10^9 \text{ s} \quad \text{and} \quad T = \frac{4.00 \times 10^9 \text{ s}}{3.16 \times 10^7 \text{ s/y}} = 127 \text{ y} \quad \lozenge$$

(b)   The total length of the end-to-end line of bills would be

$$L = (4.00 \times 10^{12} \text{ bills})(0.155 \text{ m/bill}) = 6.20 \times 10^{11} \text{ m}$$

The circumference of the Earth at the equator is

$$C = 2\pi R = 2\pi(6.378 \times 10^6 \text{ m}) = 4.01 \times 10^7 \text{ m}$$

So
$$N = \frac{L}{C} = \frac{6.20 \times 10^{11} \text{ m}}{4.01 \times 10^7 \text{ m}} = 1.55 \times 10^4 \text{ times} \quad \lozenge$$

31. One gallon of paint (volume = 3.78×10^{-3} m^3) covers an area of 25.0 m^2. What is the thickness of the paint on the wall?

Solution   We assume the paint keeps the same volume in the can and on the wall. The volume of the film on the wall is its surface area, multiplied by its thickness t: $V = At$

Therefore,
$$t = \frac{V}{A} = \frac{3.78 \times 10^{-3} \text{ m}^3}{25.0 \text{ m}^2} = 1.51 \times 10^{-4} \text{ m} \quad \lozenge$$

---

35. The diameter of our disk-shaped galaxy, the Milky Way, is about 1.0×10^5 light years. The distance to Andromeda, the galaxy nearest our own Milky Way, is about 2.0 million light years. If we represent the Milky Way by a dinner plate 25 cm in diameter, determine the distance to the next dinner plate.

Solution   The scale used in the "dinner plate" model is

$$S = \frac{1.0 \times 10^5 \text{ lightyears}}{25 \text{ cm}} = 4.00 \times 10^3 \frac{\text{lightyears}}{\text{cm}}$$

The distance to Andromeda in the dinner plate model will be

$$D = \frac{2.00 \times 10^6 \text{ lightyears}}{4.00 \times 10^3 \text{ lightyears / cm}} = 5.00 \times 10^2 \text{ cm } = 5.00 \text{ m} \quad \lozenge$$

---

39. One cubic meter (1.00 m^3) of aluminum has a mass of 2.70×10^3 kg, and 1.00 m^3 of iron has a mass of 7.86×10^3 kg. Find the radius of a solid aluminum sphere that will balance a solid iron sphere of radius 2.00 cm on an equal-arm balance.

Solution   We require equal masses: $m_{Al} = m_{Fe}$,

or
$$\rho_{Al} V_{Al} = \rho_{Fe} V_{Fe}$$

Therefore,
$$\rho_{Al}\left[\frac{4}{3}\pi r^3\right] = \rho_{Fe}\left[\frac{4}{3}\pi(2.00 \text{ cm})^3\right]$$

$$r^3 = \left(\frac{\rho_{Fe}}{\rho_{Al}}\right)(2.00 \text{ cm})^3 = \left(\frac{7.86 \text{ kg / m}^3}{2.70 \text{ kg / m}^3}\right)(2.00 \text{ cm})^3 = 23.3 \text{ cm}^3$$

and
$$r = 2.86 \text{ cm} \quad \lozenge$$

---

41. Estimate the number of Ping-Pong balls that would fit into an average-size room (without being crushed).

Solution A Ping-Pong ball has a diameter of about 3 cm and can be thought of as an object which occupies a cube of volume $(3 \times 3 \times 3)$ cm^3 = 27 cm^3. That is, V_{Ball} = 27 cm^3. A typical room has dimensions 12 ft × 15 ft × 8 ft. Using the conversion, 1 ft ≅ 30 cm, we find

$$V_{\text{Room}} = 12 \text{ ft} \times 15 \text{ ft} \times 8 \text{ ft} = 1440 \text{ ft}^3 = (1440 \text{ ft}^3)\left(\frac{30 \text{ cm}}{\text{ft}}\right)^3 \cong 3.9 \times 10^7 \text{ cm}^3$$

The number of Ping-Pong balls which can fill the room is

$$N = \frac{V_{\text{Room}}}{V_{\text{Ball}}} \cong \frac{4 \times 10^7 \text{ cm}^3}{27 \text{ cm}^3} = 1.50 \times 10^6 \text{ balls} \quad \lozenge$$

Therefore, a typical room can hold about 10^6 Ping-Pong balls. $\quad \lozenge$

---

47. A high fountain of water is located at the center of a circular pool as in Figure P1.47. A student walks around the pool and estimates its circumference to be 150 m. Next, the student stands at the edge of the pool and uses a protractor to gauge the angle of elevation of the top of the fountain to be 55°. How high is the fountain?

Solution Define a right triangle whose legs represent the height and radius of the fountain. From the dimensions of the triangle and fountain, the circumference

$$C = 2\pi r \quad \text{and} \quad \tan\theta = \frac{h}{r}$$

Figure P1.47

Therefore,

$$h = r\tan\theta = \left(\frac{C}{2\pi}\right)\tan\theta$$

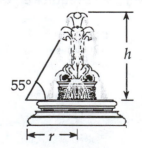

or

$$h = \frac{150 \text{ m}}{2\pi}\tan 55° = 34.1 \text{ m} \quad \lozenge$$

(The angle of elevation, 55°, limits the accuracy of our result to two significant figures.)

---

49. Estimate the number of piano tuners living in New York City. This problem was posed by the physicist Enrico Fermi, who was well known for making order-of-magnitude calculations.

Solution In this problem, assume a total population of 10^7 people. Also, let us assume 4 people per family and estimate that one family in ten owns a piano. With these assumptions, there will be 1 piano for every 40 people. In addition, assume that in one year a single piano tuner can service about 1000 pianos (about 4 per day for 250 weekdays, assuming each piano is tuned once per year. Therefore,

$$\text{The number of tuners} = \left(\frac{1 \text{ tuner}}{1000 \text{ pianos}}\right)\left(\frac{1 \text{ pianos}}{40 \text{ people}}\right)(10^7 \text{ people}) \approx 250 \approx 10^2 \text{ tuners} \quad \lozenge$$

53. If the length and width of a rectangular plate are measured to be (15.30 ± 0.05) cm and (12.80 ± 0.05) cm, respectively, find the area of the plate and the approximate uncertainty in the calculated area.

Solution   (See Example 1.10 in the textbook)

Referring to the sketch we have,

$A = Lw = (15.30 \pm 0.05)(12.80 \pm 0.05) \text{ cm}^2$

$A = [(15.30)(12.80) \pm (15.30)(0.05) \pm (0.05)(12.80)]$

$\quad = (195.8 \pm 1.4) \text{ cm}^2 \quad \lozenge$

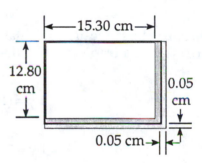

In the above result we have given the area ± an error term. However, we neglected the term $(0.05)(0.05)$ or $(\text{error})^2$. From the sketch, it is clearly insignificant.

59. Determine whether the expression $\sum_{i=1}^{4} i^2$ is equal to $\left(\sum_{i=1}^{4} i\right)^2$.

Solution   Evaluate each expression.

$$\sum_{i=1}^{4} i^2 = (1)^2 + (2)^2 + (3)^2 + (4)^2 = 30$$

$$\left(\sum_{i=1}^{4} i\right)^2 = (1+2+3+4)^2 = (10)^2 = 100$$

The two expressions are *not* equal.   ◊

_____

61.   A useful fact is that there are about $\pi \times 10^7$ s in one year. Use a calculator to find the percentage error in this approximation. *Note:* "Percentage error" is defined as (difference/true value) × 100%.

Solution   First we use calculation to evaluate the "true value." Remember that every fourth year is a leap year, and therefore there are 365.25 days in an average year.

$$1\,y = (1\,y)\left(\frac{365.25\,d}{1\,y}\right)\left(\frac{24\,h}{1\,d}\right)\left(\frac{3600\,s}{1\,h}\right)$$

$$1\,y = 3.1558 \times 10^7\,s$$

$$\text{percent error} = \frac{\left|\text{true value} - \text{approximate value}\right|}{\text{true value}} \times 100\%$$

$$\text{percent error} = \frac{(3.1558 - 3.1416) \times 10^7}{3.1558 \times 10^7} \times 100\%$$

$$\text{percent error} = 0.450\%   ◊$$

_____

62. Assume that there are 50 million passenger cars in the United Sates and that the average fuel consumption is 20 mi/gal of gasoline. If the average distance traveled by each car is 10 000 mi/yr, how much gasoline would be saved per year if average fuel consumption could be increased to 25 mi/gal?

Solution

$$\text{fuel consumed} = \frac{\text{total miles driven}}{\text{average fuel consumption rate}} \qquad \text{or symbolically,} \qquad f = \frac{s}{c_f}$$

Since the total distance is the same, we can calculate the fuel saved as

$$\Delta f = f_{20} - f_{25} = s\left(\frac{1}{c_{20}} - \frac{1}{c_{25}}\right)$$

Substituting,
$$\Delta f = \left(50 \times 10^6 \text{ cars}\right)\left(10^4 \text{ mi / yr / car}\right)\left(\frac{1}{20 \text{ mi / gal}} - \frac{1}{25 \text{ mi / gal}}\right)$$

$$\Delta f = \left(5.0 \times 10^{11} \text{ mi / yr}\right)(0.01 \text{ gal / mi})$$

Thus, we estimate $\Delta f = 5 \times 10^9$ gal/yr $\lozenge$

Chapter 2

Motion in One Dimension

MOTION IN ONE DIMENSION

INTRODUCTION

As a first step in studying mechanics, it is convenient to describe motion in terms of space and time, ignoring for the present the agents that caused that motion. This portion of mechanics is called *kinematics*. In this chapter we consider motion along a straight line, that is, one-dimensional motion. Starting with the concept of displacement discussed in the previous chapter, we first define velocity and acceleration. Then, using these concepts, we study the motion of objects traveling in one dimension under a constant acceleration.

In this and the next few chapters, we are concerned only with translational motion. In many situations, we can treat the moving object as a particle, which in mathematics is defined as a point having no size. In Chapter 4 we shall discuss the motions of objects in two dimensions.

NOTES FROM SELECTED CHAPTER SECTIONS

2.2 Instantaneous Velocity and Speed

The velocity of a particle at any instant of time (i.e. at some point on a space-time graph) is called the **instantaneous velocity**.

The **instantaneous speed** of an object, which is a scalar quantity, is defined as the magnitude of the instantaneous velocity. Hence, by definition, *speed can never be negative*.

The slope of the line tangent to the position-time curve at a point P is defined to be the **instantaneous velocity** at the corresponding time.

The area under the v vs. t curve in any time interval equals the displacement of the particle during the corresponding interval.

2.3 Acceleration

The **average acceleration** during a given time interval is defined as the change in velocity divided by the time interval during which this change occurs.

The **instantaneous acceleration** of an object at a certain time equals the slope of the tangent to the velocity-time graph at that instant of time.

2.5 Freely Falling Bodies

A freely falling body is an object moving freely under the influence of gravity only, regardless of its initial motion. Objects thrown upward or downward and those released from rest are all falling freely once they are released!

It is important to emphasize that any freely falling object experiences an *acceleration directed downward*. This is true regardless of the initial motion of the object.

Once they are in free fall, all objects have an acceleration downward equal to the acceleration due to gravity.

EQUATIONS AND CONCEPTS

The *displacement* Δx of a particle moving from position x_i to position x_f equals the final coordinate minus the initial coordinate.

$$\Delta x \equiv x_f - x_i \tag{2.1}$$

Displacement should not be confused with distance traveled. When $x_f = x_i$, the displacement is zero; however, if the particle leaves x_i, travels along a path, and returns to x_i, the distance traveled will not be zero.

The average velocity of an object during a time interval is the ratio of the total displacement to the time interval during which the displacement occurred. Note that the instantaneous velocity of the object during the time interval might have different values from instant to instant.

$$\overline{v} = \frac{x_f - x_i}{t_f - t_i} = \frac{\Delta x}{\Delta t} \tag{2.2}$$

The average speed of a particle is defined as the ratio of the total distance traveled to the time required to travel that distance.

The *instantaneous velocity* v is defined as the limit of the ratio $\Delta x / \Delta t$ as Δt approaches zero. The instantaneous can be positive, negative, or zero.

$$v \equiv \lim_{\Delta t \to 0} \frac{\Delta x}{\Delta t} = \frac{dx}{dt} \qquad (2.4)$$

The average acceleration of an object during a time interval is the ratio of the change in velocity to the time interval during which the change in velocity occurs.

$$\bar{a} = \frac{v_f - v_i}{t_f - t_i} \qquad (2.5)$$

The *instantaneous acceleration* a is defined as the limit of the ratio $\Delta v / \Delta t$ as Δt approaches zero.

$$a \equiv \lim_{\Delta t \to 0} \frac{\Delta v}{\Delta t} = \frac{dv}{dt} \qquad (2.6)$$

Equations 2.8 - 2.12 are called the equations of kinematics, and can be used to describe one-dimensional motion along the x axis with constant acceleration. Note that each equation shows a different relationship among physical quantities: initial velocity, final velocity, acceleration, time, and position.

$$v = v_0 + at \qquad (2.8)$$

$$\bar{v} = \frac{v_0 + v}{2} \qquad (2.9)$$

$$x - x_0 = \tfrac{1}{2}(v + v_0)t \qquad (2.10)$$

$$x - x_0 = v_0 t + \tfrac{1}{2}at^2 \qquad (2.11)$$

$$v^2 = v_0^2 + 2a(x - x_0) \qquad (2.12)$$

Equations 2.13 - 2.16 apply to an object in free fall ($\mathbf{a} = -g\mathbf{j}$) along the y axis.

$$v = v_0 - gt \qquad (2.13)$$

$$y - y_0 = \tfrac{1}{2}(v + v_0)t \qquad (2.14)$$

$$y - y_0 = v_0 t - \tfrac{1}{2}gt^2 \qquad (2.15)$$

$$v^2 = v_0^2 - 2g(y - y_0) \qquad (2.16)$$

SUGGESTIONS, SKILLS, AND STRATEGIES

The following procedure is recommended for solving problems involving acceleration motion:

- Make sure all the units in the problem are consistent. That is, if distances are measured in meters, be sure that velocities have units of m/s and accelerations have units of m/s^2.

- Choose a coordinate system.

- Make a list of all the quantities given in the problem and a separate list of those to be determined.

- Select from the list of kinematic equations the one or ones that will enable you to determine the unknowns.

- Construct an appropriate motion diagram and check to see if your answers are consistent with the diagram.

REVIEW CHECKLIST

▷ Define the displacement and average velocity of a particle in motion. Define the instantaneous velocity and understand how this quantity differs from average velocity.

▷ Define average acceleration and instantaneous acceleration.

▷ Construct a graph of position versus time (given a function such as $x = 5 + 3t - 2t^2$) for a particle in motion along a straight line. From this graph, you should be able to determine both average and instantaneous values of velocity by calculating the slope of the tangent to the graph.

▷ Describe what is meant by a body in *free fall* (one moving under the influence of gravity—where air resistance is neglected). Recognize that the equations of kinematics apply directly to a freely falling object and that the acceleration is then given by $a = -g$ (where $g = 9.8$ m/s^2).

▷ Apply the equations of kinematics to any situation where the motion occurs under constant acceleration.

SOLUTIONS TO SELECTED END-OF-CHAPTER PROBLEMS

3. The displacement versus time graph for a certain particle moving along the x axis is shown in Figure P2.3. Find the average velocity in the time intervals (a) 0 to 2 s, (b) 0 to 4 s, (c) 2 s to 4 s, (d) 4 s to 7 s, (e) 0 to 8 s.

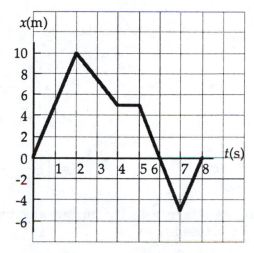

Figure P2.3

Solution   On this graph, we can tell positions to two significant figures:

(a) $x = 0$ at $t = 0$ and $x = 10.0$ m at $t = 2.0$ s

0 to 2

$$\overline{v} = \frac{\Delta x}{\Delta t} = \frac{10.0 \text{ m} - 0}{2.0 \text{ s} - 0} = 5.00 \text{ m/s} \quad \Diamond$$

(b) $x = 5.0$ m at $t = 4.0$ s

0 to 4

$$\overline{v} = \frac{\Delta x}{\Delta t} = \frac{5.0 \text{ m} - 0}{4.0 \text{ s} - 0} = 1.25 \text{ m/s} \quad \Diamond$$

(c)

2 to 4

$$\overline{v} = \frac{\Delta x}{\Delta t} = \frac{5.0 \text{ m} - 10.0 \text{ m}}{4.0 \text{ s} - 2.0 \text{ s}} = -2.50 \text{ m/s} \quad \Diamond$$

(d)

4 to 7

$$\overline{v} = \frac{\Delta x}{\Delta t} = \frac{-5.0 \text{ m} - 5.0 \text{ m}}{7.0 \text{ s} - 4.0 \text{ s}} = -3.33 \text{ m/s} \quad \Diamond$$

(e)

0 to 8

$$\overline{v} = \frac{\Delta x}{\Delta t} = \frac{0.0 \text{ m} - 0.0 \text{ m}}{8.0 \text{ s} - 0 \text{ s}} = 0 \text{ m/s} \quad \Diamond$$

---

5. A person walks first at a constant speed of 5.0 m/s along a straight line from point A to point B and then back along the line from B to A at a constant speed of 3.0 m/s. (a) What is her average speed over the entire trip? (b) Her average velocity over the entire trip?

Solution

(a)   The average speed during any time interval is equal to the total distance of travel divided by the total time:

$$\bar{v} = \frac{\Delta x}{\Delta t} = \frac{d_{AB} + d_{BA}}{t_{AB} + t_{BA}}$$

But $\qquad d_{AB} = d_{BA} = d, \qquad t_{AB} - \frac{d}{v_{AB}}, \qquad$ and $\qquad t_{BA} - \frac{d}{v_{BA}}$

so $\qquad \bar{v} = \dfrac{d+d}{\left(\dfrac{d}{v_{AB}}\right) + \left(\dfrac{d}{v_{BA}}\right)} = 2\left(\dfrac{v_{AB} \cdot v_{BA}}{v_{AB} + v_{BA}}\right)$

so $\qquad \bar{v} = 2\left[\dfrac{(5.0 \text{ m}/\text{s})(3.0 \text{ m}/\text{s})}{5.0 \text{ m}/\text{s} + 3.0 \text{ m}/\text{s}}\right] = 3.75 \text{ m}/\text{s} \quad \Diamond$

(b) Average velocity during any time interval equals total displacement divided by elapsed time.

$$\bar{v} = \frac{\Delta x}{\Delta t}$$

In this case, the walker returns to the starting point, so $\Delta x = 0$ and $\bar{v} = 0.$ $\Diamond$

7. A car makes a 200-km trip at an average speed of 40 km/h. A second car starting 1.0 h later arrives at their mutual destination at the same time. What was the average speed of the second car for the period that it was in motion?

Solution

Time of travel for the first car: $\qquad t_1 = \dfrac{d}{v_1} = \dfrac{200 \text{ km}}{40 \text{ km}/\text{h}} = 5.00 \text{ h}$

Time of travel for the second car (starting one hour later):

$$t_2 = t_1 - 1 \text{ h} = 4 \text{ h}$$

Average speed of second car: $\qquad v_2 = \dfrac{d}{t_2} = \dfrac{200 \text{ km}}{4 \text{ h}} = 50.0 \text{ km}/\text{h} \quad \Diamond$

9. The space-time graph for a particle moving along the x axis is as shown in Figure P2.9. (a) Find the average velocity in the time interval $t = 1.5$ s to $t = 4.0$ s. (b) Determine the instantaneous velocity at $t = 2.0$ s by measuring the slope of the tangent line shown in the graph. (c) At what value of t is the velocity zero?

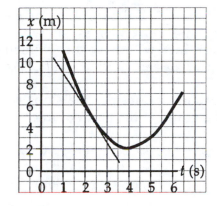

Figure P2.9

Solution

(a) From the graph:

At $t_1 = 1.5$ s, $x = x_1 = 8.0$ m

At $t_2 = 4.0$ s, $x = x_2 = 2.0$ m

Therefore,
$$\bar{v}_{1\to2} = \frac{\Delta x}{\Delta t} = \frac{2.0 \text{ m} - 8.0 \text{ m}}{4.0 \text{ s} - 1.5 \text{ s}} = -2.40 \text{ m} / \text{s} \quad \Diamond$$

(b) Choose two points along line which is tangent to the curve at $t = 2.0$ s. We will use the two points:

$(t_i = 1.0 \text{ s}, x_i = 9.0 \text{ m})$ and $(t_f = 3.5 \text{ s}, x_f = 1.0 \text{ m})$

Instantaneous velocity equals the slope of the tangent line, so

$$v = \frac{x_f - x_i}{t_f - t_i} = \frac{1.0 \text{ m} - 9.0 \text{ m}}{3.5 \text{ s} - 1.0 \text{ s}} = -3.20 \text{ m} / \text{s}$$

The negative sign indicates that the *direction* of **v** is along the negative x direction. $\Diamond$

(c) The velocity will be zero when the slope of the tangent line is zero. This occurs for the point on the graph where x has its minimum value. Therefore,

$v = 0$ at $t \cong 4$ s (visible to 1 significant figure) $\quad \Diamond$

14. The space-time graph for a particle moving along the z axis is as shown in Figure P2.14. Determine whether the velocity is positive, negative, or zero at times (a) t_1, (b) t_2, (c) t_3, (d) t_4.

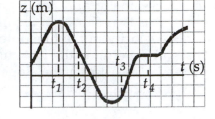

Solution   Consider the slope of the line tangent to the graph at the point of interest.

Figure P2.14

(a) $v = 0$ (slope is zero) ◊

(b) $v < 0$ (slope is negative) ◊

(c) $v > 0$ (zero is positive) ◊

(d) $v = 0$ (zero slope) ◊

---

17. A particle moves along the x axis according to the equation $x = 2t + 3t^2$, where x is in meters and t is in seconds.   Calculate the instantaneous velocity and instantaneous acceleration at $t = 3.0$ s.

Solution   With the position given by $x = (2t + 3t^2)$ m, we can use the rules of differentiation to find expressions for velocity and acceleration as functions of time:

$$v(t) = \frac{dx}{dt} = (2 + 6t) \text{ m/s}$$

$$a(t) = \frac{dv}{dt} = 6.00 \text{ m/s}^2 \quad \text{◊ (valid at all times)}$$

At $t = 3.0$ s, $v = [2 + 6(3.0)] \text{ m/s} = 20.0 \text{ m/s}$ ◊

---

21. A particle moves along the x axis according to the equation $x = 2 + 3t - t^2$, where x is in meters and t is in seconds. At $t = 3.00$ s, find (a) the position of the particle, (b) its velocity, and (c) its acceleration.

Solution   With the position given by $x = 2 + 3t - t^2$, where x is in meters and t is in seconds, we can use the rules for differentiation to write expressions for the velocity and acceleration as functions of time:

$$v = \frac{dx}{dt} = 3 - 2t$$

$$a = \frac{dv}{dt} = -2$$

Now we can evaluate x, v, and a at $t = 3.00$ s.

(a) $x = 2 + 3(3.00) - (3.00)^2 = 2.00$ m   ◊

(b) $v = 3 - 2(3.00) = -3.00$ m/s   ◊

(c) $a = -2.00$ m/s^2 ◊

---

25. A body moving with uniform acceleration has a velocity of 12.0 cm/s when its x coordinate is 3.00 cm. If its x coordinate 2.00 s later is –5.00 cm, what is the magnitude of its acceleration?

Solution   Take $t = 0$ to be the time when $x_0 = 3.00$ cm and $v_0 = 12.0$ cm/s. Also, at $t = 2.00$ s, $x = -5.00$ cm.

Use the kinematic equation $x - x_0 = v_0 t + \frac{1}{2}at^2$,   and solve for a.

$$a = \frac{2[x - x_0 - v_0 t]}{t^2}$$

$$a = \frac{2[-5.00 \text{ cm} - 3.00 \text{ cm} - (12.0 \text{ cm/s})(2.00 \text{ s})]}{(2.00 \text{ s})^2}$$

$$a = -16.0 \text{ cm/s}^2 \quad ◊$$

---

31. A jet plane lands with a velocity of 100 m/s and can accelerate at a maximum rate of −5.0 m/s² as it comes to rest. (a) From the instant it touches the runway, what is the minimum time needed before it stops? (b) Can this plane land at a small airport where the runway is 0.80 km long?

Solution

(a) Assume that the acceleration of the plane is *constant* at the maximum rate, $a = -5.0$ m/s². Given $v_0 = 100$ m/s and $v = 0$, use the equation $v = v_0 + at$ and solve for t:

$$t = \frac{v - v_0}{a} = \frac{0 - 100 \text{ m/s}}{-5.0 \text{ m/s}^2} = 20.0 \text{ s} \quad \lozenge$$

(b) Find the required stopping distance and compare this to the length of the runway. Taking x_0 to be zero, we get

$$v^2 = v_0^2 + 2ax$$

or

$$x = \frac{v^2 - v_0^2}{2a} = \frac{0 - (100 \text{ m/s})^2}{2(-5.0 \text{ m/s}^2)} = 1000 \text{ m}$$

The stopping distance is greater than the length of the runway; therefore the plane *cannot land.* $\lozenge$

====================

35. A particle starts from rest from the top of an inclined plane and slides down with constant acceleration. The inclined plane is 2.00 m long, and it takes 3.00 s for the particle to reach the bottom. Find (a) the acceleration of the particle, (b) its speed at the bottom of the incline, (c) the time it takes the particle to reach the middle of the incline, and (d) its speed at the midpoint.

Solution

(a) The particle moves with *constant* acceleration along the incline, starting from rest (with $v_0 = 0$),

$$x = v_0 t + \tfrac{1}{2}at^2 \quad \text{becomes} \quad x = \tfrac{1}{2}at^2$$

or
$$a = \frac{2x}{t^2} = \frac{2(2.00 \text{ m})}{(3.00 \text{ s})^2} = 0.444 \text{ m/s}^2 \quad \Diamond$$

(b) At the bottom of the incline, $v^2 = v_0{}^2 + 2ax;$

when $v_0 = 0$, $\quad v = \sqrt{v_0{}^2 + 2ax},$

so $\quad v = \sqrt{2(0.444 \text{ m/s}^2)(2.00 \text{ m})} = 1.33 \text{ m/s} \quad \Diamond$

(c) At the middle of the incline, $x = 1.00$ m and from the equation $x = \frac{1}{2}at^2$

find $\quad t = \sqrt{\frac{2x}{a}} = \sqrt{\frac{2(1.00 \text{ m})}{0.444 \text{ m/s}^2}} = 2.12 \text{ s} \quad \Diamond$

(d) Use the value of t from part (c) to find the speed at the midpoint of the incline:

$$v = v_0 + at = 0 + (0.444 \text{ m/s}^2)(2.12 \text{ s}) = 0.943 \text{ m/s} \quad \Diamond$$

39. A car moving at a constant speed of 30.0 m/s suddenly stalls at the bottom of a hill. The car undergoes a constant acceleration of –2.00 m/s² (opposite its motion) while ascending the hill. (a) Write equations for the position and the velocity as functions of time, taking $x = 0$ at the bottom of the hill, where $v_0 = 30.0$ m/s. (b) Determine the maximum distance traveled by the car after it stalls.

Solution

(a) Take $t_0 = 0$ at the bottom of the hill where $x_0 = 0$, $v_0 = 30.0$ m/s, and $a = -2$ m/s². Use these values in the general equation

$$x = x_0 + v_0t + \frac{1}{2}at^2$$

to find $\quad x = 0 + (30.0 \text{ m/s})t + \frac{1}{2}(-2.00 \text{ m/s}^2)t^2$

when t is in seconds

$$x = (30.0t - t^2) \text{ m} \quad \Diamond$$

To find an equation for the velocity, use

$$v = v_0 + at = 30.0 \text{ m/s} + (-2.00 \text{ m/s}^2)t$$

and when t is in seconds

$$v = (30.0 - 2.00t) \text{ m/s} \quad \Diamond$$

(b) The distance of travel x becomes a maximum, x_{max}, when $v = 0$ (turning point in the motion). Use the expressions found in part (a) for v to find the value of t when x has its maximum value.

When $v = 0$, $\qquad\qquad v = (30.0 - 2.00t) \text{ m/s} = 0$

so $\qquad\qquad\qquad\qquad t = 15.0 \text{ s}$

Now use this value in the equation for x to find x_{max}.

$$x_{max} = (30.0t - t^2) \text{ m} = (30.0)(15.0) - (15.0^2) = 225 \text{ m} \quad \Diamond$$

---

43. Until recently, the world's land speed record was held by Colonel John P. Stapp, USAF. On March 19, 1954, he rode a rocket-propelled sled that moved down the track at 632 mi/h. He and the sled were safely brought to rest in 1.4 s. Determine (a) the negative acceleration he experienced and (b) the distance he traveled during this negative acceleration.

Solution

(a) We assume the acceleration remains constant during the 1.4 s period of negative acceleration.

$$v_0 = 632 \frac{\text{mi}}{\text{h}} = 632 \frac{\text{mi}}{\text{h}} \left(\frac{1609 \text{ m}}{1 \text{ mi}} \right) \left(\frac{1 \text{ h}}{3600 \text{ s}} \right) = 282 \text{ m/s}$$

and $v = 0$. Then $v = v_0 + at$ gives

$$a = \frac{v - v_0}{t} = \frac{0 - 282 \text{ m/s}}{1.4 \text{ s}} = -200 \text{ m/s}^2 \quad (\text{ approximately } 20g \text{ !}) \quad \Diamond$$

(b) $\quad x - x_0 = \frac{1}{2}(v_0 + v)t = \frac{1}{2}(282 \text{ m/s} + 0)(1.4 \text{ s}) = 200 \text{ m} \quad \Diamond$

---

47. A student throws a set of keys vertically upward to her sorority sister in a window 4.00 m above. The keys are caught 1.50 s later by the sister's outstretched hand. (a) With what initial velocity were the keys thrown? (b) What was the velocity of the keys just before they were caught?

Solution

(a) Taking $y_0 = 0$ (at the position of the thrower) and given that $y = 4.00$ m at $t = 1.50$ s, we find (with $a = -9.80$ m/s^2)

$$y = v_0 t + \tfrac{1}{2}at^2 \quad \text{or} \quad v_0 = \frac{y - \tfrac{1}{2}at^2}{t}$$

$$v_0 = \frac{4.00 \text{ m} - \tfrac{1}{2}(-9.80 \text{ m/s}^2)(1.50 \text{ s})^2}{1.50 \text{ s}}$$

$$v_0 = 10.0 \text{ m/s} \quad \lozenge$$

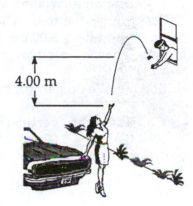

4.00 m

(b) The velocity at any time $t > 0$ is given by $v = v_0 + at$. Therefore, at $t = 1.50$ s,

$$v = 10.0 \text{ m/s} - (9.80 \text{ m/s}^2)(1.50 \text{ s}) = -4.68 \text{ m/s} \quad \lozenge$$

The negative sign means that the keys are moving *downward* just before they are caught.

---

51. A baseball is hit such that it travels straight upward after being struck by the bat. A fan observes that it requires 3.00 s for the ball to reach its maximum height. Find (a) its initial velocity and (b) its maximum height. Ignore the effects of air resistance.

Solution

(a) After leaving the bat, the ball is in free fall and has a constant acceleration, $a = -g = -9.80$ m/s^2.

Solve the equation $v = v_0 + at$ for v_0, to get $v_0 = v - at$

When $t = 3.00$ s (at maximum height), $v = 0$.

Therefore, $v_0 = 0 - (-9.80 \text{ m/s}^2)(3.00 \text{ s}) = 29.4 \text{ m/s} \quad \lozenge$

(b) $\quad y = v_0 t + \frac{1}{2} a t^2 \quad$ and $\quad y = y_{max} \quad$ when $\quad t = 3.00$ s

So, $\qquad\qquad\qquad y_{max} = (29.4 \text{ m/s})(3.00 \text{ s}) + \frac{1}{2}(-9.80 \text{ m/s}^2)(3.00 \text{ s})^2$

$\qquad\qquad\qquad y_{max} = 44.1$ m $\quad \lozenge$

---

55. A daring stunt woman sitting on a tree limb wishes to drop vertically onto a horse galloping under the tree. The speed of the horse is 10.0 m/s, and the distance from the limb to the saddle is 3.00 m. (a) What must be the horizontal distance between the saddle and limb when the woman makes her move? (b) How long is she in the air?

Solution   We do part (b) first.

(b) Consider the vertical motion of the woman after leaving the limb (with $v_0 = 0$ at $y_0 = 3.00$ m) until reaching the saddle (at $y = 0$). We find her time of fall from

$$y - y_0 = v_0 t + \frac{1}{2} a t^2$$

When $v_0 = 0$,

$$t = \sqrt{\frac{2(y - y_0)}{a}} = \sqrt{\frac{2(0 - 3.00 \text{ m})}{-9.80 \text{ m/s}^2}} = 0.782 \text{ s} \quad \lozenge$$

(a) During the time interval found in part (b), the horse travels horizontally at constant speed.

$$v_0 = v = 10.0 \text{ m/s}, \quad a = 0$$

$$x - x_0 = v_0 t = (10.0 \text{ m/s})(0.782 \text{ s}) = 7.82 \text{ m}$$

so the "cowgirl" must let go when the horse is 7.82 m from the tree. $\lozenge$

---

59. Automotive engineers refer to the time rate of change of acceleration as the "jerk." If an object moves in one dimension such that its jerk J is constant, (a) determine expressions for its acceleration $a(t)$, velocity $v(t)$, and position $x(t)$, given that its initial acceleration, speed, and position are a_0, v_0, and x_0, respectively. (b) Show that $a^2 = a_0^2 + 2J(v - v_0)$.

Solution

(a) $J = \dfrac{da}{dt} = $ constant, so $\qquad\qquad da = J\, dt$

$a = J \int dt = Jt + c_1$, but $a = a_0$ when $t = 0$, so $\qquad c_1 = a_0$.

Therefore, $\qquad\qquad\qquad\qquad\qquad\qquad\qquad a = Jt + a_0 \quad \Diamond$

$a = \dfrac{dv}{dt}$, and $dv = a\,dt$; integration yields $\qquad v = \int a\, dt = \int (Jt + a_0)\, dt = \tfrac{1}{2} Jt^2 + a_0 t + c_2$

But $v = v_0$ when $t = 0$, so $c_2 = v_0$, and $\qquad v = \tfrac{1}{2} Jt^2 + a_0 t + v_0 \quad \Diamond$

$v = \dfrac{dx}{dt}$, or $dx = v\, dt$. Again integrating, $\qquad x = \int v\, dt = \int \left(\tfrac{1}{2} Jt^2 + a_0 t + v_0 \right) dt$

and $\qquad\qquad\qquad\qquad\qquad\qquad\qquad x = \tfrac{1}{6} Jt^3 + \tfrac{1}{2} a_0 t^2 + v_0 t + c_3$

Since $x = x_0$ when $t = 0$, $c_3 = x_0$.

Therefore, $\qquad\qquad\qquad\qquad\qquad\qquad x = \tfrac{1}{6} Jt^3 + \tfrac{1}{2} a_0 t^2 + v_0 t + x_0 \quad \Diamond$

(b) $a^2 = (Jt + a_0)^2 = J^2 t^2 + a_0^2 + 2Ja_0 t = a_0^2 + (J^2 t^2 + 2Ja_0 t)$

Solving, $\qquad\qquad\qquad\qquad\qquad\qquad a^2 = a_0^2 + 2J(\tfrac{1}{2} Jt^2 + a_0 t)$

Now recall the expression for v: $\qquad\qquad v = \tfrac{1}{2} Jt^2 + a_0 t + v_0$

So $\qquad\qquad\qquad\qquad\qquad\qquad\qquad (v - v_0) = \tfrac{1}{2} Jt^2 + a_0 t$

Therefore, $\qquad\qquad\qquad a^2 = a_0^2 + 2J(v - v_0) \quad \Diamond$

63. An inquisitive physics student climbs a 50.0-m cliff that overhangs a calm pool of water. She throws two stones vertically downward 1.00 s apart and observes that they cause a single splash. The first stone has an initial velocity of 2.00 m/s. (a) At what time after release of the first stone do the two stones hit the water? (b) What initial velocity must the second stone have if they are to hit simultaneously? (c) What is the velocity of each stone at the instant they hit the water?

Solution

(a)   Find the time required for the first stone to reach the water using the equation

$$y = v_0 t + \tfrac{1}{2} a t^2$$

where $y_0 = 0$ at the top of the cliff. If we take the direction downward to be negative, $y = -50.0$ m, $v_0 = -2.00$ m/s, and $a = -9.80$ m/s^2. Using these values in the equation, we find

$$-50.0 \text{ m} = (-2.00 \text{ m}/\text{s})(t) + \tfrac{1}{2}(-9.80 \text{ m}/\text{s}^2)t^2$$

This can be written in the standard form of a quadratic equation as

$$\left(4.90 \text{ m}/\text{s}^2\right)t^2 + (2.00 \text{ m}/\text{s})t - 50.0 \text{ m} = 0$$

Use the quadratic formula and solve for t :

$$t = \frac{-2.00 \text{ m}/\text{s} \pm \sqrt{(2.00 \text{ m}/\text{s})^2 - 4\left(4.90 \text{ m}/\text{s}^2\right)(-50.0 \text{ m})}}{2\left(-4.90 \text{ m}/\text{s}^2\right)}$$

Since only the positive root describes this physical situation,

$$t = 3.00 \text{ s} \quad \lozenge$$

(b)   For the second stone, the time of travel is 3.00 s – 1.00 s = 2.00 s

From $y = v_0 t + \frac{1}{2} a t^2$,   we have

$$v_0 = \frac{y - \frac{1}{2} a t^2}{t} = \frac{-50.0 \text{ m} - \left(\frac{1}{2}\right)\left(-9.80 \text{ m}/\text{s}^2\right)(2.00 \text{ s})^2}{2.00 \text{ s}}$$

or $v_0 = -15.3 \text{ m}/\text{s}$ $\lozenge$

The negative value indicates the downward direction of the initial velocity of the second stone.

(c)   For the first stone,

$$v_1 = v_{01} + a_1 t_1 = -2.00 \text{ m}/\text{s} + (-9.80 \text{ m}/\text{s}^2)(3.00 \text{ s})$$

$$v_1 = -31.4 \text{ m}/\text{s}   \lozenge$$

For the second stone,

$$v_2 = v_{02} + a_2 t_2 = -15.2 \text{ m}/\text{s} + (-9.80 \text{ m}/\text{s}^2)(2.00 \text{ s})$$

$$v_2 = -34.8 \text{ m}/\text{s}   \lozenge$$

---

69. A young woman named Kathy Kool buys a super-deluxe sports car that can accelerate at the rate of 4.90 m/s². She decides to test the car by dragging with another speedster, Stan Speedy. Both start from rest, but experienced Stan leaves 1.00 s before Kathy. If Stan moves with a constant acceleration of 3.50 m/s² and Kathy maintains an acceleration of 4.90 m/s², find (a) the time it takes Kathy to overtake Stan, (b) the distance she travels before she catches him, and (c) the velocities of both cars at the instant she overtakes him.

Solution

(a) Let the times of travel for Kathy and Stan be t_K and t_S where $t_S = t_K + 1.00$ s. Both start from rest ($v_0 = 0$), so the expressions for the distances traveled are:

$$x_K = \tfrac{1}{2}a_K t_K^2 = \tfrac{1}{2}(4.90 \text{ m/s}^2)t_K^2$$

$$x_S = \tfrac{1}{2}a_S t_S^2 = \tfrac{1}{2}(3.50 \text{ m/s}^2)(t_K + 1.00 \text{ s})^2$$

When Kathy overtakes Stan, the two distances will be equal; setting $x_K = x_S$,

$$\tfrac{1}{2}(4.90 \text{ m/s}^2)t_K^2 = \tfrac{1}{2}(3.50 \text{ m/s}^2)(t_K + 1.00 \text{ s})^2$$

This can be simplified and written in the standard form of a quadratic as

$$t_K^2 - 5.0 t_K \text{ s} - 2.5 \text{ s}^2 = 0$$

and solved using the quadratic formula to find

$$t_K = 5.45 \text{ s} \quad \lozenge$$

(b) Use equation from part (a) for distance of travel,

$$x_K = \tfrac{1}{2}a_K t_K^2 = \tfrac{1}{2}(4.90 \text{ m/s}^2)(5.45 \text{ s})^2 = 73.0 \text{ m} \quad \lozenge$$

(c) The final velocities will be (remember $v_0 = 0$ for each car):

$$v_K = a_K t_K = (4.90 \text{ m/s}^2)(5.45 \text{ s}) = 26.7 \text{ m/s} \quad \lozenge$$

$$v_S = a_S t_S = (3.50 \text{ m/s}^2)(6.45 \text{ s}) = 22.6 \text{ m/s} \quad \lozenge$$

---

73. In 1987, Art Boileau won the Los Angeles Marathon, 26 mi and 385 yd, in 2 h, 13 min, and 9 s. (a) Find his average speed in meters per second and in miles per hour. (b) At the 21-mi marker, Boileau had a 2.50-min lead on the second-place winner, who later crossed the finish line 30.0 s after Boileau. Assume that Boileau maintained his constant average speed and that both runners were running at the same speed when Boileau passed the 21-mi marker. Find the average acceleration (in meters per second squared) that the second-place runner had during the remainder of the race after Boileau passed the 21-mi marker.

Solution Average speed, $\bar{v} = \dfrac{\text{total distance}}{\text{total time}}$

(a) Use conversion factors given in text.

$$\bar{v} = \frac{26 \text{ mi} + \dfrac{385 \text{ yd}}{1760 \text{ yd}/\text{mi}}}{2 \text{ h} + \dfrac{13 \text{ min}}{60 \text{ min}/\text{h}} + \dfrac{9 \text{ s}}{3600 \text{ s}/\text{h}}} = \frac{26.22 \text{ mi}}{2.219 \text{ h}} = 11.8 \text{ mi}/\text{h} \quad \lozenge$$

$$\bar{v} = \left(11.8 \ \frac{\text{mi}}{\text{h}}\right)\left(\frac{1 \text{ h}}{3600 \text{ s}}\right)\left(\frac{1609 \text{ m}}{\text{mi}}\right) = 5.27 \text{ m}/\text{s} \quad \lozenge$$

(b) At the 21-mi marker, the remaining distance for Boileau is 26.22 − 21.0 = 5.22 mi. At a constant speed of 11.8 mi/h, a 2.5 min lead corresponds to a distance of

$$\Delta x = \left(11.8 \ \frac{\text{mi}}{\text{h}}\right)\left(\frac{2.5 \text{ min}}{60 \text{ min}/\text{h}}\right) = 0.4920 \text{ mi}$$

Therefore, when Boileau is at the 21-mi marker, the second runner has a remaining distance of d = 5.22 mi + 0.4920 mi = 5.712 mi = 9191 m. Boileau covers the last 5.22 mi in a time,

$$t' = \frac{5.22 \text{ mi}}{11.8 \text{ mi}/\text{h}} = 0.4418 \text{ h} = 1590 \text{ s}$$

Finishing 30 s after Boileau, the second runner must cover the last 9191 m in a time of 1590 s + 30 s = 1620 s. The average speed is

$$\bar{v} = \frac{9191 \text{ m}}{1620 \text{ s}} = 5.67 \text{ m}/\text{s}$$

The initial speed of the second runner is assumed to be equal to Boileau's constant average speed which [from part (a)] is v_0 = 5.28 m/s.

Using $\bar{v} = \frac{1}{2}(v_0 + v)$ or $v = 2\bar{v} - v_0$ to find v = 2(5.67 m/s) − 5.28 m/s = 6.06 m/s

Therefore, the acceleration of the second-place runner is

$$a = \frac{v - v_0}{t} = \frac{6.06 \text{ m}/\text{s} - 5.28 \text{ m}/\text{s}}{1623 \text{ s}} = 4.83 \times 10^{-4} \text{ m}/\text{s}^2 \quad \lozenge$$

77. In a 100-m race, Maggie and Judy cross the finish line in a dead heat, both taking 10.2 s. Accelerating uniformly, Maggie takes 2.00 s and Judy 3.00 s to attain maximum speed, which they maintain for the rest of the race. (a) What is the acceleration of each sprinter? (b) What are their respective maximum speeds? (c) Which sprinter is ahead at the 6.00-s mark, and by how much?

Solution

(a) Maggie moves with constant positive acceleration a_M for 2.00 s, then with constant speed (zero acceleration) for 8.20 s, covering altogether distance $x_{M1} + x_{M2} = 100$ m.

$$x_{M1} = \tfrac{1}{2} a_M (2.00 \text{ s})^2 \quad \text{and} \quad x_{M2} = v_M (8.20 \text{ s}),$$

where v_M is her maximum speed. Since $v_M = 0 + a_M(2.00 \text{ s})$,

by substitution $\quad 100 \text{ m} = \tfrac{1}{2} a_M (2.00 \text{ s})^2 + a_M (2.00 \text{ s})(8.20 \text{ s}) = (18.4 \text{ s}^2) a_M$

$$a_M = 5.43 \text{ m/s}^2 \quad \lozenge$$

Similarly, for Judy, $100 \text{ m} = x_{J1} + x_{J2}$ with

$$x_{J1} = \tfrac{1}{2} a_J (3.00 \text{ s})^2; \quad x_{J2} = v_J (7.20 \text{ s}); \quad v_J = a_J (3.00 \text{ s})$$

$$100 \text{ m} = \tfrac{1}{2} a_J (3.00 \text{ s})^2 + a_J (3.00 \text{ s})(7.20 \text{ s}) = (26.1 \text{ s}^2) a_J$$

$$a_J = 3.83 \text{ m/s}^2 \quad \lozenge$$

(b) Their speeds after accelerating are

$$v_M = a_M (2.00 \text{ s}) = (5.43 \text{ m/s}^2)(2.00 \text{ s}) = 10.9 \text{ m/s} \quad \lozenge$$

and $\qquad\qquad v_J = a_J (3.00 \text{ s}) = (3.83 \text{ m/s}^2)(3.00 \text{ s}) = 11.5 \text{ m/s} \quad \lozenge$

(c) In the first 6.00 s, Maggie covers a distance

$$\tfrac{1}{2} a_M (2.00 \text{ s})^2 + v_M (4.00 \text{ s}) = \tfrac{1}{2}(5.43 \text{ m/s}^2)(2.00 \text{ s})^2 + (10.9 \text{ m/s})(4.00 \text{ s}) = 54.3 \text{ m}$$

and Judy has run a distance

$$\tfrac{1}{2} a_J (3.00 \text{ s})^2 + v_J (3.00 \text{ s}) = \tfrac{1}{2}(3.83 \text{ m/s}^2)(3.00 \text{ s})^2 + (11.5 \text{ m/s})(3.00 \text{ s}) = 51.7 \text{ m}$$

So Maggie is ahead by 54.3 m – 51.7 m = 2.62 m $\quad \lozenge$

---

81. Two objects A and B are connected by a rigid rod that has a length L. The objects slide along perpendicular guide rails, as shown in Figure P2.81. If A slides to the left with a constant speed v, find the velocity of B when $\alpha = 60°$.

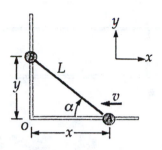

Figure P2.81

Solution The distances x and y are always related by $x^2 + y^2 = L^2$. Differentiating this equation with respect to time, we have

$$2x \frac{dx}{dt} + 2y \frac{dy}{dt} = 0$$

Now $\dfrac{dy}{dt}$ is v_B, the unknown velocity of B; and $\dfrac{dx}{dt} = -v$.

So the differentiated equation becomes

$$\frac{dy}{dt} = -\frac{x}{y}\left(\frac{dx}{dt}\right) = \left(-\frac{x}{y}\right)(-v) = v_B$$

But $\dfrac{y}{x} = \tan\alpha$, so $v_B = \left(\dfrac{1}{\tan\alpha}\right)v$

When $\alpha = 60°$,

$$v_B = \frac{v}{\tan 60°} = \frac{v\sqrt{3}}{3} = 0.577v \quad \lozenge$$

Chapter 3

Vectors

VECTORS

INTRODUCTION

Physical quantities that have both numerical and directional properties are represented by vectors. Some examples of vector quantities are force, displacement, velocity, and acceleration. This chapter is primarily concerned with vector algebra and with some general properties of vector quantities. The addition and subtraction of vector quantities are discussed, together with some common applications to physical situations. Discussion of the products of vector quantities shall be delayed until these operations are needed.

Vector quantities are used throughout this text, and it is therefore imperative that you master both their graphical and their algebraic properties.

NOTES FROM SELECTED CHAPTER SECTIONS

3.1 Coordinate Systems and Frames of Reference

A *coordinate system* used to specify locations in space consists of:

- A fixed reference point called the *origin*.

- A set of specified *axes* or directions.

- Instructions (properties of the particular coordinate system) which describe how to label a point in space relative to the origin and the axes.

Convenient coordinate systems include *cartesian* and *plane polar* coordinate systems.

3.2 Vector and Scalar Quantities

A *vector* is a physical quantity that must be specified by both magnitude and direction.

A *scalar* quantity has only magnitude.

3.3 Some Properties of Vectors

- Equality of Two Vectors—Two vectors are *equal* if they have the same magnitude and the same direction.

- Addition of Vectors—In order for two or more vectors to be added, they must have the same units; their sum is independent of the order of addition. The triangle method and the parallelogram method are graphic methods for determining the resultant or sum of two or more vectors.

- Negative of a Vector—The sum of a vector and its negative is zero. A vector and its negative have the same magnitude, but have opposite directions.

- Multiplication by a Scalar—When a vector is *multiplied* or *divided* by a positive (negative) scalar, the result is a vector in the same (opposite) direction. The magnitude of the resulting vector is equal to the product of the scalar and the magnitude of the original vector.

3.4 Components of a Vector and Unit Vectors

Any vector can be completely described by its *components*.

A *unit vector* is a dimensionless vector *one unit in length* used to *specify a given direction*.

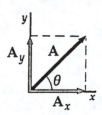

Figure 1 Components of a vector

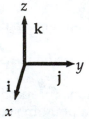

Figure 2 Unit Vectors

The unit vectors **i**, **j**, and **k** form a set of *mutually perpendicular* vectors as illustrated above.

$\mathbf{i} \equiv$ a unit vector along the x axis
$\mathbf{j} \equiv$ a unit vector along the y axis
$\mathbf{k} \equiv$ a unit vector along the z axis
where $|\mathbf{i}| = |\mathbf{j}| = |\mathbf{k}| = 1$

EQUATIONS AND CONCEPTS

The location of a point P in a plane can be specified by either cartesian coordinates, x and y, or polar coordinates, r and θ. If one set of coordinates is known, values for the other set can be calculated.

$$x = r\cos\theta \qquad (3.1)$$

$$y = r\sin\theta \qquad (3.2)$$

$$\tan\theta = \frac{y}{x} \qquad (3.3)$$

$$r = \sqrt{x^2 + y^2} \qquad (3.4)$$

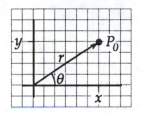

Vector quantities obey the *commutative law of addition*. In order to add vector **A** to vector **B** using the graphical method, first construct **A**, and then draw **B** such that the tail of **B** starts at the head of **A**. The sum of **A** + **B** is the vector that completes the triangle by connecting the tail of **A** to the head of **B**.

$$\mathbf{A} + \mathbf{B} = \mathbf{B} + \mathbf{A} \qquad (3.5)$$

When three or more vectors are added, the sum is independent of the manner in which the vectors are grouped. This is the *associative law* of addition.

$$\mathbf{A} + (\mathbf{B} + \mathbf{C}) = (\mathbf{A} + \mathbf{B}) + \mathbf{C} \qquad (3.6)$$

When more than two vectors are to be added, they are all connected head-to-tail in any order. The resultant or sum is the vector which joins the tail of the first vector to the head of the last vector.

$$\mathbf{R} = \mathbf{A} + \mathbf{B} + \mathbf{C} + \mathbf{D}$$

When two or more vectors are to be added, all of them must represent the same physical quantity—that is, have the same units. In the graphical or geometrical method of vector addition, the length of each vector is proportional to the magnitude of the vector. Also, each vector must be pointed along a direction which makes the proper angle relative to the others.

The operation of vector subtraction utilizes the definition of the negative of a vector. The negative of vector **A** is the vector which has a magnitude equal to the magnitude of **A**, but acts or points along a direction opposite the direction of **A**.

$$A - B = A + (-B) \tag{3.7}$$

A vector **A** in a two-dimensional coordinate system can be resolved into its components along the x and y directions. The projection of **A** onto the x axis is the x component of **A**; and the projection of **A** onto the y axis is the y component of **A**.

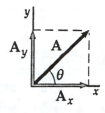

The magnitude of **A** and the angle, θ, which the vector makes with the positive x axis can be determined from the values of the x and y components of **A**.

$$A_x = A \cos \theta \tag{3.8}$$

$$A_y = A \sin \theta \tag{3.9}$$

$$A = \sqrt{A_x^2 + A_y^2} \tag{3.10}$$

$$\tan \theta = \frac{A_y}{A_x} \tag{3.11}$$

A vector **A** lying in the x-y plane, having rectangular components A_x and A_y, can be expressed in unit vector notation.

$$A = A_x \mathbf{i} + A_y \mathbf{j} \tag{3.12}$$

When two vectors are added, the resultant vector can be expressed in terms of the components of the two vectors.

If $\mathbf{R} = \mathbf{A} + \mathbf{B}$,

$$\mathbf{R} = (A_x + B_x)\mathbf{i} + (A_y + B_y)\mathbf{j} \tag{3.14}$$

SUGGESTIONS, SKILLS, AND STRATEGIES

When two or more vectors are to be added, the following step-by-step procedure is recommended:

- Select a coordinate system.

- Draw a sketch of the vectors to be added (or subtracted), with a label on each vector.

- Find the x and y components of all vectors.

- Find the resultant components (the algebraic sum of the components) in both the x and y directions.

- Use the Pythagorean theorem to find the magnitude of the resultant vector.

- Use a suitable trigonometric function to find the angle the resultant vector makes with the x axis.

To add vector **A** to vector **B** graphically, first construct **A**, and then draw **B** such that the tail of **B** starts at the head of **A**. The sum **A + B** is the vector that completes the triangle as shown in Figure 3 below. The procedure for adding more than two vectors (the polygon rule) is also illustrated.

In order to subtract two vectors *graphically*, recognize that **A − B** is equivalent to the operation **A + (−B)**. Since the vector **−B** is a vector whose magnitude is B and is opposite in direction to **B**, the construction in Figure 3 (c) below is obtained:

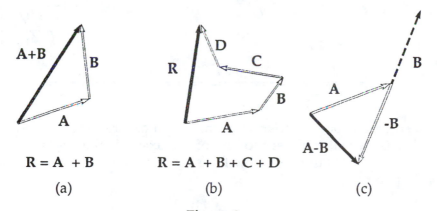

$$R = A + B \qquad\qquad R = A + B + C + D$$

(a)   (b)   (c)

Figure 3
Adding vectors by (a) the triangle rule and (b) the polygon rule. (c) Subtracting two vectors graphically.

REVIEW CHECKLIST

▷ Locate a point in space using both cartesian coordinates and polar coordinates.

▷ Describe the basic properties of vectors such as the rules of vector addition and graphical solutions for addition of two or more vectors.

▷ Resolve a vector into its rectangular components. Determine the magnitude and direction of a vector from its rectangular components.

▷ Understand the use of unit vectors to express any vector in unit vector notation.

SOLUTIONS TO SELECTED END-OF-CHAPTER PROBLEMS

3. The polar coordinates of a point are $r = 5.50$ m and $\theta = 240.0°$. What are the cartesian coordinates of this point?

Solution

When the polar coordinates (r, θ) of a point P are known, the cartesian coordinates can be found:

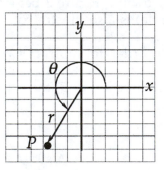

$$x = r\cos\theta, \quad y = r\sin\theta$$

$$x = (5.50 \text{ m})\cos 240° = -2.75 \text{ m} \quad \Diamond$$

$$y = (5.50 \text{ m})\sin 240° = -4.76 \text{ m} \quad \Diamond$$

5. A certain corner of a room is selected as the origin of a rectangular coordinate system. A fly is crawling on a wall adjacent to one of the axes. If the fly is located at a point having coordinates (2.00, 1.00) m, (a) how far is it from the corner of the room? (b) What is its location in polar coordinates?

Solution

(a) Assume the wall is in the x-y plane so that the coordinates are $x = 2.00$ m and $y = 1.00$ m; and the fly is located at point P. The distance between two points in the x, y plane is

$$d = \sqrt{(x_2 - x_1)^2 + (y_2 - y_1)^2}$$

$$d = \sqrt{(2.00 \text{ m} - 0)^2 + (1.00 \text{ m} - 0)^2} = 2.24 \text{ m} \quad \lozenge$$

(b) From the figure,

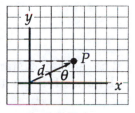

$$\theta = \tan^{-1}\left(\frac{y}{x}\right) = \tan^{-1}\left(\frac{1.00}{2.00}\right) = 26.6°$$

and $r = d$. Therefore, the polar coordinates of the point P are (2.24 m, 26.6°) $\lozenge$

---

9. A surveyor estimates the distance across a river by the following method: standing directly across from a tree on the opposite bank, she walks 100 m along the riverbank, then sights across to the tree. The angle from her baseline to the tree is 35.0°. How wide is the river?

Solution   Make a sketch of the area as viewed from above. Assume the river-banks are straight and parallel, show the location of the tree, and the original location of the surveyor. The width w between tree and surveyor must be perpendicular to the riverbanks, as the surveyor chooses to start "directly across" from the tree. Now draw the 100-meter baseline, b, and the line showing the "line of sight" to the tree.

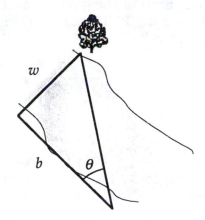

From the figure, $\tan\theta = \dfrac{w}{b}$

and $w = b\tan\theta = (100 \text{ m}) \tan 35.0° = 70.0 \text{ m} \quad \lozenge$

---

13. A person walks along a circular path of radius 5.00 m, around one half of the circle. (a) Find the magnitude of the displacement vector. (b) How far did the person walk? (c) What is the magnitude of the displacement if the person walks all the way around the circle?

Solution   See the sketch on the right.

(a) $|\mathbf{d}| = |-10.0\mathbf{i}| = 10.00$ m   ◊

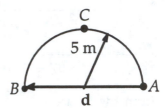

since the displacement is a straight line from point A to point B.

(b)   The actual distance walked is not equal to the straight-line displacement. The distance follows the curved path of the semicircle (ACB).

$$s = \left(\tfrac{1}{2}\right)(2\pi r) = 5.00\pi = 15.7 \text{ m} \quad ◊$$

(c)   If the circle is complete, $\mathbf{d}$ begins and ends at point A. Hence, $|\mathbf{d}| = 0$.   ◊

―――――――――――――――――――

15. Each of the displacement vectors **A** and **B** shown in Figure P3.15 has a magnitude of 3.00 m. Find graphically (a) **A** + **B**, (b) **A** − **B**, (c) **B** − **A**, (d) **A** − 2**B**.

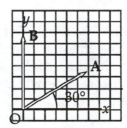

Figure P3.15

Solution   To find these vector expressions graphically, we draw each set of vectors as indicated by the drawings on the next page. Measurements of the results can be taken using a ruler and protractor.

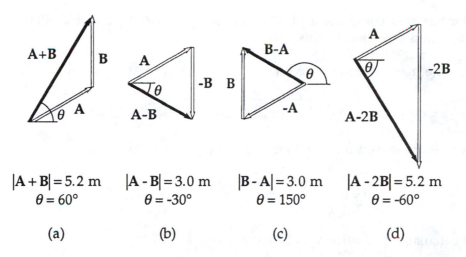

$|A+B| = 5.2$ m $|A-B| = 3.0$ m $|B-A| = 3.0$ m $|A-2B| = 5.2$ m
$\theta = 60°$ $\theta = -30°$ $\theta = 150°$ $\theta = -60°$

(a)   (b)   (c)   (d)

Remember: when adding vectors graphically, they are connected "head-to-tail," represented by arrows whose lengths correspond to their magnitudes. Also, the relative directions of the vectors must be maintained. When subtracting vectors, remember also that $A - B = A + (-B)$.

17. A roller coaster moves 200 ft horizontally, then rises 135 ft at an angle of 30.0° above the horizontal. It then travels 135 ft at an angle of 40.0° downward. What is its displacement from its starting point? Use graphical techniques.

Solution Your sketch when drawn to scale should look somewhat like the one below. The distance R and the angle θ can be measured to give, upon use of your scale factor, the values of:

$R = 423$ ft at about 2.63° below the horizontal. ◊
(You will probably only be able to obtain a measurement to 1-2 significant figures)

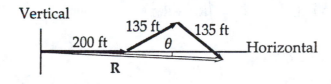

23. A vector has an x component of -25.0 units and a y component of 40.0 units. Find the magnitude and direction of this vector.

Solution   The magnitude of a vector, r, in terms of the x and y components is

$$r = \sqrt{x^2 + y^2} = \sqrt{(-25.0 \text{ units})^2 + (40.0 \text{ units})^2} = 47.2 \text{ units} \quad \Diamond$$

The direction, θ, measured relative to the positive-going x axis is

$$\theta = \tan^{-1}\left(\frac{y}{x}\right) = \tan^{-1}\left(\frac{40.0}{-25.0}\right) = \tan^{-1}(-1.60)$$

Using your calculaor, this value will be shown as $\theta = -58.0°$.

However, since the x component is negative and the y component is positive, the angle relative to the positive x axis is
$$\theta = 180° - 58.0° = 122° \quad \Diamond$$

---

29. Consider two vectors $\mathbf{A} = 3\mathbf{i} - 2\mathbf{j}$ and $\mathbf{B} = -\mathbf{i} - 4\mathbf{j}$. Calculate (a) $\mathbf{A} + \mathbf{B}$, (b) $\mathbf{A} - \mathbf{B}$, (c) $|\mathbf{A} + \mathbf{B}|$, (d) $|\mathbf{A} - \mathbf{B}|$, (e) the direction of $\mathbf{A} + \mathbf{B}$ and $\mathbf{A} - \mathbf{B}$.

Solution   Use the property of vector addition that states if

$$\mathbf{R} = \mathbf{A} + \mathbf{B} \quad \text{then} \quad R_x = A_x + B_x \quad \text{and} \quad R_y = A_y + B_y$$

(a) $\mathbf{A} + \mathbf{B} = (3\mathbf{i} - 2\mathbf{j}) + (-\mathbf{i} - 4\mathbf{j}) = 2\mathbf{i} - 6\mathbf{j} \quad \Diamond$

(b) $\mathbf{A} - \mathbf{B} = (3\mathbf{i} - 2\mathbf{j}) - (-\mathbf{i} - 4\mathbf{j}) = 4\mathbf{i} + 2\mathbf{j} \quad \Diamond$

For a vector $\mathbf{R} = R_x\mathbf{i} + R_y\mathbf{j}$, $|\mathbf{R}| = \sqrt{R_x^2 + R_y^2}$

(c) $|\mathbf{A} + \mathbf{B}| = \sqrt{2^2 + (-6)^2} = 6.32 \quad \Diamond$

(d) $|\mathbf{A} - \mathbf{B}| = \sqrt{4^2 + 2^2} = 4.47 \quad \Diamond$

The direction of a vector relative to the positive x axis is $\theta = \tan^{-1}\left(\frac{R_y}{R_x}\right)$

(e)   For **A** + **B**, $\theta = \tan^{-1}\left(\dfrac{-6}{2}\right) = -71.6° = 288°$ ◊

   For **A** − **B**, $\theta = \tan^{-1}\left(\dfrac{2}{4}\right) = 26.6°$ ◊

31. Obtain expressions for the position vectors with polar coordinates (a) 12.8 m, 150°; (b) 3.30 cm, 60.0°; (c) 22.0 in., 215°.

Solution

Find the x and y components of each vector using $x = R\cos\theta$ and $y = R\sin\theta$. In unit vector notation, $\mathbf{R} = R_x\mathbf{i} + R_y\mathbf{j}$.

(a) $x = (12.8 \text{ m}) \cos 150° = -11.1 \text{ m}$

   $y = (12.8 \text{ m}) \sin 150° = 6.40 \text{ m}$

   $\mathbf{R} = (-11.1\mathbf{i} + 6.40\mathbf{j}) \text{ m}$ ◊

(b) $x = (3.30 \text{ cm}) \cos 60.0 = 1.65 \text{ cm}$

   $y = (3.30 \text{ cm}) \sin 60.0 = 2.86 \text{ cm}$

   $\mathbf{R} = (1.65\mathbf{i} + 2.86\mathbf{j}) \text{ cm}$ ◊

(c) $x = (22.0 \text{ in}) \cos 215° = -18.0 \text{ in}$

   $y = (22.0 \text{ in}) \sin 215° = -12.6 \text{ in}$

   $\mathbf{R} = (-18.0\mathbf{i} - 12.6\mathbf{j}) \text{ in}$ ◊

33. A particle undergoes the following consecutive displacements: 3.50 m south, 8.20 m northeast, and 15.0 m west. What is the resultant displacement?

Solution Take the direction east to be along +i. The three displacements can be written as:

$$d_1 = -3.50j \text{ m}$$

$$d_2 = (8.20 \cos 45.0°)i \text{ m} + (8.20 \sin 45.0°)j \text{ m} = 5.80i \text{ m} + 5.80j \text{ m}$$

$$d_3 = -15.0i \text{ m}$$

The resultant, $R = d_1 + d_2 + d_3$

$$R = (0 + 5.80 - 15.0)i \text{ m} + (-3.50 + 5.80 + 0)j \text{ m}$$

$$R = (-9.20i + 2.30j) \text{ m} \quad ◊$$

The magnitude of the resultant displacement is

$$|R| = \sqrt{R_x^2 + R_y^2} = \sqrt{(-9.20 \text{ m})^2 + (2.30 \text{ m})^2} = 9.48 \text{ m} \quad ◊$$

The direction of the resultant vector is given by

$$\tan^{-1}\left(\frac{R_y}{R_x}\right) = \tan^{-1}\left(\frac{2.30}{-9.20}\right) = -14.0°$$

or relative to the positive x axis, $\quad \theta = 180° - 14.0° = 166° \quad ◊$

---

39. The vector **A** has x, y, and z components of 8, 12, and –4 units, respectively. (a) Write a vector expression for **A** in unit-vector notation. (b) Obtain a unit-vector expression for a vector **B** one fourth the length of **A** pointing in the same direction as **A**. (c) Obtain a unit-vector expression for a vector **C** three times the length of **A** pointing in the direction opposite the direction of **A**.

Solution

(a) $\quad A = A_x i + A_y j + A_z k$

$\quad A = 8i + 12j - 4k \quad ◊$

(b) $B = \dfrac{A}{4}$

$B = 2i + 3j - k$ ◊

(c) $C = -3A$

$C = -24i - 36j + 12k$ ◊

---

43. Vector **A** has a negative x component 3.00 units in length and a positive y component 2.00 units in length. (a) Determine an expression for **A** in unit-vector notation. (b) Determine the magnitude and direction of **A**. (c) What vector **B** when added to **A** gives a resultant vector with no x component and a negative y component 4.00 units in length?

Solution

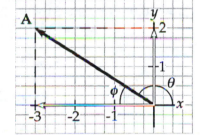

$A_x = -3.00$ units, $A_y = 2.00$ units

(a) $\mathbf{A} = A_x\mathbf{i} + A_y\mathbf{j} = -3.00\mathbf{i} + 2.00\mathbf{j}$ units ◊

(b) $|\mathbf{A}| = \sqrt{A_x{}^2 + A_y{}^2} = \sqrt{(-3.00)^2 + (2.00)^2} = 3.61$ units

$\tan\phi = \left|\dfrac{A_y}{A_x}\right| = \left|\dfrac{2.00}{-3.00}\right| = 0.667$ so $\phi = 33.7°$ (relative to the $-x$ axis)

A is in the *second quadrant*: $\theta = 180° - \phi = 146°$ ◊

(c) $R_x = 0,$ $R_y = -4.00$

Since $\mathbf{R} = \mathbf{A} + \mathbf{B},$ $\mathbf{B} = \mathbf{R} - \mathbf{A}$

$B_x = R_x - A_x = 0 - (-3.00) = 3.00$

$B_y = R_y - A_y = -4.00 - 2.00 = -6.00$

Therefore, $\mathbf{B} = B_x\mathbf{i} + B_y\mathbf{j} = (3.00\mathbf{i} - 6.00\mathbf{j})$ units ◊

---

47. Three vectors are oriented as shown in Figure P3.47, where $|\mathbf{A}| = 20.0$ units, $|\mathbf{B}| = 40.0$ units, and $|\mathbf{C}| = 30.0$ units. Find (a) the x and y components of the resultant vector and (b) the magnitude and direction of the resultant vector.

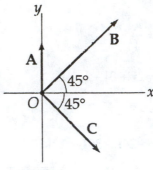

Solution

(a) $A_x = (20.0 \text{ units}) \cos 90° = 0$

$A_y = (20.0 \text{ units}) \sin 90° = 20.0 \text{ units}$

Figure P3.47

$B_x = (40.0 \text{ units}) \cos 45° = 28.3 \text{ units}$

$B_y = (40.0 \text{ units}) \sin 45° = 28.3 \text{ units}$

$C_x = (30.0 \text{ units}) \cos 315° = 21.2 \text{ units}$

$C_y = (30.0 \text{ units}) \sin 315° = -21.2 \text{ units}$

$$R_x = A_x + B_x + C_x = (0 + 28.3 + 21.2) \text{ units}$$

$$= 49.5 \text{ units} \quad \lozenge$$

$$R_y = A_y + B_y + C_y = (20 + 28.3 - 21.2) \text{ units}$$

$$= 27.1 \text{ units} \quad \lozenge$$

(b) $$|\mathbf{R}| = \sqrt{R_x{}^2 + R_y{}^2} = \sqrt{(49.5 \text{ units})^2 + (27.1 \text{ units})^2} = 56.4 \text{ units} \quad \lozenge$$

$$\theta = \tan^{-1}\left(\frac{R_y}{R_x}\right) = \tan^{-1}\left(\frac{27.1}{49.5}\right) = 28.7° \quad \lozenge$$

49. A person going for a walk follows the path shown in Figure P3.49. The total trip consists of four straight-line paths. At the end of the walk, what is the person's resultant displacement measured from the starting point?

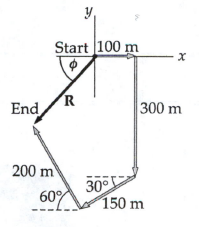

Figure P3.49

Solution The resultant displacement **R** is equal to the sum of the four individual displacements.

$\mathbf{d_1} = 100\mathbf{i}$ m

$\mathbf{d_2} = -300\mathbf{j}$ m

$\mathbf{d_3} = (-150\cos30°)\mathbf{i}$ m $+ (-150\sin30°)\mathbf{j}$ m

$\quad = -130\mathbf{i}$ m $- 75\mathbf{j}$ m

$\mathbf{d_4} = (-200\cos60°)\mathbf{i}$ m $+ (200\sin60°)\mathbf{j}$ m

$\quad\quad = -100\mathbf{i}$ m $+ 173\mathbf{j}$ m

$\mathbf{R} = \mathbf{d_1} + \mathbf{d_2} + \mathbf{d_3} + \mathbf{d_4}$

$R_x = d_{1x} + d_{2x} + d_{3x} + d_{4x} = (100 + 0 - 130 - 100)$ m

$R_x = -130$ m

$R_y = d_{1y} + d_{2y} + d_{3y} + d_{4y} = (0 - 300 - 75 + 173)$ m

$R_y = -202$ m

$\mathbf{R} = -130\mathbf{i}$ m $- 202\mathbf{j}$ m $\quad \Diamond$

$|\mathbf{R}| = \sqrt{R_x{}^2 + R_y{}^2} = \sqrt{(-130 \text{ m})^2 + (-202 \text{ m})^2} = 240$ m $\quad \Diamond$

$\phi = \tan^{-1}\left(\dfrac{R_y}{R_x}\right) = \tan^{-1}\left(\dfrac{-202}{-130}\right) = 57.2°$ or $\theta = 57.2° + 180° = 237°$ relative to $+x$ axis $\Diamond$

Chapter 4

Motion in Two Dimensions

MOTION IN TWO DIMENSIONS

INTRODUCTION

In this chapter we deal with the kinematics of a particle moving in a plane, which is two-dimensional motion. Common examples of motion in a plane include the motion of projectiles and satellites in a gravitational field, and the motion of charged particles in uniform electric fields.

We begin by showing that displacement, velocity, and acceleration are vector quantities. As in the case of one-dimensional motion, we derive the kinematic equations for two-dimensional motion from the fundamental definitions of displacement, velocity, and acceleration. As special cases of motion in two dimensions, we then treat constant-acceleration motion in a plane (projectile motion) and uniform circular motion.

NOTES FROM SELECTED CHAPTER SECTIONS

4.1 Displacement, Velocity, and Acceleration Vectors

The *displacement vector* is the final position vector minus the initial position vector. Note that the magnitude of the displacement vector is in general less than the distance measured along the actual path of travel.

It is important to recognize that a particle experiences an *acceleration* when:

- the magnitude of the velocity (speed) changes while the direction remains constant (e.g., a sphere rolling down an inclined plane);

- the magnitude of the velocity remains constant while the direction of the velocity changes (e.g., a particle moving at constant speed around a circle of constant radius);

- both the magnitude and direction of the velocity change (e.g., a mass vibrating up and down on the end of a spring).

4.3 Projectile Motion

Projectile motion is a common example of motion in two dimensions under constant acceleration. *Provided air resistance is negligible*, the characteristics of projectile motion can be summarized as follows:

- The horizontal component of velocity, v_x, remains constant since there is no horizontal component of acceleration.

- The vertical component of acceleration is equal to the acceleration due to gravity, g.

- The vertical component of velocity, v_y, and the displacement in the y direction change in time in a manner identical to that of a freely-falling body.

- Projectile motion can be described as a superposition, or vector addition, of the two motions in the x and y directions.

The trajectory of a projectile is a *parabola* as shown in Figure 4.2. We choose the motion to be in the x-y plane, and take the *initial* velocity of the projectile to have a magnitude of v_0, directed at an angle θ_0 with the horizontal. The parabolic path of travel is completely determined when the magnitude, v_0, and the direction, θ_0, of the initial velocity vector are given. The actual motion of the projectile is a superposition of two motions: motion of a freely-falling body with constant acceleration in the vertical direction, $-g$, and motion in the horizontal direction with constant velocity, $v_0 \cos \theta_0$.

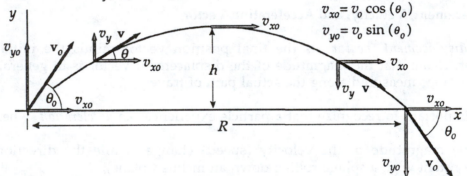

$$v_{xo} = v_o \cos (\theta_o)$$
$$v_{yo} = v_o \sin (\theta_o)$$

Figure 4.2

The *maximum height h* and *horizontal range R* are *special* coordinates defined in Figure 4.3. These can be obtained from Equations 4.10 - 4.13.

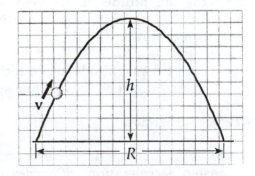

Figure 4.3

4.4 Uniform Circular Motion

The *centripetal* acceleration is the acceleration experienced by a mass which moves uniformly in a circular path of constant radius. The *direction* of the centripetal acceleration is always toward the center of the circular path.

Uniform circular motion is the motion of an object moving in a circular path with *constant linear* speed. The velocity vector is always tangent to the path of the moving body and therefore *perpendicular* to the radius.

4.5 Tangential and Radial Acceleration

Tangential acceleration of a particle moving in a circular path is due to a change in the speed (magnitude of the velocity vector) of the particle. The direction of the tangential acceleration at any instant is tangent to the circular path (perpendicular to the radius).

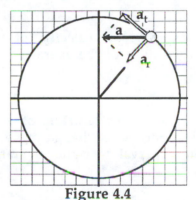

If a particle moves in a circle such that the speed v is *not* constant, the components of acceleration and the total acceleration at some instant are as shown in Figure 4.4.

Figure 4.4
Components of acceleration when
$|v|$ is not constant

4.6 Relative Velocity and Relative Acceleration

Observers in different frames of reference may make different measurements of a moving particle's displacement, velocity, and acceleration. That is, measurements taken by two observers moving with respect to each other generally do not agree.

Although observers in two different reference frames will measure different velocities for the particles, they will measure the *same acceleration* when the relative velocity of one frame with respect to the other is constant.

EQUATIONS AND CONCEPTS

A particle whose position vector changes from $\mathbf{r}_i$ to $\mathbf{r}_f$ undergoes a *displacement* $\Delta\mathbf{r} \equiv \mathbf{r}_f - \mathbf{r}_i$.

$$\Delta\mathbf{r} \equiv \mathbf{r}_f - \mathbf{r}_i \qquad (4.1)$$

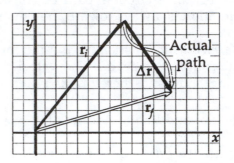

The *average velocity* of a particle which undergoes a displacement $\Delta\mathbf{r}$ in a time interval Δt equals the ratio $\Delta\mathbf{r}/\Delta t$.

$$\overline{\mathbf{v}} \equiv \frac{\Delta\mathbf{r}}{\Delta t} \qquad (4.2)$$

The *instantaneous velocity* of a particle equals the limit of the average velocity as $\Delta t \to 0$.

$$\mathbf{v} \equiv \lim_{\Delta t \to 0} \frac{\Delta\mathbf{r}}{\Delta t} = \frac{d\mathbf{r}}{dt} \qquad (4.3)$$

The *average acceleration* of a particle which undergoes a change in velocity $\Delta\mathbf{v}$ in a time interval Δt equals the ratio $\Delta\mathbf{v}/\Delta t$.

$$\overline{\mathbf{a}} \equiv \frac{\Delta\mathbf{v}}{\Delta t} \qquad (4.4)$$

The *instantaneous acceleration* is defined as the limit of the average velocity as $\Delta t \to 0$.

$$\mathbf{a} \equiv \lim_{\Delta t \to 0} \frac{\Delta\mathbf{v}}{\Delta t} = \frac{d\mathbf{v}}{dt} \qquad (4.5)$$

The *velocity* of a particle as a function of time moving with *constant acceleration* : (at $t = 0$, the velocity is $\mathbf{v}_0$).

$$\mathbf{v} = \mathbf{v}_0 + \mathbf{a}t \qquad (4.8)$$

The *position vector* as a function of time for a particle moving with *constant acceleration* : (at $t = 0$, the position vector is $\mathbf{r}_0$ and the velocity is $\mathbf{v}_0$).

$$\mathbf{r} = \mathbf{r}_0 + \mathbf{v}_0 t + \tfrac{1}{2}\mathbf{a}t^2 \qquad (4.9)$$

Equations 4.10 and 4.11 give the x and y components of velocity versus time for a projectile.

$$v_x = v_0 \cos \theta_0 = \text{constant} \tag{4.10}$$

$$v_y = v_0 \sin \theta_0 - gt \tag{4.11}$$

Equations 4.12 and 4.13 give the x and y position coordinates for a projectile, as a function of time.

$$x = (v_0 \cos \theta_0)t \tag{4.12}$$

$$y = (v_0 \sin \theta_0)t - \tfrac{1}{2}gt^2 \tag{4.13}$$

The *maximum height* of a projectile can be written in terms of v_0 and θ_0.

$$h = \frac{v_0^2 \sin^2 \theta_0}{2g} \tag{4.17}$$

Likewise, the *horizontal range* of a projectile can also be stated in terms of v_0 and θ_0.

$$R = \frac{v_0^2 \sin 2\theta_0}{g} \tag{4.13}$$

A particle moving in a circle of radius r with speed v undergoes a *centripetal acceleration a_r* equal in magnitude to v^2/r.

$$a_r = \frac{v^2}{r} \tag{4.19}$$

If a particle which moves with constant speed v in a circular path, the period is equal to the total distance traveled in one revolution, divided by the speed.

$$T \equiv \frac{2\pi r}{v}$$

In general, a particle moving on a curved path can have both a centripetal component and tangential component of acceleration, where $\mathbf{a}_r$ is directed towards the center of curvature and $\mathbf{a}_t$ is tangent to the path.

$$\mathbf{a} = \mathbf{a}_r + \mathbf{a}_t \tag{4.20}$$

If the speed of a particle moving on a curved path changes with time, the particle has a *tangential* component of acceleration equal in magnitude to dv/dt.

$$a_t = \frac{d|\mathbf{v}|}{dt} \tag{4.21}$$

Galilean transformation equations relate the position and velocity of a particle measured by an observer in a moving frame of reference to those values measured by an observer in a fixed frame (moving with the object).

$$\mathbf{r}' = \mathbf{r} - \mathbf{u}t \qquad\qquad (4.24)$$

$$\mathbf{v}' = \mathbf{v} - \mathbf{u} \qquad\qquad (4.25)$$

SUGGESTIONS, SKILLS, AND STRATEGIES

- You should be familiar with the mathematical expression for a *parabola*. In particular, the equation which describes the trajectory of a projectile moving under the influence of gravity is given by

$$y = Ax - Bx^2$$

where $\qquad A = \tan\theta_0 \qquad$ and $\qquad B = \dfrac{g}{2v_0^2\,\cos^2\theta_0}$

Note that this expression for y assumes that the particle leaves the origin at $t = 0$, with a velocity $\mathbf{v}_0$. A sketch of y versus x for this situation is shown in Figure 4.1.

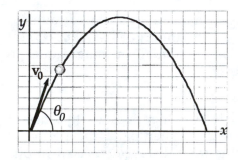

Figure 4.1

- If you are given v_0 and θ_0, you should be able to make a point-by-point plot of the trajectory using the expressions for $x(t)$ and $y(t)$. Furthermore, you should know how to calculate the velocity component v_y at any time t. (Note that the component $v_x = v_{x0} = v_0\cos\theta_0 = \text{constant}$, since $a_x = 0$.)

- Assuming that you have values for x and y at any time $t > 0$, you should be able to write an expression for the position vector $\mathbf{r}$ at that time using the relation $\mathbf{r} = x\mathbf{i} + y\mathbf{j}$. From this you can find the *displacement* r, where $r = \sqrt{x^2 + y^2}$. Likewise, if v_x and v_y are known at any time $t > 0$, you can express the velocity vector $\mathbf{v}$ in the formula $\mathbf{v} = v_x\mathbf{i} + v_y\mathbf{j}$. From this, you can find the *speed* at any time, since $v = \sqrt{v_x^2 + v_y^2}$.

Problem-Solving Strategy: Projectile Motion

We suggest that you use the following approach to solving projectile motion problems:

- Select a coordinate system.

- Resolve the initial velocity vector into x and y components.

- Treat the horizontal motion and the vertical motion independently.

- Follow the techniques for solving problems with constant velocity to analyze the horizontal motion of the projectile.

- Follow the techniques for solving problems with constant acceleration to analyze the vertical motion of the projectile.

REVIEW CHECKLIST

▷ Describe the displacement, velocity, and acceleration of a particle moving in the x-y plane.

▷ Recognize that two-dimensional motion in the x-y plane with constant acceleration is equivalent to two independent motions along the x and y directions with constant acceleration components a_x and a_y.

▷ Recognize the fact that if the initial speed v_0 and initial angle θ_0 of a projectile are known at a given point at $t = 0$, the velocity components and coordinates can be found at any later time t. Furthermore, one can also calculate the horizontal range R and maximum height h if v_0 and θ_0 are known.

▷ Understand the nature of the acceleration of a particle moving in a circle with constant speed. In this situation, note that although $|\mathbf{v}|$ = constant, the *direction* of $\mathbf{v}$ varies in time, the result of which is the radial, or centripetal acceleration.

▷ Describe the components of acceleration for a particle moving on a curved path, where both the magnitude and direction of $\mathbf{v}$ are changing with time. In this case, the particle has a tangential component of acceleration and a radial component of acceleration.

▷ Realize that the outcome of a measurement of the motion of a particle (its position, velocity, and acceleration) depends on the frame of reference of the observer.

SOLUTIONS TO SELECTED END-OF-CHAPTER PROBLEMS

3. A motorist drives south at 20.0 m/s for 3.00 min, then turns west and travels at 25.0 m/s for 2.00 min, and finally travels northwest at 30.0 m/s for 1.00 min. For this 6.00-min trip, find (a) the resultant vector displacement, (b) the average speed, and (c) the average velocity.

Solution

(a) Her displacements are

$$\Delta r = (20.0 \text{ m}/\text{s})(180 \text{ s}) \text{ south} + (25.0 \text{ m}/\text{s})(120 \text{ s}) \text{ west} + (30.0 \text{ m}/\text{s})(60.0 \text{ s}) \text{ northwest}$$

Choosing i = east and j = north, we have

$$\Delta r = (3.60 \text{ km})(-j) + (3.00 \text{ km})(-i) + (1.80 \text{ km cos } 45.0°)(-i) + (1.80 \text{ km sin } 45.0°)(j)$$

$$= (3.00 + 1.27) \text{ km } (-i) + (1.27 - 3.60) \text{ km } j$$

$$= (-4.27i - 2.33j) \text{ km } \Diamond$$

The answer can also be written as

$$\Delta r = \sqrt{(-4.27 \text{ km})^2 + (-2.33 \text{ km})^2} \text{ at } \tan^{-1}\left(\frac{2.33}{4.27}\right) \text{ south of west}$$

$$= 4.87 \text{ km at } 209° \text{ from east } \Diamond$$

(b) The total distance or path-length traveled is

$$(3.60 + 3.00 + 1.80) \text{ km} = 8.40 \text{ km}$$

$$\text{So average speed} = \frac{8.40 \text{ km}}{6.00 \text{ min}}\left(\frac{1.00 \text{ min}}{60.0 \text{ s}}\right)\left(\frac{1000 \text{ m}}{\text{km}}\right) = 23.4 \text{ m}/\text{s } \Diamond$$

(c) $v_{av} = \dfrac{\Delta r}{t} = \dfrac{4.87 \text{ km}}{360 \text{ s}} = 13.5 \text{ m}/\text{s at } 209° \Diamond$

$$\text{or} \quad v_{av} = \frac{\Delta r}{t} = \frac{(-4.27 \text{ east} - 2.33 \text{ north}) \text{ km}}{360 \text{ s}}$$

$$v_{av} = (11.9 \text{ west} + 6.47 \text{ south}) \text{ m}/\text{s } \Diamond$$

7. A fish swimming in a horizontal plane has velocity $\mathbf{v}_0 = (4.0\mathbf{i} + 1.0\mathbf{j})$ m/s at a point in the ocean whose position vector is $\mathbf{r}_0 = (10.0\mathbf{i} - 4.0\mathbf{j})$ m relative to a stationary rock at the shore. After the fish swims with constant acceleration for 20.0 s, its velocity is $\mathbf{v} = (20.0\mathbf{i} - 5.0\mathbf{j})$ m/s. (a) What are the components of the acceleration? (b) What is the direction of the acceleration with respect to the fixed x axis? (c) Where is the fish at $t = 25.0$ s and in what direction is it moving?

Solution

At $t = 0$, $\qquad\qquad \mathbf{v}_0 = (4.0\mathbf{i} + 1.0\mathbf{j})$ m/s $\quad$ and $\quad$ $\mathbf{r}_0 = (10.0\mathbf{i} - 4.0\mathbf{j})$ m

At $t = 20.0$ s, $\qquad \mathbf{v} = (20.0\mathbf{i} - 5.0\mathbf{j})$ m/s

(a) $\quad a_x = \dfrac{\Delta v_x}{\Delta t} = \dfrac{20.0 \text{ m/s} - 4.0 \text{ m/s}}{20.0 \text{ s}} = 0.800 \text{ m/s}^2$ ◊

$\quad a_y = \dfrac{\Delta v_y}{\Delta t} = \dfrac{-5.0 \text{ m/s} - 1.0 \text{ m/s}}{20.0 \text{ s}} = -0.300 \text{ m/s}^2$ ◊

(b) $\quad \theta = \tan^{-1}\left(\dfrac{a_y}{a_x}\right) = \tan^{-1}\left(\dfrac{-0.30 \text{ m/s}^2}{0.80 \text{ m/s}^2}\right) = -20.6°$ $\quad$ or $\quad$ 339° from the $+x$ axis ◊

(c) $\quad$ At $t = 25.0$ s, its coordinates are

$\quad x = x_0 + v_{x0}t + \tfrac{1}{2}a_x t^2 = 10.0 \text{ m} + (4.0 \text{ m/s})(25.0 \text{ s}) + \tfrac{1}{2}(0.800 \text{ m/s}^2)(25.0 \text{ s})^2 = 360 \text{ m}$ ◊

$\quad y = y_0 + v_{y0}t + \tfrac{1}{2}a_y t^2 = -4.0 \text{ m} + (1.0 \text{ m/s})(25.0 \text{ s}) + \tfrac{1}{2}(-0.30 \text{ m/s}^2)(25.0 \text{ s})^2 = -72.8 \text{ m}$ ◊

$\quad v_y = v_{y0} + a_y t = (1.0 \text{ m/s}) - (0.30 \text{ m/s}^2)(25.0 \text{ s}) = -6.5 \text{ m/s}$

$\quad v_x = v_{x0} + a_x t = (4.0 \text{ m/s}) + (0.800 \text{ m/s}^2)(25.0 \text{ s}) = 24 \text{ m/s}$

Therefore,

$$\theta = \tan^{-1}\left(\dfrac{v_y}{v_x}\right) = \tan^{-1}\left(\dfrac{-6.5 \text{ m/s}}{24 \text{ m/s}}\right) = -15.0° = 345 ° \text{ from the } +x \text{ axis.} \quad ◊$$

11. In a local bar, a customer slides an empty beer mug on the counter for a refill. The bartender is momentarily distracted and does not see the mug, which slides off the counter and strikes the floor 1.40 m from the base of the counter. If the height of the counter is 0.860 m, (a) with what speed did the mug leave the counter and (b) what was the direction of the mug's velocity just before it hit the floor?

Solution

Choose the origin as the point where the mug just leaves the counter. Here, its motion is horizontal, with $v_{y0} = 0$. Choose the final point just before it hits the floor.

Here, $x = 1.40$ m, and $y = -0.860$ m; v_x and v_y are both unknown.

For the horizontal motion: $x = 1.40$ m, $a_x = 0$

For the vertical motion: $y = -0.860$ m, $v_{y0} = 0$, $a_y = -9.80$ m/s^2

(a)   To find v_x using $x = v_x t$, we need first to know the time of flight. Since it is the one quantity in common between horizontal and vertical motions, we find it from

$$y = v_{y0}t + \tfrac{1}{2}a_y t^2$$

Solving, $-0.860 \text{ m} = 0 + \tfrac{1}{2}(-9.80 \text{ m/s}^2)t^2$

and $t = 0.419$ s

Then, $v_x = \dfrac{x}{t} = \dfrac{1.40 \text{ m}}{0.419 \text{ s}} = 3.34 \text{ m/s} \lozenge$

(b)   As the mug hits the floor,

$v_y = v_{y0} + a_y t = 0 - (9.80 \text{ m/s}^2)(0.419 \text{ s}) = -4.11 \text{ m/s}$

The impact angle, $\theta = \tan^{-1}\left(\dfrac{v_y}{v_x}\right) = \tan^{-1}\dfrac{-4.11 \text{ m/s}}{3.34 \text{ m/s}} = -50.9° \lozenge$

17. A place kicker must kick a football from a point 36.0 m (about 40 yards) from the goal, and the ball must clear the crossbar, which is 3.05 m high. When kicked, the ball leaves the ground with a speed of 20.0 m/s at an angle of 53.0° to the horizontal. (a) By how much does the ball clear or fall short of clearing the crossbar? (b) Does the ball approach the crossbar while still rising or while falling?

Solution

(a) To find the actual height of the football when it reaches the goal line, use the equation

$$y = x \tan \theta_0 - \frac{gx^2}{2v_0^2 \cos^2 \theta_0}$$

where $x = 36.0$ m, $v_0 = 20.0$ m/s, and $\theta_0 = 53.0°$

So, $y = (36.0 \text{ m})(\tan 53.0°) - \dfrac{(9.80 \text{ m/s}^2)(36.0 \text{ m})^2}{(2)(20.0 \text{ m/s})^2 \cos^2 53.0°} = 47.774 - 43.834 = 3.940 \text{ m}$

The ball clears the bar by (3.940 − 3.050) m = 0.890 m. ◊

(b) The time the ball takes to reach the maximum height ($v_y = 0$) is

$$t_1 = \frac{(v_0 \sin \theta_0) - v_y}{g} = \frac{(20.0 \text{ m/s})(\sin 53.0°) - 0}{9.80 \text{ m/s}^2} = 1.63 \text{ s}$$

The time to travel 36.0 m horizontally is $t_2 = \dfrac{x}{v_{0x}}$

$$t_2 = \frac{36.0 \text{ m}}{(20.0 \text{ m/s})(\cos 53.0°)} = 3.00 \text{ s}$$

Since $t_2 > t_1$, the ball clears the goal on its way down. ◊

21. During World War I, the Germans had a gun called Big Bertha that was used to shell Paris. The shell had an initial speed of 1.70 km/s at an initial inclination of 55.0° to the horizontal. In order to hit the target, adjustments were made for air resistance and other effects. If we ignore those effects, (a) how far away did the shell hit? (b) How long was it in the air?

Solution

(a) The range R is given by

$$R = \frac{v_0^2}{g} \sin 2\theta_0 = \frac{(1.70 \times 10^3 \text{ m/s}^2)}{9.80 \text{ m/s}^2} \sin 110.0° = 2.77 \times 10^5 \text{ m} = 277 \text{ km} \quad \lozenge$$

(b) If the time in the air is t, then t is given from $v_{0x} = v_0 \cos \theta_0 = \dfrac{R}{t}$

$$t = \frac{R}{v_0 \cos \theta_0} = \frac{2.77 \times 10^5 \text{ m}}{(1700 \text{ m/s}) \cos 55.0°} = 284 \text{ s} = 4.74 \text{ min} \quad \lozenge$$

---

23. A projectile is fired in such a way that its horizontal range is equal to three times its maximum height. What is the angle of projection?

Solution In this problem, we want to find θ_0 such that $R = 3h$. We can use:

$$R = \frac{v_0^2 \sin (2\theta_0)}{g} \qquad \text{and} \qquad h = \frac{v_0^2 \sin^2 \theta_0}{2g}$$

Since we require $R = 3h$,

$$\frac{v_0^2 \sin (2\theta_0)}{g} = \frac{3v_0^2 \sin^2 \theta_0}{2g}, \qquad \text{or} \qquad \frac{2}{3} = \frac{\sin^2 \theta_0}{\sin (2\theta_0)}$$

But $\sin(2\theta_0) = 2(\sin\theta_0)(\cos\theta_0)$, so

$$\frac{\sin^2 \theta_0}{\sin (2\theta_0)} = \frac{\sin^2 \theta_0}{2 \sin \theta_0 \cos \theta_0} = \frac{\tan \theta_0}{2}$$

Substituting and solving for θ_0,

$$\theta_0 = \tan^{-1}\left(\frac{4}{3}\right) = 53.1° \quad \lozenge$$

66

29. An athlete rotates a 1.00-kg discus along a circular path of radius 1.06 m. The maximum speed of the discus is 20.0 m/s. Determine the magnitude of its maximum radial acceleration.

Solution   The maximum radial acceleration occurs when maximum tangential speed is attained, just before the discus is released.

Here,
$$a_r = \frac{v^2}{r} = \frac{(20.0 \text{ m/s})^2}{(1.06 \text{ m})} = 377 \text{ m/s}^2 \quad \lozenge$$

---

35. Figure P4.35 represents, at a given instant, the total acceleration of a particle moving clockwise in a circle of radius 2.50 m. At this instant of time, find (a) the centripetal acceleration, (b) the speed of the particle, and (c) its tangential acceleration.

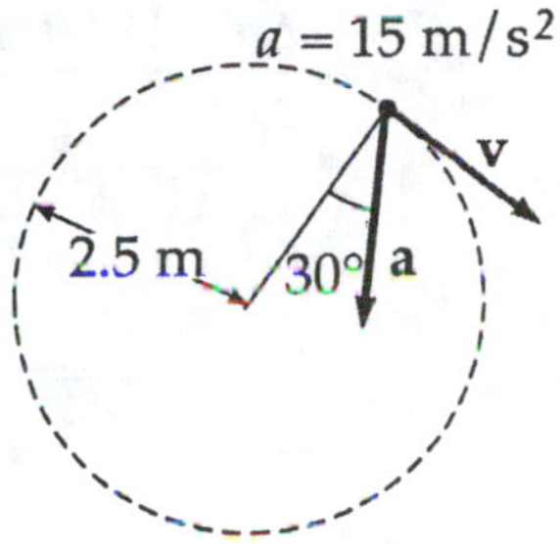

Figure P4.35

Solution

(a)   The acceleration has an inward radial component:

$$a_r = a \cos 30.0° = (15.0 \text{ m/s}^2) \cos 30.0° = 13.0 \text{ m/s}^2 \quad \lozenge$$

(b)   The speed at the instant shown can be found by using

$$a_r = \frac{v^2}{r} \qquad \text{or} \qquad v = \sqrt{a_r r}$$

$$v = \sqrt{(13.0 \text{ m/s}^2)(2.50 \text{ m})} = 5.70 \text{ m/s} \quad \lozenge$$

(c)   The acceleration also has a tangential component:

$$a_t = a \sin 30.0° = (15.0 \text{ m/s}^2) \sin 30.0° = 7.50 \text{ m/s}^2 \quad \lozenge$$

---

37. A train slows down as it rounds a sharp, level turn, slowing from 90.0 km/h to 50.0 km/h in the 15.0 s that it takes to round the bend. The radius of the curve is 150 m. Compute the acceleration at the moment the train speed reaches 50.0 km/h.

Solution
$$50.0 \text{ km/h} = \left(50.0 \frac{\text{km}}{\text{h}}\right)\left(10^3 \frac{\text{m}}{\text{km}}\right)\left(\frac{1\text{ h}}{3600\text{ s}}\right) = 13.89 \text{ m/s}$$

$$90.0 \text{ km/h} = \left(90.0 \frac{\text{km}}{\text{h}}\right)\left(10^3 \frac{\text{m}}{\text{km}}\right)\left(\frac{1\text{ h}}{3600\text{ s}}\right) = 25.0 \text{ m/s}$$

Therefore, when $v = 13.89$ m/s,

$$a_r = \frac{v^2}{r} = \frac{(13.89 \text{ m/s})^2}{150 \text{ m}} = 1.29 \text{ m/s}^2$$

$$a_t = \frac{\Delta v}{\Delta t} = \frac{13.89 \text{ m/s} - 25.0 \text{ m/s}}{15.0 \text{ s}} = -0.741 \text{ m/s}^2$$

$$a = \sqrt{a_r^2 + a_t^2} = \sqrt{(1.29 \text{ m/s}^2)^2 + (-0.741 \text{ m/s}^2)^2} = 1.48 \text{ m/s}^2 \quad \Diamond$$

---

43. A river has a steady speed of 0.500 m/s. A student swims upstream a distance of 1.00 km and returns to the starting point. If the student can swim at a speed of 1.20 m/s in still water, how long does the trip take? Compare this with the time the trip would take if the water were still.

Solution Total time in river equals time upstream (against current) plus time downstream (with current).

$$t_{up} = \frac{1000 \text{ m}}{1.20 \text{ m/s} - 0.500 \text{ m/s}} = 1429 \text{ s}$$

$$t_{dn} = \frac{1000 \text{ m}}{1.20 \text{ m/s} + 0.500 \text{ m/s}} = 588 \text{ s}$$

$$t_R = t_{up} + t_{dn} = 2.02 \times 10^3 \text{ s} \quad \Diamond$$

In still water, $\quad t_{still} = \dfrac{d}{v} = \dfrac{2000 \text{ m}}{1.20 \text{ m/s}} = 1.67 \times 10^3 \text{ s} \quad \Diamond \quad$ or $\quad t_{still} = 0.827 t_R \quad \Diamond$

---

47. The pilot of an airplane notes that the compass indicates a heading due west. The airplane's speed relative to the air is 150 km/h. If there is a wind of 30.0 km/h toward the north, find the velocity of the airplane relative to the ground.

Solution   The velocity of the plane relative to the ground, v_{pg}, is the sum of two vectors:

v_{pa}, velocity of the plane relative to the air
and v_{ag}, velocity of the air relative to the ground

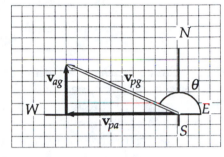

So, $\mathbf{v}_{pg} = \mathbf{v}_{pa} + \mathbf{v}_{ag}$

$\mathbf{v}_{pg} = (-150\mathbf{i} + 30.0\mathbf{j})\ km/h$ ◊

or $\left|\mathbf{v}_{pg}\right| = \sqrt{(-150\ km/h)^2 + (30.0\ km/h)^2} = 153\ km/h$

and $\tan^{-1}\left(\dfrac{30.0}{-150}\right) = -11.3°$

or $\theta = 180° - 11.3° = 169°$ from east   ◊

---

49. A car travels due east with a speed of 50.0 km/h. Rain is falling vertically relative to the Earth. The traces of the rain on the side windows of the car make an angle of 60.0° with the vertical. Find the velocity of the rain relative to (a) the car and (b) the Earth.

Solution

Let $\mathbf{v}_{rg}$ = velocity of the rain relative to ground,

$\mathbf{v}_{rc}$ = velocity of rain relative to the car,

and $\mathbf{v}_{cg}$ = velocity of the car relative to ground.

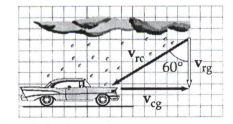

These vectors are related as shown in the figure:

$\mathbf{v}_{rg} = \mathbf{v}_{rc} + \mathbf{v}_{cg}$

Note that the vectors are in a vertical plane.

(a) From the figure,

$$v_{rc} = \frac{v_{cg}}{\sin 60.0°} = \frac{50.0 \text{ km/h}}{\sin 60.0°} = 57.7 \text{ km/h at } 60.0° \text{ relative to the vertical} \lozenge$$

(b) $v_{rg} = v_{cg} \tan 30.0° = (50.0 \text{ km/h})\tan 30.0° = 28.9 \text{ km/h downward} \lozenge$

---

52. A science student is riding on a flatcar of a train traveling along a straight horizontal track at a constant speed of 10.0 m/s. The student throws a ball into the air along a path that he judges to make an initial angle of 60.0° with the horizontal and to be in line with the track. The student's professor, who is standing on the ground nearby, observes the ball to rise vertically. How high does she see the ball rise?

Solution   Shown on the right, $\mathbf{v}_{be} = \mathbf{v}_{bc} + \mathbf{v}_{ce}$

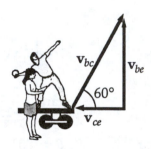

where $\mathbf{v}_{bc}$ = the velocity of the ball relative to the car

$\mathbf{v}_{be}$ = the velocity of the ball relative to the Earth

and $\mathbf{v}_{ce}$ = velocity of the car relative to the Earth = 10.0 m/s

From the figure, we have $v_{ce} = v_{bc} \cos 60.0°$

So $v_{bc} = \frac{10.0 \text{ m/s}}{\cos 60.0°} = 20.0 \text{ m/s}$

Again from the figure,

$$v_{be} = v_{bc} \sin 60.0° + 0 = (20.0 \text{ m/s})(0.866) = 17.3 \text{ m/s}$$

This is the initial velocity of the ball relative to the Earth.

Now from $v_y^2 = v_{0y}^2 + 2ah$, we have $0 = (17.3 \text{ m/s})^2 + 2(-9.8 \text{ m/s}^2)h$

to give $h = 15.3$ m, the maximum height the ball rises. $\lozenge$

---

55. A car is parked overlooking the ocean on an incline that makes an angle of 37.0° with the horizontal. The distance from where the car is parked to the bottom of the incline is 50.0 m, and the incline terminates at a cliff that is 30.0 m above the ocean surface. The negligent driver leaves the car in neutral, and the parking brakes are defective. If the car rolls from rest down the incline with a constant acceleration of 4.00 m/s^2, find (a) the speed of the car just as it reaches the cliff and the time it takes to get there, (b) the velocity of the car just as it lands in the ocean, (c) the total time the car is in motion, and (d) the position of the car relative to the base of the cliff just as it lands in the ocean.

Solution

(a)   While on the incline,

$$v^2 = v_0^2 + 2a(\Delta x)$$

In this case, $v_0 = 0$

and $v = \sqrt{(2)(4.00 \text{ m}/\text{s}^2)(50.0 \text{ m})}$

$= 20.0 \text{ m}/\text{s} \quad \lozenge$

Also, $v = v_0 + at$

so $t = \dfrac{v - v_0}{a} = \dfrac{20.0 \text{ m/s} - 0}{4.00 \text{ m/s}^2} = 5.00 \text{ s} \quad \lozenge$

(b)   Initial "free-flight" conditions give us

$$v_{x0} = (20.0 \text{ m}/\text{s}) \cos 37.0° = 16.0 \text{ m}/\text{s}$$

at the edge of the cliff

$$v_{y0} = (-20.0 \text{ m}/\text{s}) \sin 37.0° = -12.0 \text{ m}/\text{s}$$

$v_x = v_{x0}$ since $a_x = 0$

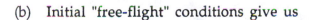

$$v_y = \pm\sqrt{2a_y(\Delta y) + v_{y0}^2} = \pm\sqrt{2(-9.80 \text{ m}/\text{s}^2)(-30.0 \text{ m}) + (-12.0 \text{ m}/\text{s})^2} = -27.1 \text{ m}/\text{s}$$

Since the velocity of the car is downward, the negative root is the physically meaningful solution.

$$v = \sqrt{v_x^2 + v_y^2} = \sqrt{(16.0 \text{ m}/\text{s})^2 + (-27.1 \text{ m}/\text{s})^2} = 31.4 \text{ m}/\text{s} \quad \lozenge$$

(c) $t_1 = 5.00$ s

$$t_2 = \frac{v_y - v_{y0}}{a_y} = \frac{(-27.1 \text{ m/s}) - (-12.0 \text{ m/s})}{(-9.80 \text{ m/s}^2)} = 1.54 \text{ s}$$

$t = t_1 + t_2 = 6.54$ s ◊

(d) $\Delta x = v_{x0}t_2 = (16.0 \text{ m/s})(1.54 \text{ s}) = 24.6 \text{ m}$ ◊

---

57. A batter hits a pitched baseball 1.00 m above the ground, imparting to the ball a speed of 40.0 m/s. The resulting line drive is caught on the fly by the left fielder 60.0 m from home plate with his glove 1.00 m above the ground. If the shortstop, 45.0 m from home plate and in line with the drive, were to jump straight up to make the catch instead of allowing the left fielder to make the play, how high above the ground would his glove have to be?

Solution   We define o and f to be points at the batter and the left fielder, respectively, and set y to be equal to the difference in height between points o and f.

We will use the equation of motion to find the initial angle of flight:

$$y = 0; \qquad v_{y0} = (40.0 \text{ m/s}) \sin \theta_0; \qquad a_y = -9.80 \text{ m/s}^2$$
$$x = 60.0 \text{ m}; \qquad v_{x0} = (40.0 \text{ m/s}) \cos \theta_0; \qquad a_x = 0$$

$y = v_{y0}t + \frac{1}{2}a_y t^2$ gives $0 = (40.0 \text{ m/s}) \sin \theta_0 t + \frac{1}{2}(-9.80 \text{ m/s}^2)t^2$

The root $t = 0$ represents when the ball left the bat; the time of flight is the other root,

$$t = \frac{2(40.0 \text{ m/s}) \sin \theta_0}{9.80 \text{ m/s}^2}$$

Substituting this into

$$x = v_{x0}t = (40.0 \text{ m/s}) \cos \theta_0 \frac{(2)(40.0 \text{ m/s}) \sin \theta_0}{9.80 \text{ m/s}^2} = 60.0 \text{ m}$$

gives $$2(\cos \theta_0)(\sin \theta_0) = \sin 2\theta_0 = \frac{(60.0 \text{ m})(9.80 \text{ m/s}^2)}{(40.0 \text{ m/s})^2} = 0.368$$

$$2\theta_0 = 21.6°; \qquad \theta_0 = 10.8°$$

Now, we redefine points o and f to be at the batter and above the shortstop:

$$v_{y0} = (40.0 \text{ m/s}) \sin (10.8°) \qquad a_y = -9.80 \text{ m/s}^2$$

and $\qquad x = 45.0 \text{ m} \qquad v_{x0} = (40.0 \text{ m/s}) \cos (10.8°) \qquad a_x = 0$

From $x = v_{x0}t$, $\qquad t = \dfrac{x}{v_{x0}} = \dfrac{45.0 \text{ m}}{(40.0 \text{ m/s}) \cos (10.8°)} = 1.15 \text{ s}$

So, $\qquad y = v_{y0}t + \frac{1}{2}a_y t^2$

or $\qquad y = (40.0 \text{ m/s})(\sin 10.8°)(1.15 \text{ s}) + \frac{1}{2}(-9.80 \text{ m/s}^2)(1.15 \text{ s})^2 = 2.14 \text{ m}$

The height above the ground would be $h = 3.14 \text{ m}$ ◊

---

63. A dart gun is fired while being held horizontally at a height of 1.00 m above ground level. With the gun at rest relative to the ground, the dart from the gun travels a horizontal distance of 5.00 m. A child holds the same gun in a horizontal position while sliding down a 45.0° incline at a constant speed of 2.00 m/s. How far will the dart travel if the gun is fired when it is 1.00 m above the ground?

Solution Find the initial velocity of the dart when shot horizontally, from one meter above the ground.

$$v_{y0} = 0; \quad y = -1.00 \text{ m}; \quad x_r = 5.00 \text{ m}$$

$$y = v_{y0}t - \frac{1}{2}gt^2 \quad \text{thus} \quad t = \sqrt{-2y/g} \quad \text{and} \quad x = v_{x0}t$$

Thus, $\qquad v_{x0} = \dfrac{x}{t} = \dfrac{x}{\sqrt{-2y/g}} = x\sqrt{\dfrac{g}{-2y}} = 5\sqrt{\dfrac{9.80 \text{ m/s}^2}{2(1.00 \text{ m})}} = 11.1 \text{ m/s}$

In the second case, there are additional horizontal and vertical velocity components of $[2.00 \cos 45.0°]$ m / s and $[2.00 \sin 45.0°]$ m / s, respectively:

$$v_{y0} = (2.00 \cos 45.0°) \text{ m / s} = -1.41 \text{ m / s}$$

$$v_{x0} = (2.00 \sin 45.0°) \text{ m / s} + 11.1 \text{ m / s} = 12.5 \text{ m / s}$$

The time of flight is calculated from $y = v_{y0}t - \frac{1}{2}gt^2$:

$$-1.00 \text{ m} = (-1.41 \text{ m / s}) - \frac{1}{2}(9.80 \text{ m / s}^2)t^2$$

or
$$(4.90 \text{ m / s}^2)t^2 + (1.41 \text{ m / s})t - 1.00 \text{ m} = 0$$

This can be solved by using the quadratic formula:

$$t = \frac{-1.41 \text{ m / s} \pm \sqrt{(1.41 \text{ m / s})^2 - (4)(4.90 \text{ m / s}^2)(-1.00 \text{ m})}}{(2)(4.90 \text{ m / s}^2)}$$

$$t = 0.330 \text{ s}$$

Finally, we calculate the horizontal distance of travel to be

$$x = v_{x0}t = (12.5 \text{ m / s})(0.330 \text{ s}) = 4.13 \text{ m} \quad \Diamond$$

---

66. A home run in a baseball game is hit in such a way that the ball just clears a wall 21.0 m high, located 130 m from home plate. The ball is hit at an angle of 35.0° to the horizontal, and air resistance is negligible. Find (a) the initial speed of the ball, (b) the time it takes the ball to reach the wall, and (c) the velocity components and the speed of the ball when it reaches the wall. (Assume the ball is hit at a height of 1.00 m above the ground.)

Solution

Let the initial speed of the ball be v_0.

Also define:

$x_0 = 0$ $y_0 = 1.00$ m $\theta_0 = 35.0°$

When $x = 130$ m, $y = 21.0$ m

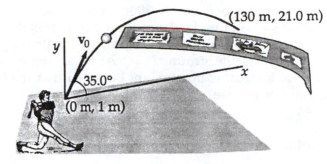

(130 m, 21.0 m)

$$x = x_0 + v_{x0}t = v_0 t \ \cos 35.0°$$

$$v_0 t \ \cos 35.0° = 130 \ \text{m}$$

$$v_{x0} = v_0 \ \cos 35.0°$$
$$v_{y0} = v_0 \ \sin 35.0°$$

and $$v_0 t = \frac{130 \ \text{m}}{\cos \ 35.0°} = 158.7 \ \text{m}$$

Next, $$y = y_0 + v_{y0}t - \tfrac{1}{2}gt^2 = 1.00 \ \text{m} + (v_0 \ \sin \ 35.0°)t - \left(\tfrac{1}{2}\right)(9.80 \ \text{m/s}^2)t^2$$

Substituting for $v_0 t$ in the second term,

$$20 \ \text{m} = (158.7 \ \text{m})(\sin 35.0°) - \left(\tfrac{1}{2}\right)(9.80 \ \text{m/s}^2)t^2$$

and   (a) $$t = \sqrt{\frac{71.0 \ \text{m}}{4.90 \ \text{m/s}^2}} = 3.81 \ \text{s} \quad ◊$$

Therefore,   (b) $$v_0 = \frac{158.7 \ \text{m}}{3.81 \ \text{s}} = 41.7 \ \text{m/s} \quad ◊$$

(c) $v_x = v_{x0} = v_0 \ \cos \ 35.0° = (41.7 \ \text{m/s}) \ \cos \ 35.0° = 34.1 \ \text{m/s} \quad ◊$

$v_y = v_{y0} - gt = v_0 \ \sin \ 35.0° - gt = (23.9 \ \text{m/s}) - \left(9.80 \ \text{m/s}^2\right)t$

At $t = 3.81$ s $v_y = -13.4$ m/s   ◊

and $$|v| = \sqrt{v_x^2 + v_y^2} = \sqrt{34.1^2 + (-13.4)^2} = 36.6 \ \text{m/s} \quad ◊$$

69. A bomber flies horizontally with a speed of 275 m/s relative to the ground. The altitude of the bomber is 3000 m and the terrain is level. Neglect the effects of air resistance. (a) How far from the point vertically under the point of release does a bomb hit the ground? (b) If the plane maintains its original course and speed, where is it when the bomb hits the ground? (c) At what angle from the vertical at the point of release must the telescopic bomb sight be set so that the bomb hits the target seen in the sight at the time of release?

Solution

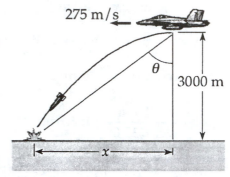

275 m/s

(a)   In horizontal flight, $v_{0y} = 0$ and $v_{0x} = 275$ m/s

Therefore, (taking $y_0 = 0$), $\qquad y = -\tfrac{1}{2}gt^2$

$\qquad\qquad$ and $\qquad\quad x = v_0 t$

Eliminate t between these two equations to find

$$x = v_0\sqrt{\frac{-2y}{g}} = (275 \text{ m/s})\sqrt{\frac{(-2)(-3000 \text{ m})}{9.80 \text{ m/s}^2}} = 6800 \text{ m} \quad \Diamond$$

(b)   The plane and the bomb have the same constant *horizontal* velocity. Therefore, the plane will be 3000 m above the bomb at impact, and 6800 m from the point of release. $\Diamond$

(c)   If θ is the angle between the vertical and the direct line of sight to the target at the time of release,

$$\theta = \tan^{-1}\left(\frac{x}{y}\right) = \tan^{-1}\left(\frac{6800}{3000}\right) = 66.2° \quad \Diamond$$

73. A rifle has a maximum range of 500 m. (a) For what angles of elevation would the range be 350 m? What is the range when the bullet leaves the rifle (b) at 14.0°? (c) at 76.0°?

Solution

(a) Let v_0 be the initial speed of the bullet. The range is given by $R = \dfrac{v_0{}^2 \sin 2\theta_0}{g}$,

so the maximum range will be $R_{max} = \dfrac{v_0^2}{g}$

So, $v_0 = \sqrt{R_{max}g} = \sqrt{(500 \text{ m})(9.80 \text{ m/s}^2)} = 70.0 \text{ m/s}$

If $R = 350$ m and the angle of projection of the bullet is θ_0, then

$$R = \frac{v_0^2}{g} \sin 2\theta_0$$

$$\sin 2\theta_0 = \frac{gR}{v_0^2} = \frac{(9.80 \text{ m/s}^2)(350 \text{ m})}{(70.0 \text{ m/s})^2} = 0.700$$

$$2\theta_0 = 44.4° \quad \text{and} \quad \theta_0 = 22.2° \quad \lozenge$$

or $$2\theta_0' = 180° - 44.4° \quad \text{and} \quad \theta_0' = 67.8° \quad \lozenge$$

(Note: *complementary angles* give the same range.)

(b) $R = \dfrac{(70.0)^2}{g} \sin 28.0° = 235 \text{ m} \quad \lozenge$

(c) $R = \dfrac{(70.0)^2}{g} \sin 152° = 235 \text{ m} \text{ (same)} \quad \lozenge$

77. Two soccer players, Mary and Jane, begin running from approximately the same point at the same time. Mary runs in an easterly direction at 4.0 m/s, while Jane takes off in a direction 60.0° north of east at 5.4 m/s. (a) How long is it before they are 25 m apart? (b) What is the velocity of Jane relative to Mary? (c) How far apart are they after 4.0 s?

Solution

Define **i** to be directed East, and **j** to be directed North.

According to the figure, set

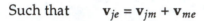

$\mathbf{v}_{je}$ = velocity of Jane, relative to the Earth

$\mathbf{v}_{me}$ = velocity of Mary, relative to the Earth

$\mathbf{v}_{jm}$ = velocity of Jane, relative to Mary,

Such that $\mathbf{v}_{je} = \mathbf{v}_{jm} + \mathbf{v}_{me}$

Solve for part (b) first. By the figure,

$$\mathbf{v}_{je} = [5.4(\cos 60.0°)\mathbf{i} + 5.4(\sin 60.0°)\mathbf{j}]\ \text{m/s}$$

$$= (2.7\mathbf{i} + 4.68\mathbf{j})\ \text{m/s}$$

and $\mathbf{v}_{me} = 4.0\mathbf{i}\ \text{m/s}$

So, $\mathbf{v}_{jm} = (-1.30\mathbf{i} + 4.68\mathbf{j})\ \text{m/s}$ ◊

The distance between the two players increases at a rate of $|\mathbf{v}_{jm}|$:

$$\left|\mathbf{v}_{jm}\right| = \sqrt{(1.30)^2 + (4.68)^2}\ \text{m/s} = 4.86\ \text{m/s}$$

(a) Therefore, $$t = \frac{d}{v_{jm}} = \frac{25\ \text{m}}{4.86\ \text{m/s}} = 5.15\ \text{s}$$ ◊

(c) After 4 s, $$d = v_{jm}t = (4.86\ \text{m/s})(4.0\ \text{s}) = 19.4\ \text{m apart}$$ ◊

79. A skier leaves the ramp of a ski jump with a velocity of 10.0 m/s, 15.0° above the horizontal, as in Figure P4.79. The slope is inclined at 50.0°, and air resistance is negligible. Find (a) the distance that the jumper lands down the slope and (b) the velocity components just before landing. (How do you think the results might be affected if air resistance were included? Note that jumpers lean forward in the shape of an airfoil with their hands at their sides to increase their distance. Why does this work?)

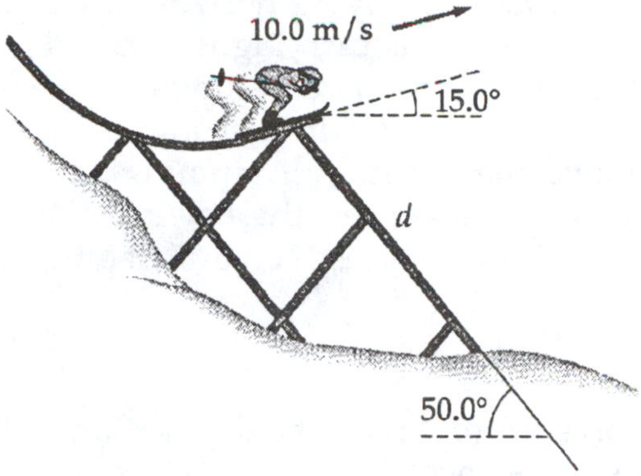

10.0 m/s

15.0°

d

50.0°

$$x = v_{x0}t \qquad\qquad v_x = v_{x0}$$
$$y = v_{y0}t - \tfrac{1}{2}gt^2 \qquad v_y = v_{y0} - gt$$

Figure P4.79

Solution

$v_0 = 10.0$ m/s at 15.0°

$v_{x0} = v_0 \cos \theta_0 = (10.0 \text{ m/s}) \cos 15.0° = 9.66$ m/s

$v_{y0} = v_0 \sin \theta_0 = (10.0 \text{ m/s}) \sin 15.0° = 2.59$ m/s

(a) $x = v_{x0}t = (9.66 \text{ m/s})t$

$y = v_{y0}t - \tfrac{1}{2}gt^2 = (2.59 \text{ m/s})t - (4.90 \text{ m/s}^2)t^2$

The skier hits the slope when $\dfrac{y}{x} = \tan(-50.0°) = -\tan 50.0°$

So, $y = -x \tan 50.0°$

or $(2.59 \text{ m/s})t - (4.90 \text{ m/s}^2)t^2 = (9.66 \text{ m/s})t(-1.19)$

Simplifying, $-4.90t^2 + 14.1t = 0$

so $t = \dfrac{14.1 \text{ m/s}}{4.90 \text{ m/s}^2} = 2.88$ s, and $x = (9.66 \text{ m/s})t = 27.8$ m.

By Figure P4.79, $d = \dfrac{x}{\cos 50.0°} = \dfrac{27.8 \text{ m}}{\cos 50.0°} = 43.2$ m ◊

(b) $v_x = v_{x0} = 9.66$ m/s ◊

$v_y = v_{y0} - gt = 2.59 \text{ m/s} - (9.8 \text{ m/s}^2)t$

When $t = 2.88$ s, $v_x = 9.66$ m/s ◊ $v_y = -25.6$ m/s ◊

83. A hawk is flying horizontally at 10.0 m/s in a straight line 200 m above the ground. A mouse it was carrying is released from its grasp. The hawk continues on its path at the same speed for two seconds before attempting to retrieve its prey. To accomplish the retrieval, it dives in a straight line at constant speed and recaptures the mouse 3.0 m above the ground. Assuming no air resistance (a) find the diving speed of the hawk. (b) What angle did the hawk make with the horizontal during its descent? (c) For how long did the mouse "enjoy" free flight?

Solution

Solve for (c) first. The mouse falls with an initial speed $v_{y0} = 0$ and acceleration $a_y = g$, for a distance of

$$h = 200 \text{ m} - 3.00 \text{ m} = 197 \text{ m}$$

Solving the equation $h = v_{y0} + \frac{1}{2}a_y t^2$ for t, we have:

$$t = \sqrt{\frac{2(197 \text{ m})}{9.80 \text{ m}/\text{s}^2}} = 6.34 \text{ s} \qquad ◊$$

To find the diving speed of the hawk, we must first calculate the horizontal distance that the mouse travels:

$$x_m = v_{x0}t = (10.0 \text{ m}/\text{s})(6.34 \text{ s}) = 63.4 \text{ m}$$

and by the hawk in the first 2.00 sec: $\quad x_{h0} = (2.00 \text{ s})(10.0 \text{ m}/\text{s}) = 20.0 \text{ m}$

Now, the hawk must dive in the remaining 4.34 seconds a vertical distance of 197 m, and a horizontal distance of

$$d_x = x_m - x_{h0} = 63.4 \text{ m} - 20.0 \text{ m} = 43.4 \text{ m}$$

for a total distance of $\qquad d = \sqrt{h^2 + d_x^2} = 202 \text{ m}$

at a speed of $\qquad v_x = \frac{d}{t - 2.00 \text{ s}} = 46.5 \text{ m}/\text{s} \qquad ◊$

and an angle of $\qquad \theta = \tan^{-1}\left[\frac{h}{d_x}\right] = 77.6° \text{ below the horizontal.} \qquad ◊$

Chapter 5

The Laws of Motion

THE LAWS OF MOTION

INTRODUCTION

In Chapters 2 and 4, we described the motion of particles based on the definitions of displacement, velocity, and acceleration. In this chapter, we use the concepts of force and mass to describe the change in motion of particles. We then discuss the three basic laws of motion, which are based on experimental observations and were formulated more than three centuries ago by Newton.

Classical mechanics describes the relationship between the motion of a body and the forces acting on that body. Classical mechanics deals only with objects that (a) are large compared with the dimensions of atoms and (b) move at speeds that are much less than the speed of light.

We learn in this chapter how it is possible to describe an object's acceleration in terms of the resultant force acting on the object and its mass. This force represents the interaction of the object with its environment. Mass is a measure of the object's inertia, that is, its tendency to resist an acceleration when a force acts on it.

We also discuss force laws, which describe the quantitative method of calculating the force on an object if its environment is known. These force laws, together with the laws of motion, are the foundations of classical mechanics.

NOTES FROM SELECTED CHAPTER SECTIONS

5.1 The Concept of Force

Equilibrium is the condition under which the *net force* (vector sum of all forces) acting on an object is zero. An object in equilibrium has a zero acceleration (velocity is constant or equals zero).

Fundamental forces in nature are: (1) gravitational (attractive forces between objects due to their masses), (2) electromagnetic forces (between electric charges at rest or in motion), (3) strong nuclear forces (between subatomic particles), and (4) weak nuclear forces (accompanying the process of radioactive decay).

Classical physics is concerned with contact forces (which are the result of physical contact between two or more objects) and field forces (which act through empty space and do not involve physical contact).

5.2 Newton's First Law and Inertial Frames

Newton's first law is called the *law of inertia* and states that an object at rest will remain at rest and an object in motion will remain in motion with a constant velocity unless acted on by a *net external force*.

Mass and *weight* are two different physical quantities. The weight of a body is equal to the *force of gravity* acting on the body and varies with location in the Earth's gravitational field. Mass is an *inherent property* of a body and is a measure of the body's inertia (resistance to change in its state of motion). The SI unit of mass is the *kilogram* and the unit of weight is the *newton*.

5.3 Inertial Mass

Mass and *weight* are two different physical quantities. The weight of a body is equal to the *force of gravity* acting on the body and varies with location in the Earth's gravitational field. Mass is an *inherent property* of a body and is a measure of the body's inertia (resistance to change in its state of motion). The SI unit of mass is the *kilogram* and the unit of weight is the *newton*.

5.4 Newton's Second Law

Newton's second law, the *law of acceleration*, states that the acceleration of an object is directly proportional to the resultant force acting on it and inversely proportional to its mass. The direction of the acceleration is the direction of the net force.

5.6 Newton's Third Law

Newton's third law, the *law of action-reaction*, states that when two bodies interact, the force which body "A" exerts on body "B" (the *action force*) is equal in magnitude and opposite in direction to the force which body "B" exerts on body "A" (the *reaction force*). A consequence of the third law is that forces occur in *pairs*. Remember that the action force and the reaction force act on *different objects*.

5.7 Some Applications of Newton's Laws

Construction of a *free-body diagram* is an important step in the application of Newton's laws of motion to the solution of problems involving bodies in equilibrium or accelerating under the action of external forces. The diagram should include an arrow labeled to identify each of the external forces acting on the body whose motion (or condition of equilibrium) is to be studied. Forces which are the *reactions* to these external forces must *not* be included. When a system consists of more than one body or mass, you must construct a free-body diagram for each mass.

5.8 Forces of Friction

Experimentally, one finds that, to a good approximation, both $f_{s,\max}$ and f_k are *proportional to the normal force acting on an object.* The experimental observations can be summarized by the following empirical laws of friction:

- The direction of the force of static friction between any two surfaces in contact is opposite the direction of any applied force and can have values

$$f_s \leq \mu_s n \tag{5.9}$$

where the dimensionless constant μ_s is called the coefficient of static friction and n is the magnitude of the normal force. The equality in Equation 5.9 holds when the block is on the verge of slipping, that is, when $f_s = f_{s,\max} = \mu_s n$. The inequality holds when the applied force is less than this value.

- The direction of the force of kinetic friction acting on an object is opposite the direction of its motion and is given by

$$f_k = \mu_k n \tag{5.10}$$

where μ_k is the coefficient of kinetic friction.

- The values of μ_k and μ_s depend on the nature of the surfaces, but μ_k is generally less than μ_s.

- The coefficients of friction are nearly independent of the area of contact between the surfaces.

EQUATIONS AND CONCEPTS

A quantitative measurement of mass (the term used to measure inertia) can be made by comparing the accelerations that a given force will produce on different bodies. If a given force acting on a body of mass m_1 produces an acceleration a_1 and the same force acting on a body of mass m_2 produces an acceleration a_2, the ratio of the two masses equals the inverse of the ratio of the two accelerations.

$$\frac{m_1}{m_2} \equiv \frac{a_2}{a_1} \tag{5.1}$$

The acceleration of an object is proportional to the resultant force acting on it and inversely proportional to its mass. This is a statement of Newton's second law.

$$\Sigma\mathbf{F} = m\mathbf{a} \quad \text{or} \quad \mathbf{a} = \frac{\Sigma\mathbf{F}}{m} \tag{5.2}$$

When several forces act on an object, it is often convenient to write the equation expressing Newton's second law as component equations. The orientation of the coordinate system can often be chosen so that the object has a nonzero acceleration along only one direction.

$$\Sigma F_x = ma_x \tag{5.3}$$

$$\Sigma F_y = ma_y$$

$$\Sigma F_z = ma_z$$

Calculations with Equation 5.4 must be made using a consistent set of units for the quantities' force, mass, and acceleration. The SI unit of force is the newton (N), defined as the force that, when acting on a 1-kg mass, produces an acceleration of 1 m/s^2.

$$1 \text{ N} \equiv 1 \text{ kg·m/s}^2 \tag{5.4}$$

$$1 \text{ dyne} \equiv 1 \text{ g·cm/s}^2 \tag{5.5}$$

$$1 \text{ lb} \equiv 4.448 \text{ N} \tag{5.6}$$

Weight is not an inherent property of a body, but depends on the local value of g and varies with location.

$$w = mg \tag{5.7}$$

Newton's third law states that the force exerted on body 1 by body 2 is equal in magnitude and opposite in direction to the force exerted on body 2 by body 1.

$$\mathbf{F}_{12} = -\mathbf{F}_{21} \tag{5.8}$$

The force of static friction between two surfaces in contact but not in motion, relative to each other, cannot be greater than $\mu_s n$, where n is the normal (perpendicular) force between the two surfaces and μ_s (coefficient of static friction) is a dimensionless constant which depends on the nature of the pair of surfaces.

$$f_s \leq \mu_s n \tag{5.9}$$

When two surfaces are in relative motion, the force of kinetic friction on each body is directed opposite to the direction of motion of the body.

$$f_k = \mu_k n \qquad\qquad (5.10)$$

SUGGESTIONS, SKILLS, AND STRATEGIES

The following procedure is recommended for problems involving objects in equilibrium:

- Make a sketch of the object under consideration.

- Draw a free-body diagram and label all external forces acting on the object. Try to guess the correct direction for each force. If you select a direction that leads to a negative sign in your solution for a force, do not be alarmed; this merely means that the direction of the force is the opposite of what you assumed.

- Resolve all forces into x and y components, choosing a convenient coordinate system.

- Use the equations $\Sigma F_x = 0$ and $\Sigma F_y = 0$. Remember to keep track of the signs of the various force components.

- Application of Step 4 above leads to a set of equations with several unknowns. All that is left is to solve the simultaneous equations for the unknowns in terms of the known quantities.

The following procedure is recommended when dealing with problems involving the application of Newton's second law:

- Draw a simple, neat diagram of the system.

- Isolate the object of interest whose motion is being analyzed. Draw a free-body diagram for this object; that is, a diagram showing *all external forces acting on the object.* For systems containing more than one object, draw *separate* diagrams for each object. Do not include forces that the object exerts on its surroundings.

- Establish convenient coordinate axes for each body and find the components of the forces along these axes.

- Apply Newton's second law, $\Sigma \mathbf{F} = m\mathbf{a}$, in the x and y directions for each object under consideration.

- Solve the component equations for the unknowns. Remember that you must have as many independent equations as you have unknowns in order to obtain a complete solution.

- Often in solving such problems, one must also use the equations of kinematics (motion with constant acceleration) to find all the unknowns.

REVIEW CHECKLIST

▷ State in your own words a description of Newton's laws of motion, recall physical examples of each law, and identify the action-reaction force pairs in a multiple-body interaction problem as specified by Newton's third law.

▷ Express the normal force in terms of other forces acting on an object and write out the equation which relates the coefficient of friction, force of friction and normal force between an object and surface on which it rests or moves.

▷ Apply Newton's laws of motion to various mechanical systems using the recommended procedure discussed in Section 5.7. Most important, you should identify all external forces acting on the system, draw the *correct* free-body diagrams which apply to each body of the system, and apply Newton's second law, $F = ma$, in *component* form.

▷ Apply the equations of kinematics (which involve the quantities' displacement, velocity, and acceleration) as described in Chapter 2 along with those methods and equations of Chapter 5 (involving mass, force, and acceleration) to the solutions of problems where *both* the kinematical and dynamic aspects are present.

▷ Be familiar with solving several linear equations simultaneously for the unknown quantities. Recall that you must have as many *independent* equations as you have unknowns.

SOLUTIONS TO SELECTED END-OF-CHAPTER PROBLEMS

7. A 3.0-kg mass undergoes an acceleration given by $a = (2.0i + 5.0j)$ m/s^2. Find the resultant force, F, and its magnitude.

Solution $F = ma = (3.0 \text{ kg})(2.0i + 5.0j) \text{ m/s}^2$; $\qquad$ $F = (6.00 \text{ i} + 15.0 \text{ j}) \text{ N}$ ◊

$$|F| = \sqrt{(F_x)^2 + (F_y)^2} = \sqrt{(6.0 \text{ N})^2 + (15.0 \text{ N})^2} \; ; \qquad |F| = 16.2 \text{ N} \qquad ◊$$

13. On planet X, an object weighs 12 N. On planet B where the magnitude of the free-fall acceleration is $1.6g$, the object weighs 27 N. What is the mass of the object and what is the free-fall acceleration (in m/s^2) on planet X?

Solution $w = mg$ at any location

On planet B, $w_B = mg_B$

$$m = \frac{w_B}{g_B} = \frac{27\ N}{(1.6)(9.80\ m/s^2)} = 1.72\ kg \quad \Diamond$$

On planet X, $w_x = mg_x$

$$g_x = \frac{w_x}{m} = \frac{12\ N}{1.72\ kg} = 6.98\ m/s^2 \quad \Diamond$$

---

18. Two forces F_1 and F_2 act on a 5.00-kg mass. If $F_1 = 20.0$ N and $F_2 = 15.0$ N, find the acceleration in (a) and (b) of Figure P5.18.

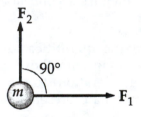

Figure P5.18 (a)

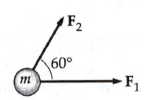

Figure P5.18 (b)

Solution $m = 5.00$ kg

(a) $\Sigma F = F_1 + F_2 = (20.0i + 15.0j)$ N

$$a = \frac{\Sigma F}{m} = (4.00i + 3.00j)\ m/s^2 \quad \Diamond; \qquad |a| = \sqrt{\left(4.00\ m/s^2\right)^2 + \left(3.00\ m/s^2\right)^2} = 5.00\ m/s^2 \quad \Diamond$$

(b) $\Sigma F = F_1 + F_2 = [20.0i + (15.0\cos 60°\ i + 15.0\sin 60°\ j)]$ N $= (27.5i + 13.0j)$ N

$$a = \frac{\Sigma F}{m} = (5.50i + 2.60j)\ m/s^2 \quad \Diamond; \qquad a = \sqrt{\left(5.50\ m/s^2\right)^2 + \left(2.60\ m/s^2\right)^2} = 6.08\ m/s^2 \quad \Diamond$$

---

21. A 4.0-kg object has a velocity of 3.0i m/s at one instant. Eight seconds later, its velocity is (8.0i + 10.0j) m/s. Assuming the object was subject to a constant net force, find (a) the components of the force and (b) its magnitude.

Solution

We are given $m = 4.0$ kg, $\mathbf{v_o} = 3.0$i m/s, $\mathbf{v} = (8$i $+ 10$j$)$ m/s, $t = 8.0$ s

The object's acceleration is

$$\mathbf{a} = \frac{\Delta \mathbf{v}}{t} = \frac{\mathbf{v} - \mathbf{v_o}}{t} = \frac{(8.0\mathbf{i} + 10.0\mathbf{j} - 3.0\mathbf{i})\ m/s}{8.0} = \frac{(5.0\mathbf{i} + 10.0\mathbf{j})}{8.0\ s}\ m/s^2$$

(a)   So the total force on it must be

$$\Sigma \mathbf{F} = m\mathbf{a} = \sum \mathbf{F} = ma = (4.0\ kg)\frac{(5.0\mathbf{i} + 10.0\mathbf{j})}{8.0}\ m/s^2 = (2.5\mathbf{i} + 5.0\mathbf{j})\ N \quad \Diamond$$

(b)   With magnitude, $\left|\sum \mathbf{F}\right| = \sqrt{(2.5\ N)^2 + (5.0\ N)^2} = 5.59\ N \quad \Diamond$

---

24. An electron of mass 9.1×10^{-31} kg has an initial speed of 3.0×10^5 m/s. It travels in a straight line, and its speed increases to 7.0×10^5 m/s in a distance of 5.0 cm. Assuming its acceleration is constant, (a) determine the force on the electron and (b) compare this force with the weight of the electron.

Solution $F = ma$ and $v^2 = v_o^2 + 2ax$ or $a = \dfrac{(v^2 - v_o^2)}{2x}$

(a)   Therefore,

$$F = \frac{m(v^2 - v_o^2)}{2x} = \frac{(9.1 \times 10^{-31}\ kg)\left[(7.0 \times 10^5\ m/s)^2 - (3.0 \times 10^5\ m/s)^2\right]}{(2)(0.050\ m)} = 3.64 \times 10^{-18}\ N \quad \Diamond$$

(b) The weight of the electron is

$$W = mg = \left(9.1 \times 10^{-31} \text{ kg}\right)\left(9.80 \text{ m/s}^2\right) = 8.92 \times 10^{-30} \text{ N}$$

The accelerating force is approximately 10^{11} times the weight of the electron. ◊

---

27. A 2.0-kg mass accelerates at 8.5 m/s² in a direction 30.0° north of east (Fig. P5.27). One of the two forces acting on the mass has a magnitude of 11 N and is directed north. Determine the magnitude of the second force.

Solution   Choose directions east along **i** and north along **j**.

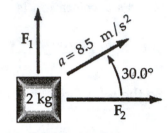

Figure P5.27

$a = [(8.5 \cos 30°)\mathbf{i} + (8.5 \sin 30°)\mathbf{j}] \text{ m/s}^2$

$a = (7.36\mathbf{i} + 4.25\mathbf{j}) \text{ m/s}^2$

$\sum F = m\,a = (14.7\mathbf{i} + 8.50\mathbf{j}) \text{ N}$

$\sum F = F_1 + F_2$

$F_2 = \sum F - F_1 = (14.7\mathbf{i} + 8.50\mathbf{j} - 11.0\mathbf{j}) \text{ N}$

$F_2 = (14.7\mathbf{i} - 2.50\mathbf{j}) \text{ N}$

$|F_2| = \sqrt{(14.7 \text{ N})^2 + (2.50 \text{ N})^2} = 14.9 \text{ N}$ ◊

---

31. A bag of cement hangs from three wires as shown in Figure P5.31. Two of the wires make angles θ_1 and θ_2 with the horizontal. If the system is in equilibrium, (a) show that

$$T_1 = \frac{w \cos \theta_2}{\sin (\theta_1 + \theta_2)}$$

(b) Given that $w = 325$ N, $\theta_1 = 10°$ and $\theta_2 = 25°$, find the tensions T_1, T_2, and T_3 in the wires.

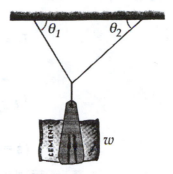

Figure P5.31

Solution

(a) Draw a free-body diagram for the point where the three ropes are joined. Choose the x axis to be horizontal and apply Newton's second law in component form.

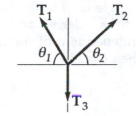

$$\Sigma F_x = 0; \quad T_2 \cos \theta_2 - T_1 \cos \theta_1 = 0 \qquad (1)$$

$$\Sigma F_y = 0; \quad T_2 \sin \theta_2 + T_1 \sin \theta_1 - w = 0 \qquad (2)$$

Solve equation (1) for $T_2 = \dfrac{T_1 \cos \theta_1}{\cos \theta_2}$

Substitute this expression for T_2 into Equation (2):

$$\left(\frac{T_1 \cos \theta_1}{\cos \theta_2} \right) \sin \theta_2 + T_1 \sin \theta_1 = w$$

Solve for $T_1 = \dfrac{w \cos \theta_2}{\cos \theta_1 \sin \theta_2 + \sin \theta_1 \cos \theta_2}$

Use trigonometric identity, $\sin (\theta_1 + \theta_2) = \cos \theta_1 \sin \theta_2 + \sin \theta_1 \cos \theta_2$, to find

$$T_1 = \frac{w \cos \theta_2}{\sin (\theta_1 + \theta_2)} \qquad \Diamond$$

(b)   Use the given values to find

$$T_1 = \frac{(325 \text{ N}) \cos 25°}{\sin (10° + 25°)} = 514 \text{ N} \quad \lozenge$$

$$T_2 = T_1 \left(\frac{\cos \theta_1}{\cos \theta_2} \right) = 514 \text{ N} \left(\frac{\cos 10°}{\cos 25°} \right) = 558 \text{ N} \quad \lozenge$$

$$T_3 = w = 325 \text{ N} \quad \lozenge$$

---

35. A simple accelerometer is constructed by suspending a mass m from a string of length L that is tied to the top of a cart. As the cart is accelerated, the string-mass system makes an angle of θ with the vertical. (a) Assuming that the string mass is negligible compared to m, derive an expression for the cart's acceleration in terms of θ and show that it is independent of the mass m and the length L. (b) Determine the acceleration of the cart when $\theta = 23°$.

Solution

(a) Use a free-body diagram to write out Newton's second law in component form.

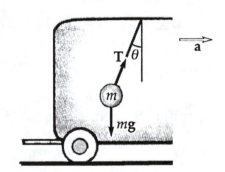

$$\Sigma F_x = T \sin \theta = ma$$

$$\Sigma F_y = T \cos \theta - mg = 0$$

Eliminate T between the two equations to find

$$\tan \theta = \frac{a}{g} \quad \text{or} \quad a = g \tan \theta \quad \lozenge$$

(b) When $\theta = 23°$,

$$a = (9.8 \text{ m/s}^2)(\tan 23°) = 4.16 \text{ m/s}^2 \quad \lozenge$$

---

37. In the system shown in Figure P5.37, a horizontal force F_x acts on the 8.00-kg mass. (a) For what values of F_x does the 2.00-kg mass accelerate upward? (b) For what values of F_x is the tension in the cord zero? (c) Plot the acceleration of the 8.00-kg mass versus F_x. Include values of F_x from -100 N to $+100$ N.

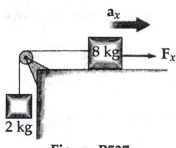

Figure P537

Solution The separate weights of the blocks are:

$$w = mg = (8.00 \text{ kg})(9.80 \text{ m/s}^2) = 78.4 \text{ N}$$

and $\quad (2.00 \text{ kg})(9.80 \text{ m/s}^2) = 19.6 \text{ N}$

We solve for part (b) first:

If $T = 0$, the 2.00-kg mass will accelerate downwards at 9.80 m/s². Likewise, the 8.00-kg mass must also accelerate to the left at 9.80 m/s².

We draw a free-body diagram for the 8.00-kg mass, taking the $+x$ direction to be directed to the right.

From $\Sigma F_x = ma_x$, $\qquad -T + F_x = (8.00 \text{ kg}) a_x$

If $T = 0$, and $a_x \leq -9.80$ m/s²,

(b) $\qquad\qquad F_x \leq -78.4 \text{ N} \quad \Diamond$

Note that:

$$a_x = \frac{F_x}{8.00 \text{ kg}} \quad \text{for all} \quad F_x \leq -78.4 \text{ N} \qquad (1)$$

(a) Now, we draw a free-body diagram for the 2.00-kg mass.

$\Sigma F_y = ma_y$, so $\qquad T - 19.6 \text{ N} = (2.00 \text{ kg}) a_y$

and $\quad T = (2.00 \text{ kg}) a_y + 19.6 \text{ N}$

Likewise, from part (b) above,

$\Sigma F_x = ma_x$ yields $\qquad\qquad F_x - T = (8.00\ \text{kg})\ a_x$

But if $T > 0$, then $a_x = a_y$.

Substituting for a_y and T, $\qquad F_x - [\ (2.00\ \text{kg})\ a_x + 19.6\ \text{N}] = (8.00\ \text{kg})\ a_x$

$$F_x = (10.00\ \text{kg})\ a_x + 19.6\ \text{N} \qquad\qquad\qquad (2)$$

If $a_x \geq 0$, then $\qquad\qquad F_x \geq 19.6\ \text{N} \qquad \Diamond$

(c) From equation (2), we calculate

$$a_x = \frac{F_x}{10.0\ \text{kg}} - 1.96\ \text{m}/\text{s}^2 \quad \text{for all} \quad F_x \geq -78.4\ \text{N} \qquad (3)$$

We can now graph equations (1) and (3) below:

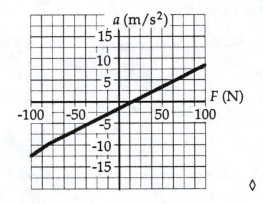

Some sample values are:

F_x	–50.0 N	0	50.0 N	100 N
a_x	–6.96 m/s^2	–1.96 m/s^2	3.04 m/s^2	8.04 m/s^2

43. A 72-kg man stands on a spring scale in an elevator. Starting from rest, the elevator ascends, attaining its maximum speed of 1.2 m/s in 0.80 s. It travels with this constant speed for the next 5.0 s. The elevator then undergoes a uniform acceleration in the negative y direction for 1.5 s and comes to rest. What does the spring scale register (a) before the elevator starts to move? (b) during the first 0.80 s? (c) while the elevator is traveling at constant speed? (d) during the time it is slowing down?

Solution Let S be the weight registered by the scale. In each case draw a free-body diagram showing forces acting on the man and apply Newton's second law.

(a) Before the elevator starts to move:

$$a = 0 \quad \text{and} \quad \Sigma F_y = S - mg = 0$$

$$S = mg = (72 \text{ kg})(9.80 \text{ m/s}^2) = 706 \text{ N} = 710 \text{ N} \quad \lozenge$$

(b) $a = \dfrac{v - v_0}{t} = \dfrac{1.2 \text{ m/s} - 0}{0.80 \text{ s}} = 1.5 \text{ m/s}^2$

During upward acceleration, the scale reading is $S = mg + ma$, because the man's feet are pressed against the floor by the elevator's upward acceleration, as well as by his weight. We also find this is true by applying the force equation, with S equal to the normal force:

$$\Sigma F_g = S - mg = ma$$

$$S = m(a + g) = (72 \text{ kg})(1.5 \text{ m/s}^2 + 9.80 \text{ m/s}^2) = 814 \text{ N} = 810 \text{ N} \quad \lozenge$$

(c) While traveling at constant speed, $a = 0$; and therefore the force registered by the scale will be the same as in part (a):

$$S = mg = 706 \text{ N} = 710 \text{ N} \quad \lozenge$$

(d) While slowing down:

$$a = \frac{v - v_0}{t} = \frac{0 - 1.2 \text{ m/s}}{1.5 \text{ s}} = -0.80 \text{ m/s}^2$$

so $\Sigma F = S - mg = ma$

$$S = m(a + g) = 72 \text{ kg}(-0.80 \text{ m/s}^2 + 9.80 \text{ m/s}^2) = 648 \text{ N} = 650 \text{ N} \quad \lozenge$$

51. An ice skater moving at 12 m/s coasts to a halt in 95 m on an ice surface. What is the coefficient of friction between ice and skates?

Solution There are no forces on the skater in the forward direction. The acceleration is negative due to the force of friction opposite the direction of motion. We can first find the acceleration:

$$v_0 = 12 \text{ m/s} \qquad v = 0 \qquad x - x_0 = 95 \text{ m}$$

$$v^2 - v_0^2 = 2a(x - x_0)$$

$$a = \frac{v^2 - v_0^2}{2(x - x_0)} = \frac{0 - (12 \text{ m/s})^2}{2(95 \text{ m})} = -0.758 \text{ m/s}^2$$

We also know $a_y = 0$, since there is no motion in the vertical direction. The free-body diagram shown gives

$$\Sigma F_y = ma_y ; \qquad n - mg = 0 \qquad \text{or} \qquad n = mg$$

$$\Sigma F_x = ma_x \qquad \text{or} \qquad -f_k = m(-0.758 \text{ m/s}^2); \qquad \text{Further,} \quad f_k = \mu_k n,$$

so we substitute to find

$$m(0.758 \text{ m/s}^2) = \mu_k mg = \mu_k m(9.80 \text{ m/s}^2) \qquad \text{or} \qquad \mu_k = (0.758/9.80) = 0.0773 \quad \Diamond$$

---

55. Two blocks connected by a massless rope are being dragged by a horizontal force **F** (Fig. P5.38). Suppose that $F = 68$ N, $m_1 = 12$ kg, $m_2 = 18$ kg, and the coefficient of kinetic friction between each block and the surface is 0.10. (a) Draw a free-body diagram for each block. (b) Determine the tension, T, and the magnitude of the acceleration of the system.

Figure P5.38

Solution

(a) The free-body diagrams for m_1 and m_2 are:

(b) Use the free-body diagrams to apply Newton's second law.

For m_1: $\qquad \Sigma F_x = T - f_1 = m_1 a \qquad$ or $\qquad T = m_1 a + f_1 \qquad\qquad$ (1)

$\qquad\qquad\qquad \Sigma F_y = n_1 - m_1 g = 0 \qquad$ or $\qquad n_1 = m_1 g$

Also, $\qquad\qquad f_1 = \mu_1 n_1 = (0.10)(12 \text{ kg})(9.8 \text{ m}/\text{s}^2) = 11.76 \text{ N}$

For m_2: $\qquad \Sigma F_x = F - T - f_2 = m_2 a \qquad$ or $\qquad T = F - m_2 a - f_2 \qquad$ (2)

$\qquad\qquad\qquad \Sigma F_y = n_2 - m_2 g = 0 \qquad$ or $\qquad n_2 = m_2 g$

Also, $\qquad\qquad f_2 = \mu n_2 = (0.10)(18 \text{ kg})(9.8 \text{ m}/\text{s}^2) = 17.64 \text{ N}$

Substitute T from equation (1) into equation (2).

$$m_1 a + f_1 = F - m_2 a - f_2$$

Solving for a. $\qquad a = \dfrac{F - f_1 - f_2}{m_1 + m_2} = \dfrac{(68 - 11.76 - 17.64) \text{ N}}{(12 + 18) \text{ kg}} = 1.29 \text{ m}/\text{s}^2 \quad \Diamond$

Use equation (1) to calculate value for T.

$$T = m_1 a + f_1 = \left(12 \text{ kg}\right)\left(1.29 \text{ m}/\text{s}^2\right) + 11.76 \text{ N}$$

$$T = 27.2 \text{ N} \quad \Diamond$$

57. A 3.0-kg block starts from rest at the top of a 30.0° incline and slides 2.0 m down the incline in 1.5 s. Find (a) the magnitude of the acceleration of the block, (b) the coefficient of kinetic friction between block and plane, (c) the frictional force acting on the block, and (d) the speed of the block after it has slid 2.0 m.

Solution

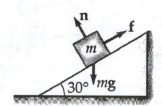

(a)   At constant acceleration, $x = v_0 t + \frac{1}{2}at^2$

So, $a = \dfrac{2(x - v_0 t)}{t^2} = \dfrac{2(2.0 \text{ m} - 0)}{(1.5 \text{ s})^2} = 1.78 \text{ m/s}^2$ ◊

Solving for answer (c) next:

(c)   Choose the x axis parallel to the incline, take the positive direction down the incline (in the direction of the acceleration) and apply the second law.

$$\Sigma F_x = mg \sin \theta - f = ma$$

or $f = m(g \sin \theta - a)$

$$f = (3.0 \text{ kg})\left[9.8 \text{ m/s}^2(\sin 30.0°) - 1.78 \text{ m/s}^2\right] = 9.36 \text{ N} \quad ◊$$

(b) $\Sigma F_y = n - mg \cos \theta = 0;$ $n = mg \cos \theta$

Also, $f = \mu n$ so $\mu = \dfrac{f}{mg \cos \theta}$

$$\mu = \dfrac{9.36 \text{ N}}{(3.0 \text{ kg})(9.80 \text{ m/s}^2)\cos 30.0°} = 0.368 \quad ◊$$

(d) $v^2 = v_0^2 + 2a(x - x_0)$

$$v = \sqrt{0 + (2)(1.78 \text{ m/s}^2)(2.0 \text{ m} - 0)} = 2.67 \text{ m/s} \quad ◊$$

61. A block is placed on a rough plane inclined at 35° relative to the horizontal. If the block slides down the plane with an acceleration of magnitude $g/3$, determine the coefficient of kinetic friction between block and plane.

Solution Take the positive x axis down the incline. Then

$$\sum F_x = mg \sin \theta - f = ma$$

$$\sum F_y = n - mg \cos \theta = 0$$

and $\qquad f = \mu n = \mu mg \cos \theta$

Therefore, when $a = g/3$,

$$mg \sin \theta - \mu mg \cos \theta = m(g/3)$$

So,

$$\mu = \frac{\left(\sin \theta - \dfrac{1}{3}\right)}{\cos \theta} = \frac{\left(\sin 35° - \dfrac{1}{3}\right)}{\cos 35°} = 0.293 \quad \Diamond$$

65. A mass m is held in place by an applied force F and a pulley system as shown in Figure P5.65. The pulleys are massless and frictionless. Find (a) the tension in each section of rope, T_1, T_2, T_3, T_4, and T_5, and (b) the magnitude of **F**.

Solution

Draw free-body diagrams and apply Newton's 2nd law. (All forces are along the y axis.)

For m,

$$\sum F = 0 = T_5 - mg, \qquad T_5 = mg \quad \Diamond$$

Assume frictionless pulleys. The tension is constant throughout a continuous rope.

Therefore, $\qquad T_1 = T_2 = T_3$

For the bottom pulley,

$$\sum F = 0 = T_2 + T_3 - T_5$$

$$2T_2 = T_5 \qquad T_2 = \frac{T_5}{2} = \frac{mg}{2}$$

$$T_1 = T_2 = T_3 = \frac{mg}{2} \quad \lozenge$$

$$F_A = T_1 = \frac{mg}{2} \quad \lozenge$$

For the top pulley,

$$\sum F = 0 = T_4 - T_1 - T_2 - T_3$$

$$T_4 = T_1 + T_2 + T_3 = \frac{3mg}{2} \quad \lozenge$$

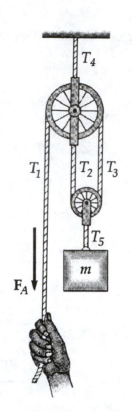

71. Three baggage carts of masses m_1, m_2, and m_3 are towed by a tractor of mass M along an airport apron. The wheels of the tractor exert a total frictional force **F** on the ground as shown (Fig. P5.71). In the following, express your answers in terms of F, M, m_1, m_2, m_3, and g. (a) What are the magnitude and direction of the horizontal force exerted on the tractor by the ground? (b) What is the smallest value of the coefficient of static friction that will prevent the wheels from slipping? Assume that each of the two drive wheels on the tractor bears 1/3 of the tractor's weight. (c) What is the acceleration, **a**, of the system (tractor plus baggage carts)? (d) What are the tensions T_1, T_2, and T_3 in the connecting cables? (e) What is the net force on the cart of mass m_2?

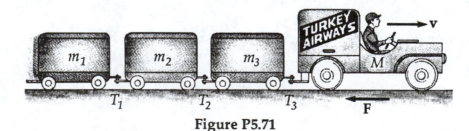

Figure P5.71

Solution

(a) The two drive wheels each exert a static friction force **F** backward on the ground, so the ground exerts static friction force **F** forward on the drive wheels. ◊

(b) Apply $f_s \le \mu_s n$ to the contact of one drive wheel with the ground:

$$\frac{F}{2} \le \frac{\mu_s M g}{3} \qquad \mu_s \ge \frac{3F}{2Mg} \quad ◊$$

(c) Applying $\Sigma F_x = m a_x$ to each of the four vehicles separately gives

$$T_1 = m_1 a$$

$$T_2 - T_1 = m_2 a$$

$$T_3 - T_2 = m_3 a$$

$$F - T_3 = M a$$

Add these equations: $F = (M + m_1 + m_2 + m_3)a$

Then $$a = \frac{F}{M + m_1 + m_2 + m_3} \quad ◊$$

(d) Now $$T_1 = m_1 a = \frac{m_1 F}{M + m_1 + m_2 + m_3} \quad ◊$$

$$T_2 = T_1 + m_2 a = \frac{(m_1 + m_2)F}{M + m_1 + m_2 + m_3} \quad ◊$$

$$T_3 = T_2 + m_3 a = \frac{(m_1 + m_2 + m_3)F}{M + m_1 + m_2 + m_3} \quad ◊$$

(e) Total force on $m_2 = m_2 a = \dfrac{m_2 F}{M + m_1 + m_2 + m_3} \quad ◊$

73. A 2.0-kg block is placed on top of a 5.0-kg block as in Figure P5.73. The coefficient of kinetic friction between the 5.0-kg block and the surface is 0.20. A horizontal force F is applied to the 5.0-kg block. (a) Draw a free-body diagram for each block. What force accelerates the 2.0-kg block? (b) Calculate the magnitude of the force necessary to pull both blocks to the right with an acceleration of 3.0 m/s². (c) Find the minimum coefficient of static friction between the blocks such that the 2.0-kg block does not slip under an acceleration of 3.0 m/s².

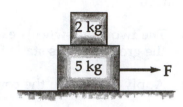

Figure P5.73

Solution

(a)

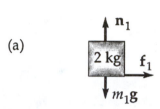
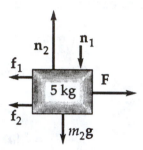

The force of static friction between the blocks accelerates the 2.0-kg block. ◊

(b) $\sum F = ma$, $\qquad F - \mu n_2 = ma$,

$F - (0.20)[(5.0 \text{ kg} + 2.0 \text{ kg})(9.80 \text{ m/s}^2)] = (5.0 \text{ kg} + 2.0 \text{ kg})(3 \text{ m/s}^2)$

Therefore, $\qquad F = 34.7 \text{ N}$ ◊

(c) For the 2-kg block,

$$f = \mu_1 n_1 = \mu_1(2.0 \text{ kg})(9.80 \text{ m/s}^2) = m_1 a = (2.0 \text{ kg})(3.0 \text{ m/s}^2)$$

Therefore, $\qquad \mu_1 = 0.306$ ◊

74. A 5.0-kg block is placed on top of a 10-kg block (Fig. P5.74). A horizontal force of 45 N is applied to the 10-kg block, and the 5.0-kg block is tied to the wall. The coefficient of kinetic friction between the moving surfaces is 0.20. (a) Draw a free-body diagram for each block and identify the action-reaction forces between the blocks. (b) Determine the tension in the string and the magnitude of the acceleration of the 10-kg block.

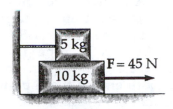

Figure P5.74

Solution

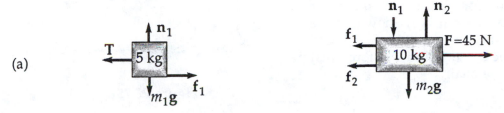

(a)

(b) For the 5.0-kg block (m_1):

$$\sum F_x = f_1 - T = 0 \qquad\qquad T = f_1$$

$$\sum F_y = n_1 - m_1 g = 0 \qquad\qquad n_1 = m_1 g$$

So, $T = f_1 = \mu n_1 = \mu m_1 g = (0.20)(5.0\ \text{kg})(9.80\ \text{m}/\text{s}^2) = 9.80\ \text{N}$ ◊

For the 10-kg block (m_2):

$$\sum F_x = 45\ \text{N} - f_1 - f_2 = m_2 a$$

$$\sum F_y = n_2 - n_1 - m_2 g = 0 \qquad \text{or} \qquad n_2 = g(m_1 + m_2)$$

So, $a = \dfrac{45\ \text{N} - f_1 - f_2}{m_2} = \dfrac{45\ \text{N} - \mu m_1 g - \mu(m_1 + m_2)g}{m_2}$

$$a = \dfrac{45\ \text{N} - (0.20)\big[5.0\ \text{kg} + (5.0 + 10)\ \text{kg}\big]\big(9.8\ \text{m}/\text{s}^2\big)}{10\ \text{kg}} = 0.580\ \text{m}/\text{s}^2 \quad ◊$$

75. An inventive child named Brian wants to reach an apple in a tree without climbing the tree. Sitting in a chair connected to a rope that passes over a frictionless pulley (Fig. P5.75), he pulls on the loose end of the rope with such a force that the spring scale reads 250 N. His true weight is 320 N, and the chair weighs 160 N. (a) Draw free-body diagrams for Brian and the chair considered as separate systems, and another diagram for Brian and the chair considered as one system. (b) Show that the acceleration of the system is upward and find its magnitude. (c) Find the force that Brian exerts on the chair.

Figure P5.75

Solution

(a)

 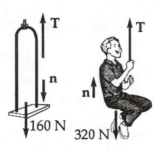 

(b) First consider Brian and the chair as the system. Note that *two* ropes support the system, and $T = 250$ N in each rope.

$$\sum F = ma$$

$$2T - (160 + 320) \text{ N} = ma \qquad \text{where} \qquad m = \frac{480 \text{ N}}{9.80 \text{ m/s}^2} = 49.0 \text{ kg}$$

Solving for a gives $\qquad a = \frac{(500 - 480) \text{ N}}{49.0 \text{ kg}} = 0.408 \text{ m/s}^2 \; \lozenge$

(c) $\sum F \text{ (on Brian)} = n + T - 320 \text{ N} = ma \qquad \text{where} \qquad m = \frac{320 \text{ N}}{9.80 \text{ m/s}^2} = 32.7 \text{ kg}$

$n = ma + 320 \text{ N} - T = 32.7 \text{ kg} (0.408 \text{ m/s}^2) + 320 \text{ N} - 250 \text{ N} = 83.3 \text{ N} \; \lozenge$

79. A wire ABC supports a body of weight w as shown in Figure P5.79. The wire passes over a fixed pulley at B and is firmly attached to a vertical wall at A. The line AB makes an angle ϕ with the vertical, and the pulley at B exerts on the wire a force of magnitude F inclined at angle θ with the horizontal. (a) Show that if the system is in equilibrium, $\theta = \phi/2$. (b) Show that $F = 2w \sin(\phi/2)$. (c) Sketch a graph of F as ϕ increases from $0°$ to $180°$.

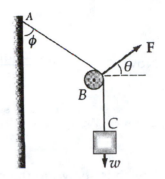

Figure P5.79

Solution

(a) For the hanging mass to be in equilibrium, the wire tension must be equal to its weight. This same tension exists on both sides of the low-friction pulley. Now take the light pulley as object. We draw its free-body diagram twice, before and after taking components, to see

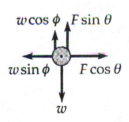

From $\Sigma F_y = ma_y$:

$$w \cos \phi + F \sin \theta = w$$

From $\Sigma F_x = ma_x$:

$$- w \sin \phi + F \cos \theta = 0$$

Therefore,
$$F = \frac{w \sin \phi}{\cos \theta}$$

or,
$$w \cos \phi + \frac{w \sin \phi \sin \theta}{\cos \theta} = w$$

Thus
$$\cos \phi \cos \theta + \sin \phi \sin \theta = \cos \theta$$

Since, from trig identities, $\cos(\phi - \theta) = \cos \phi \cos \theta + \sin \phi \sin \theta$,

we simplify the left side: $\cos(\phi - \theta) = \cos \theta$

Thus, $(\phi - \theta) = 2\theta$, abd $\theta = \phi/2$ $\Diamond$

(b) Now, $F = \dfrac{w \sin \phi}{\cos \theta} = \dfrac{w\, 2 \sin (\phi/2) \cos (\phi/2)}{\cos (\phi/2)}$

$F = 2\, w \sin (\phi/2)$ ◊

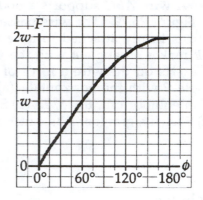

(c) Sketch is shown to the right   ◊

83. What horizontal force must be applied to the cart shown in Figure P5.83 in order that the blocks remain stationary relative to the cart? Assume all surfaces, wheels, and pulley are frictionless. (*Hint:* Note that the force exerted by the string accelerates m_1.)

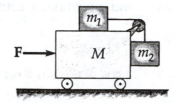

Figure P5.83

Solution

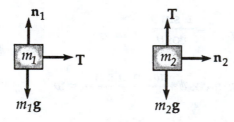

Note that m_2 should be in *contact* with the cart. Use $\Sigma F = ma$ and the free-body diagrams above.

For m_2, $T - m_2 g = 0$ or $T = m_2 g$

For m_1, $T = m_1 a$ or $a = \dfrac{T}{m_1}$

Substituting for T, we have $\qquad a = \dfrac{m_2 g}{m_1}$

For all 3 blocks, $\quad F = (M + m_1 + m_2)a.$

Therefore, $\qquad F = (M + m_1 + m_2)\left(\dfrac{m_2 g}{m_1}\right) \quad \Diamond$

---

86. In Figure P5.86, the coefficient of kinetic friction between the 2.00-kg and 3.00-kg blocks is 0.300. The horizontal surface and the pulleys are frictionless, and the masses are released from rest. (a) Draw a free-body diagram for each block. (b) Determine the acceleration of each block. (c) Find the tension in the strings.

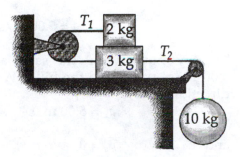

Figure P5.86

Solution

(a)

(b) and (c) $\qquad \Sigma F = ma,\quad$ therefore

(1) $\quad T_1 - \mu m_1 g = m_1 a \qquad$ or $\qquad T_1 - (0.300)(2.00 \text{ kg})(9.80 \text{ m} / \text{s}^2) = (2.00 \text{ m} / \text{s}^2)a$

(2) $\quad T_2 - T_1 - \mu m_1 g = m_2 a \quad$ or $\quad T_2 - T_1 - (0.300)(2.00 \text{ kg})(9.80 \text{ m} / \text{s}^2) = (3.00 \text{ kg})a$

(3) $\quad T_2 - m_3 g = -m_3 a \qquad$ or $\qquad T_2 - (10.0 \text{ kg})(9.80 \text{ m} / \text{s}^2) = -(10.0 \text{ kg})a$

Solving the above gives: $\qquad T_1 = 17.4 \text{ N}, \quad T_2 = 40.5 \text{ N}, \quad$ and $\quad a = 5.75 \text{ m} / \text{s}^2 \quad \Diamond$

---

89. A van accelerates down a hill (Fig. P5.89), going from rest to 30.0 m/s in 6.00 s. During the acceleration, a toy (m = 100 g) hangs by a string from the ceiling. The acceleration is such that the string remains perpendicular to the ceiling. Determine (a) the angle θ and (b) the tension in the string.

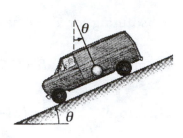

Solution   The acceleration is obtained from $v = v_0 + at$:

$$30.0 \text{ m/s} = 0 + a \ (6.00 \text{ s})$$

Figure P5.89

$$a = 5.00 \text{ m/s}^2$$

The toy moves with the same acceleration as the van, 5.00 m/s² parallel to the hill. We take the x axis in this direction, so

$$a_x = 5.00 \text{ m/s}^2$$

and $a_y = 0$

The only forces on the toy are the string tension in the y direction and its weight, as shown in the free-body diagram.

$$w = mg = (0.100 \text{ kg}) (9.80 \text{ m/s}^2) = 0.980 \text{ N}$$

Now $\Sigma F_y = ma_y$ reads: $+T - (0.980 \text{ N}) \cos \theta = 0$

and $\Sigma F_x = ma_x$ says $(0.980 \text{ N}) \sin \theta = (0.100 \text{ kg})(5.00 \text{ m/s}^2)$

(a) $\sin \theta = \dfrac{0.500}{0.980}$, so $\theta = 30.7° \quad \Diamond$

(b) $T = (0.980 \text{ N}) \cos 30.7° = 0.843 \text{ N} \qquad \Diamond$

Chapter 6

Circular Motion and Other
Applications of Newton's Laws

Chapter 6

CIRCULAR MOTION AND OTHER
APPLICATIONS OF NEWTON'S LAWS

INTRODUCTION

In the previous chapter we introduced Newton's laws of motion and applied them to situations involving linear motion. In this chapter we shall apply Newton's laws of motion to circular motion. We shall also discuss the motion of an object when observed in an accelerated, or noninertial, frame of reference and the motion of an object through a viscous medium. Finally, we conclude this chapter with a brief discussion of the fundamental forces in nature.

NOTES FROM SELECTED CHAPTER SECTIONS

6.1 Newton's Second Law Applied to Uniform Circular Motion

If a particle moves in a circle of radius r with *constant speed*, it undergoes a centripetal acceleration v^2/r directed toward the center of rotation. (Recall that in this case, the centripetal acceleration arises from the change in *direction* of $\mathbf{v}$.) Newton's second law applied to the motion says that the centripetal acceleration arises from some external, centripetal force acting toward the center of rotation. That is:

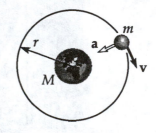

Figure 6.1

$$\Sigma \mathbf{F} \text{ (along } \hat{\mathbf{r}}) = m\mathbf{a}_r = -\frac{mv^2}{r}\hat{\mathbf{r}}$$

where $\hat{\mathbf{r}}$ is a unit vector pointing *radially outward* from the center.

The *universal gravitational constant*, G, is not to be confused with the acceleration due to gravity. The gravitational force is always a force of attraction and, as shown in Figure 6.2 , the force on m_1 due to m_2 is equal and opposite the force on m_2 due to m_1 (Newton's third law).

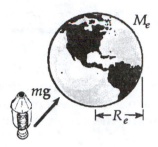

Figure 6.2

Figure 6.3

The gravitational force exerted by a spherically symmetric mass distribution on a particle outside the sphere is the *same* as if the entire mass of the sphere were concentrated at its center. The gravitational force of attraction on a mass *m* near the surface of the Earth is shown in Figure 6.3.

6.2 Nonuniform Circular Motion

If a particle moves in a circular path such that its *speed changes in time,* the particle also has a *tangential* component of acceleration a_t, whose magnitude is dv/dt. In this case, the *total* acceleration **a** is the vector sum of a_r and a_t.

$$a = a_r + a_t$$

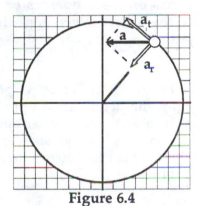

Figure 6.4

6.3 Motion in Accelerated Frames

Observers in noninertial (accelerated) frames of reference must introduce fictitious forces in order to apply Newton's laws in those frames. If these fictitious forces are properly defined, the description of the motion in an accelerated frame of reference will be equivalent to the description in the inertial frame.

6.4 Motion in the Presence of Resistive Forces

A body moving through a gas or liquid experiences a resistive force (opposing its motion) which can have a complicated velocity dependence. A falling body reaches a terminal velocity (maximum velocity) when the downward force of gravity is balanced by the upward resistive force. That is, when $\Sigma F = 0$, **a** = 0, and **v** = constant.

EQUATIONS AND CONCEPTS

When an object of mass m moves uniformly in a circular path, the net force acting on the object is a centripetal force (directed toward the center of the circular path.)

$$F_r = ma_r = m\frac{v^2}{r} \qquad (6.1)$$

When Newton's second law is applied to the motion of an object falling vertically through a viscous medium, the motion can be described by a differential equation. The constant, b, has a value which depends on the properties of the medium and the dimensions and shape of the object.

$$\frac{dv}{dt} = g - \frac{b}{m}v \qquad (6.3)$$

As the resistive force approaches the weight, the acceleration approaches zero, and the object reaches a terminal speed, v_t. Equation 6.4 gives the speed as a function of time, where the object is released from rest at $t = 0$.

$$v = v_t\left(1 - e^{-t/\tau}\right) \qquad (6.4)$$

where $\qquad v_t = \dfrac{mg}{b}$

and $\qquad \tau = \dfrac{m}{b}$

SUGGESTIONS, SKILLS, AND STRATEGIES

Section 6.4 deals with the motion of a body through a gas or liquid. If you covered this section in class, the following solution to Equation 6.3 (when the resistive force $\mathbf{R} = -b\mathbf{v}$) may be useful to know:

$$\frac{dv}{dt} = g - \frac{b}{m}v \qquad (6.3)$$

In order to solve this equation, it is convenient to change variables. If we let $y = g - \dfrac{b}{m}v$,

it follows that $dy = -\dfrac{b}{m}dv$. With these substitutions, Equation 6.3 becomes

$$-\left(\frac{m}{b}\right)\frac{dy}{dt} = y \qquad \text{or} \qquad \frac{dy}{y} = -\frac{b}{m}dt$$

Integrating this expression (now that the variables are separated) gives

$$\int \frac{dy}{y} = -\frac{b}{m}\int dt \qquad \text{or} \qquad \ln y = -\frac{b}{m}t + \text{const.}$$

This is equivalent to $y = (\text{const})\, e^{-bt/m} = g - \dfrac{b}{m}v.$ Taking $v = 0$ at $t = 0$, we see const $= g$.

so

$$v = \frac{mg}{b}(1 - e^{-bt/m}) = v_t(1 - e^{-t/\tau}) \tag{6.4}$$

where $\tau = m/b$.

REVIEW CHECKLIST

▷ Discuss Newton's universal law of gravity (the inverse-square law), and understand that it is an *attractive* force between two *particles* separated by a distance r.

▷ Discuss the nature of the fundamental forces in nature (gravitational, electromagnetic, and nuclear) and characterize the properties and relative strengths of these forces.

▷ Apply Newton's second law to uniform and nonuniform circular motion.

▷ Remember that Newton's laws are only valid in constant velocity (inertial) frames of reference. When motion is described by an observer in an accelerated (noninertial) frame, the observer must invent fictitious forces which arise due to the acceleration of the reference frame.

▷ Recognize that motion of an object through a liquid or gas can involve resistive forces which have a complicated velocity dependence.

SOLUTIONS TO SELECTED END-OF-CHAPTER PROBLEMS

5. A 3.00-kg mass attached to a light string rotates on a horizontal, frictionless table. The radius of the circle is 0.800 m, and the string can support a mass of 25.0 kg before breaking. What range of speeds can the mass have before the string breaks?

Solution   The string will break if the tension T exceeds the weight

$$mg = (25.0 \text{ kg})(9.80 \text{ m/s}^2) = 245 \text{ N}$$

As the 3.00-kg mass rotates in a horizontal circle, the tension provides the central force.

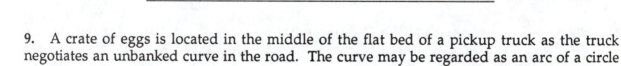

From $\Sigma F = ma$, $T = \dfrac{mv^2}{r} = \dfrac{(3.00 \text{ kg})v^2}{(0.800 \text{ m})}$

Then,

$$v^2 = \frac{rT}{m} = \frac{(0.800 \text{ m})T}{(3.00 \text{ kg})} \le \frac{(0.800 \text{ m})}{(3.00 \text{ kg})} T_{max}$$

Substituting $T_{max} = 245$ N, we find $v^2 \le 65.3 \text{ m}^2 / \text{s}^2$

or $0 < v < 8.08 \text{ m/s}$ ◊

---

9. A crate of eggs is located in the middle of the flat bed of a pickup truck as the truck negotiates an unbanked curve in the road. The curve may be regarded as an arc of a circle of radius 35 m. If the coefficient of static friction between crate and truck is 0.60, what must be the maximum speed of the truck if the crate is not to slide?

Solution   Call the mass of the egg crate m. The forces on it are its weight mg vertically down, the normal force n of the truck bed vertically up, and static friction $\mathbf{f}_s$ directed to oppose relative sliding motion of the crate over the truck bed. The friction force is directed radially inward. It is the only horizontal force on the crate, so it must provide the centripetal acceleration.

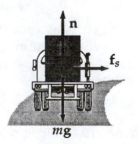

When the truck has maximum speed, friction f_s will have its maximum value when

$$f_s = \mu_s n$$

$\Sigma F_y = ma_y$ gives $n - mg = 0$ or $n = mg$

$\Sigma F_x = ma_x$ gives $f_s = ma_r$

From these two equations, $\mu_s n = mv^2/r$

$$\mu_s mg = mv^2/r$$

The mass divides out, leaving

$$v = \sqrt{\mu_s gr} = \sqrt{(0.60)(9.80 \text{ m}/\text{s}^2)(35 \text{ m})}$$

$$v = 14.3 \text{ m}/\text{s} \quad \lozenge$$

---

13. A coin placed 30.0 cm from the center of a rotating, horizontal turntable slips when its speed is 50.0 cm/s. (a) What provides the central force when the coin is stationary relative to the turntable? (b) What is the coefficient of static friction between coin and turntable?

Solution

(a) The central force is provided by the force of static friction.

(b) The motion of the coin is shown in the upper diagram. The forces on the coin, shown in the free-body diagram below, are the normal force, the weight, and the force of static friction. The only force in the radial direction is **f**.

Therefore, $f = m\dfrac{v^2}{r}$

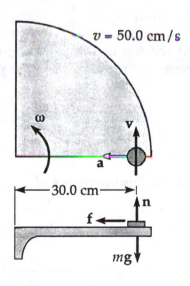

$v = 50.0 \text{ cm}/s$

ω

v

a

—30.0 cm—

n

f

mg

Since the normal force balances the weight,

$n - mg = 0$, or $\qquad\qquad\qquad\qquad\qquad\qquad n = mg$

Dividing these two equations gives $\qquad\qquad\qquad \dfrac{f}{n} = \dfrac{v^2}{rg} < \mu_s$

The coin starts to slip when $\dfrac{v^2}{rg} = \mu_s$

Taking $r = 30.0$ cm, $v = 50.0$ cm/s, and $g = 980$ cm/s²

$$\mu_s = \frac{(50.0 \text{ cm/s})^2}{(30.0 \text{ cm})(980 \text{ cm/s}^2)} = 0.0850 \quad \lozenge$$

---

15. A pail of water is rotated in a vertical circle of radius 1.00 m. What is the minimum speed of the pail at the top of the circle if no water is to spill out?

Solution

At the top of the vertical circle, the speed must be great enough so the central force is equal to or greater than the weight, mg, of the water. That is,

$$\frac{mv^2}{r} \geq mg \qquad \text{or} \qquad v^2 \geq rg$$

At the minimum speed, we have $\qquad v^2_{min} = rg \qquad$ or

$$v_{min} = \sqrt{rg} = \sqrt{(1.00 \text{ m})(9.80 \text{ m/s}^2)} = 3.13 \text{ m/s} \quad \lozenge$$

116

17. A 40.0-kg child sits in a swing supported by two chains, each 3.00 m long. If the tension in each chain at the lowest point is 350 N, find (a) the child's speed at the lowest point and (b) the force of the seat on the child at the lowest point. (Neglect the mass of the seat.)

Solution

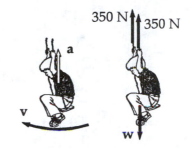

(a)   Choose the seat and child together as one object. The forces acting on the composite object are:

weight $w = mg = (40.0 \text{ kg})(9.80 \text{ m/s}^2) = 392$ N down, and two 350 N forces up.

Now $\Sigma F_y = ma_y$ becomes

$+350 \text{ N} + 350 \text{ N} - 392 \text{ N} = (40.0 \text{ kg})(v^2/3.00 \text{ m})$

$$v = \sqrt{\frac{(308 \text{ N})(3.00 \text{ m})}{40.0 \text{ kg}}} = 4.81 \text{ m/s} \quad \lozenge$$

(b)   The child feels an upward normal force **n** exerted by the seat, and the downward force of gravity, of magnitude $w = 392$ N.

$$\Sigma F_y = ma_y = \frac{mv^2}{r}$$

$$+n - 392 \text{ N} = \frac{(40.0 \text{ kg})(4.81 \text{ m/s})^2}{3.00 \text{ m}}$$

$$n = 700 \text{ N} \quad \lozenge$$

21. Tarzan ($m = 85.0$ kg) tries to cross a river by swinging from a vine. The vine is 10.0 m long, and his speed at the bottom of the swing (as he just clears the water) is 8.00 m/s. Tarzan doesn't know that the vine has a breaking strength of 1000 N. Does he make it safely across the river?

Solution   The forces acting on Tarzan are the force of gravity m**g** and the force of tension **T**. At the lowest point in his motion, **T** is upward and m**g** is downward as in the free-body diagram.

Thus, Newton's second law gives

$$T - mg = \frac{mv^2}{r}$$

Solving for T, with $v = 8.00$ m/s, $r = 10.0$ m, and $m = 85.0$ kg, gives

$$T = m\left(g + \frac{v^2}{r}\right)$$

$$= (85.0 \text{ kg})\left[9.80 \text{ m/s}^2 + \frac{(8.00 \text{ m/s})^2}{10.0 \text{ m}}\right]$$

$$= 1.38 \times 10^3 \text{ N} \quad \lozenge$$

Since T *exceeds* the breaking strength of the vine (1000 N), **Tarzan** *doesn't make it!* The vine breaks *before* he reaches the bottom of the swing.

---

25. A 0.500-kg object is suspended from the ceiling of an accelerating boxcar as in Figure 6.11. If $a = 3.00$ m/s^2, find (a) the angle that the string makes with the vertical and (b) the tension in the string.

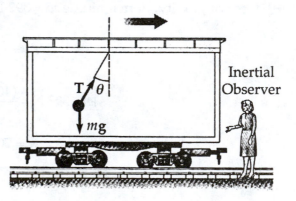

Figure 6.11

Solution

The only forces acting on the suspended object are the force of gravity $m\mathbf{g}$ and the force of tension $\mathbf{T}$, as in the free-body diagram.

Applying Newton's second law in the x and y directions,

$$\Sigma F_x = T \sin\theta = ma \qquad (1)$$

$$\Sigma F_y = T \cos\theta - mg = 0 \qquad (2)$$

or $\qquad T \cos\theta = mg \qquad (2)$

118

(a) Dividing (1) by (2) gives $\qquad \tan\theta = \dfrac{a}{g} = \dfrac{3.00 \text{ m/s}^2}{9.80 \text{ m/s}^2} = 0.306$

$$\theta = 17.0° \quad \lozenge$$

(b) From Equation (1), $\qquad T = \dfrac{ma}{\sin\theta} = \dfrac{(0.500 \text{ kg})(3.00 \text{ m/s}^2)}{\sin(17.0°)} = 5.13 \text{ N} \quad \lozenge$

---

27. A person stands on a scale in an elevator. The maximum and minimum scale readings are 591 N and 391 N, respectively. Assume the magnitude of the acceleration is the same during starting and stopping, and determine (a) the weight of the person, (b) the person's mass, and (c) the acceleration of the elevator.

Solution The scale reads the upward normal force exerted by the floor on the passenger. The maximum force occurs during upward acceleration (when starting an upward trip or ending a downward trip). The minimum normal force occurs with downward acceleration. For each respective situation,

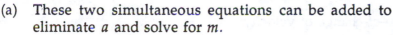

$\Sigma F_y = ma_y \quad$ reads $\quad +591 \text{ N} - mg = +ma$

$\qquad\qquad$ and $\quad +391 \text{ N} - mg = -ma$

(a) These two simultaneous equations can be added to eliminate a and solve for m.

$$+591 \text{ N} - mg + 391 \text{ N} - mg = 0 \quad \text{or} \quad 982 \text{ N} - 2mg = 0$$

$$w = mg = \dfrac{982 \text{ N}}{2} = 491 \text{ N} \quad \lozenge$$

(b) $\qquad m = \dfrac{w}{g} = \dfrac{491 \text{ N}}{9.80 \text{ m/s}^2} = 50.1 \text{ kg} \quad \lozenge$

(c) Substituting back gives $\qquad +591 \text{ N} - 491 \text{ N} = (50.1 \text{ kg})a$

$$a = \dfrac{100 \text{ N}}{50.1 \text{ kg}} = 2.00 \text{ m/s}^2 \quad \lozenge$$

---

31. A motor boat cuts its engine when its speed is 10.0 m/s and coasts to rest. The equation governing the motion of the motorboat during this period is $v = v_0 e^{-ct}$, where v is the speed at time t, v_0 is the initial speed, and c is a constant. At $t = 20.0$ s, the speed is 5.00 m/s. (a) Find the constant c. (b) What is the speed at $t = 40.0$ s? (c) Differentiate the expression for $v(t)$ and thus show that the acceleration of the boat is proportional to the speed at any time.

Solution

(a) We must fit the equation $v = v_0 e^{-ct}$ to the two data points: at $t = 0$, $v = 10.0$ m/s.

Substitution gives $10.0 \text{ m/s} = v_0 e^0 = v_0 \times 1$ so $v_0 = 10.0$ m/s.

And at $t = 20.0$ s, $v = 5.00$ m/s. Substitution gives 5.00 m/s = (10.0 m/s) $e^{-c(20.0\text{ s})}$

$$0.500 = e^{-c(20.0\text{ s})}$$

$$\ln 0.500 = (-c)(20.0 \text{ s})$$

$$c = \frac{-\ln 0.500}{20.0 \text{ s}} = 0.0347 \text{ s}^{-1} \quad \lozenge$$

(b) At all times $v = (10.0 \text{ m/s}) e^{-0.0347t}$

At $t = 40.0$ s, $v = (10.0 \text{ m/s}) e^{-0.0347 \times 40.0} = 2.50$ m/s $\lozenge$

(c) The acceleration is the rate-of-change of velocity:

$$a = \frac{dv}{dt} = \frac{d}{dt} v_0 e^{-ct} = v_0 \left(e^{-ct} \right)(-c)$$

$$= -c \left(v_0 e^{-ct} \right) = -cv = \left(-0.0347 \text{ s}^{-1} \right) v$$

Thus, the acceleration is a negative constant times the speed. $\lozenge$

33. A small, spherical bead of mass 3.00 g is released from rest at $t = 0$ in a bottle of liquid shampoo. The terminal speed is observed to be 2.00 cm/s. Find (a) the value of the constant b in Equation 6.4, (b) the time, t, it takes to reach $0.630v_t$, and (c) the value of the resistive force when the bead reaches terminal speed.

Solution

(a) The speed v varies with time according to Equation 6.4,

$$v = \frac{mg}{b}\left(1 - e^{-bt/m}\right) = v_t\left(1 - e^{-bt/m}\right)$$

where $v_t = \frac{mg}{b}$ is the terminal speed.

Hence, $\qquad b = \frac{mg}{v_t} = \frac{\left(3.00 \times 10^{-3} \text{ kg}\right)\left(9.80 \text{ m/s}^2\right)}{2.00 \times 10^{-2} \text{ m/s}} = 1.47 \text{ N} \cdot \text{s/m}$ ◊

(b) To find the time for v to reach $0.630v_t$, we substitute $v = 0.630v_t$ into Equation 6.4, giving

$$0.630v_t = v_t\left(1 - e^{-bt/m}\right) \qquad \text{or} \qquad 0.370 = e^{-(1.47t/0.00300)}$$

Solve for t by taking the log of each side of the equation:

$$\ln(0.370) = -\frac{1.47t}{3.00 \times 10^{-3}} \qquad \text{and} \qquad t = 2.03 \times 10^{-3} \text{ s}$$ ◊

(c) At terminal speed, $R = v_t b = mg$

Therefore, $R = \left(3.00 \times 10^{-3} \text{ kg}\right)\left(9.80 \text{ m/s}^2\right) = 2.94 \times 10^{-2} \text{ N}$ ◊

39. Because the Earth rotates about its axis, a point on the equator experiences a centripetal acceleration of 0.0340 m/s², while a point at the poles experiences no centripetal acceleration. (a) Show that at the equator the gravitational force on an object (the true weight) must exceed the object's apparent weight. (b) What is the apparent weight at the equator and at the poles of a person having a mass of 75.0 kg? (Assume the Earth is a uniform sphere and take g = 9.800 N/kg.)

Solution

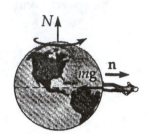

(a) Let **n** represent the force exerted on the object by the scale, which is the "apparent weight." The true weight is $m\mathbf{g}$. Summing up forces on the object in the direction towards the Earth's center gives

$$mg - n = ma_c \qquad (1)$$

where
$$a_c = \frac{v^2}{R_e}$$

is the centripetal acceleration directed toward the center of the Earth.

Thus, we see that
$$n = m(g - a_c) < mg$$

or
$$mg = n + ma_c > n \quad \Diamond \qquad (2)$$

(b) Taking m = 75.0 kg, a_c = 0.0340 m/s², and g = 9.800 m/s²,

at Equator: $n = m(g - a_c) = (75.0 \text{ kg})(9.800 \text{ m/s}^2 - 0.0340 \text{ m/s}^2) = 732.45 \text{ N} \quad \Diamond$

at Poles: $n = mg = (75.0 \text{ kg})(9.800 \text{ m/s}^2) = 735.00 \text{ N} \quad \Diamond \qquad (a_c = 0)$

43. A model airplane of mass 0.75 kg flies in a horizontal circle at the end of a 60-m control wire, with a speed of 35 m/s. Compute the tension in the wire if it makes a constant angle of 20° with the horizontal. The forces exerted on the airplane are the pull of the control wire, its own weight, and aerodynamic lift, which acts at 20° inward from the vertical as shown in Figure P6.43.

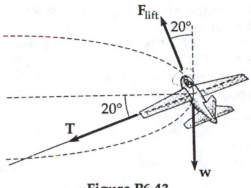

Figure P6.43

Solution

The plane's acceleration is toward the center of the circle of motion, so it is horizontal. The radius of the circle of motion is (60 m) cos 20° = 56.4 m, and the acceleration is

$$a_c = \frac{v^2}{r} = \frac{(35 \text{ m/s})^2}{56.4 \text{ m}} = 21.7 \text{ m/s}^2$$

We can also calculate the weight of the airplane:

$$w = mg = (0.75 \text{ kg})(9.80 \text{ m/s}^2) = 7.35 \text{ N}$$

We define our axes for convenience. In this case, two of the forces--one of them our force of interest-- are directed along the 20° lines.

We define the x-axis to be directed in the (+**T**) direction, and the y-axis to be directed in the direction of lift.

With these definitions, the x component of the centripetal acceleration is $a_{cx} = a_c \cos(20°)$,

and $\Sigma F_x = ma_x$ yields $\qquad\qquad T + w \sin(20°) = ma_{cx}$

Solving for T, $\qquad\qquad\qquad T = ma_{cx} - w \sin(20°)$

$$T = (0.75 \text{ kg})(21.7 \text{ m}/\text{s}^2)(\cos 20°) - (7.35 \text{ N})\sin 20°$$

and $\qquad\qquad\qquad T = (15.3 \text{ N}) - (2.5 \text{ N}) = 12.8 \text{ N} \quad ◊$

45. The pilot of an airplane executes a constant-speed loop-the-loop maneuver in a vertical plane. The speed of the airplane is 300 mi/h, and the radius of the circle is 1200 ft. (a) What is the pilot's apparent weight at the lowest point if his true weight is 160 lb? (b) What is his apparent weight at the highest point? (c) Describe how the pilot could experience weightlessness if both the radius and the speed can be varied. (*Note:* His apparent weight is equal to the force that the seat exerts on his body.)

Solution

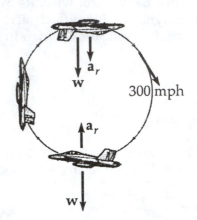

(a) $v = (300 \text{ mph})\left(\dfrac{88 \text{ ft/s}}{60 \text{ mph}}\right) = 440 \text{ ft/s}$

At the lowest point, his seat pushes up toward him and his weight seems to increase.

Since $n - w = ma$, and $n = w' = $ "apparent weight,"

$$w' = w + ma = 160 \text{ lb} + \left(\dfrac{160 \text{ lb}}{32 \text{ ft/s}^2}\right)\left(\dfrac{440 \text{ ft/s}}{1200 \text{ ft}}\right)^2 = 967 \text{ lb} \quad \lozenge$$

(b) At the highest point, the force of the seat on the pilot is directed away from the pilot (downward), and we find

$$w' = w - ma = -647 \text{ lb} \quad \lozenge$$

Since $w' < 0$, the pilot must be strapped down.

(c) When $w' = 0$, then $w = ma = \dfrac{mv^2}{R}$.

If we vary the radius of the loop, R, and the aircraft speed, v, such that this is true, then the pilot feels weightless. $\lozenge$

49. An amusement park ride consists of a large vertical cylinder that spins about its axis fast enough that any person inside is held up against the wall when the floor drops away (Fig. P6.49). The coefficient of static friction between person and wall is μ_s, and the radius of the cylinder is R. (a) Show that the maximum period of revolution necessary to keep the person from falling is $T = (4\pi^2 R \mu_s / g)^{1/2}$. (b) Obtain a numerical value for T if R = 4.00 m and μ_s = 0.400. How many revolutions per minute does the cylinder make?

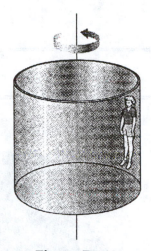

Solution

(a) The normal force of the wall pushes inward:

$$n = \frac{mv^2}{R} = \frac{m}{R}\left(\frac{2\pi R}{T}\right)^2 = \frac{4\pi^2 Rm}{T^2}$$

Figure P6.49

and the maximum force of friction balances the weight:

$$f_s = \mu_s n = mg$$

Therefore, with $\mu_s n = mg$, $\mu_s n = \mu_s \dfrac{4\pi^2 Rm}{T^2} = mg,$

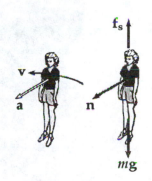

Solving, $T^2 = \dfrac{4\pi^2 R \mu_s}{g}$ gives $T = \sqrt{\dfrac{4\pi^2 R \mu_s}{g}}$ ◊

(b) $T = \left[\dfrac{4\pi^2 (4.00 \text{ m})(0.400)}{9.80 \text{ m}/\text{s}^2}\right]^{1/2} = 2.54$ s ◊

The angular speed is $\left(\dfrac{1 \text{ rev}}{2.54 \text{ s}}\right)\left(\dfrac{60 \text{ s}}{\text{min}}\right) = 23.6$ rev / min ◊

Related Why is the normal force inward? The wall is vertical, and on the outside.
Questions Why is there no upward normal force? Because there is no floor.
and Why is the frictional force directed upward? The friction opposes the
Answers: possible relative motion of the person sliding down the wall.
Why is it not kinetic friction? Because person and wall are moving together, stationary with respect to each other.

Chapter 7

Work and Energy

WORK AND ENERGY

INTRODUCTION

The transformation of energy from one form to another is an essential part of the study of physics, engineering, chemistry, biology, geology, and astronomy. When energy is changed from one form to another, its total amount remains the same. Conservation of energy says that if an isolated system loses energy in some form, it gains an equal amount of energy in other forms.

In this chapter, we introduce the concepts of work and energy. Work is done by a force acting on an object when the point of application of that force moves through some distance and the force has a component along the line of motion. Kinetic energy is energy associated with the motion of an object. The concepts of work and energy can be applied to the dynamics of a mechanical system without resorting to Newton's laws. However, it is important to note that the work-energy concepts are based upon Newton's laws and therefore do not involve any new physical principles. In a complex situation, the "energy approach" can often provide a much simpler analysis than the direct application of Newton's second law.

NOTES FROM SELECTED CHAPTER SECTIONS

7.1   Work Done By a Constant Force

Work done by a constant force is defined as the product of the component of the force in the direction of the displacement and the magnitude of the displacement.

The *unit of work* in the SI system is the newton·meter, N·m.

$$1 \text{ newton·meter} = 1 \text{ joule (J)}.$$

7.2   The Scalar Product of Two Vectors

The *scalar product* or dot product of any two vectors is a scalar quantity equal to the product of the magnitudes of the two vectors and the cosine of the angle included between the directions of the two vectors.

7.3   Work Done By a Varying Force

Work done by a varying force is equal to the area under the force-displacement curve.

7.4 Kinetic Energy and the Work-Energy Theorem

The work done by a force **F** in displacing a particle equals the change in the kinetic energy of the particle. This is known as the *work-energy theorem*.

7.5 Power

Power is the time rate of doing work or expending energy. The SI unit of power is the watt, W. $1\ W = 1\ J/s$.

EQUATIONS AND CONCEPTS

The work done on a body by a *constant* force **F** is defined to be the product of the displacement and the component of force in the direction of the displacement.

$$W = Fs \cos \theta \tag{7.1}$$

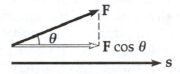

It is convenient to express the work done by a constant force as the *dot product* (scalar product) **F·s**.

$$W = \mathbf{F \cdot s} = Fs \cos \theta \tag{7.2}$$

The dot product of any two vectors **A** and **B** is defined to be a scalar quantity whose magnitude is $AB \cos \theta$. (The product of the magnitudes of the two vectors and the cosine of the angle that is included between their directions.)

$$\mathbf{A \cdot B} \equiv AB \cos \theta \tag{7.3}$$

Note that work done by a force can be positive, negative, or zero depending on the value of θ. Work done by a force is positive if **F** has a component in the direction of **s** $(0 \le \theta < 90°)$; W is negative if the projection of **F** onto **s** is opposite to **s** $(90° < \theta < 180°)$. Finally, W is zero if **F** is perpendicular to **s** $(\theta = \pm 90°)$. Work is a *scalar* quantity which has SI units of joules (J), where $1\ J = 1\ N \cdot m$.

The scalar product of two vectors **A** and **B** can be expressed in terms of the x, y, and z components of the two vectors.

$$\mathbf{A} \cdot \mathbf{B} = A_x B_x + A_y B_y + A_z B_z \qquad (7.6)$$

If a force acting along x varies with position, and the body is displaced from x_i to x_f, the *work done by that force* is given by an integral expression. Graphically, the work done equals the area under the F_x versus x curve.

$$W = \int_{x_i}^{x_f} F_x \, dx \qquad (7.7)$$

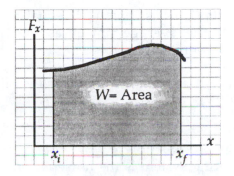

If a mass is connected to a spring of force constant k, *the work done by the spring force* ($-kx$) *as the mass undergoes an arbitrary displacement from x_i to x_f* is given by Equation 7.11.

$$W_s = \tfrac{1}{2}kx_i^2 - \tfrac{1}{2}kx_f^2 \qquad (7.11)$$

Kinetic energy, K, is energy associated with the motion of an object.

$$K \equiv \tfrac{1}{2}mv^2 \qquad (7.14)$$

The *work-energy theorem* states that the work done by the resultant force on a body equals the *change* in kinetic energy of the body.

$$W_{net} = K_f - K_i = \Delta K \qquad (7.15)$$

$$W_{net} = \tfrac{1}{2}mv_f^2 - \tfrac{1}{2}mv_i^2 \qquad (7.16)$$

Note that if W_{net} is positive, the kinetic energy increases; if W_{net} is negative, the kinetic energy decreases. Therefore, the speed of a body will only change if there is net work done on it. (When $W_{net} = 0$, the resultant work is zero, and $\Delta K = 0$.)

$$\tfrac{1}{2}mv_i^2 + W_{net} = \tfrac{1}{2}mv_f^2$$

The loss in kinetic energy due to friction corresponds to the energy dissipated by the force of kinetic friction.

$$\Delta K = -fs \qquad (7.17)$$

The *average power* supplied by a force is the ratio of the work done by that force to the time interval over which it acts.

$$\overline{P} \equiv \frac{W}{\Delta t}$$

The *instantaneous power* is equal to the limit of the average power as the time interval approaches zero.

$$P = \frac{dW}{dt} = \mathbf{F} \cdot \mathbf{v} \qquad (7.18)$$

The SI unit of power is J/s, which is called a watt (W).

$$1\ W = 1\ J/s$$

$$1\ kW \cdot h = 3.60 \times 10^6\ J$$

The unit of power in the British engineering system is the horsepower.

$$1\ hp = 746\ W$$

When particle speeds are comparable to the speed of light, the equations of Newtonian mechanics must be replaced by more general equations. In particular, the relativistic form of the kinetic energy equation must be used.

$$K = mc^2\left(\frac{1}{\sqrt{1-(v/c)^2}} - 1\right) \qquad (7.20)$$

SUGGESTIONS, SKILLS, AND STRATEGIES

There are two new mathematical skills you must learn. The first is the definition of the scalar (or dot) product, $\mathbf{A} \cdot \mathbf{B} \equiv AB \cos\theta$, where θ is the angle between $\mathbf{A}$ and $\mathbf{B}$. Since $\mathbf{A} \cdot \mathbf{B}$ is a scalar, then the order of product can be interchanged. That is, $\mathbf{A} \cdot \mathbf{B} = \mathbf{B} \cdot \mathbf{A}$. Furthermore, $\mathbf{A} \cdot \mathbf{B}$ can be positive, negative, or zero depending on the value of θ. (That is, $\cos\theta$ varies from −1 to +1.) If vectors are expressed in unit vector form, then the dot product is conveniently carried out using the multiplication table for unit vectors:

$$\mathbf{i} \cdot \mathbf{i} = \mathbf{j} \cdot \mathbf{j} = \mathbf{k} \cdot \mathbf{k} = 1; \quad \mathbf{i} \cdot \mathbf{j} = \mathbf{i} \cdot \mathbf{k} = \mathbf{k} \cdot \mathbf{j} = 0$$

The second operation introduced in this chapter is the definite integral. In Section 7.4, it is shown that the work done by a *variable* force F_x in displacing a particle a small distance Δx is given by

$$\Delta W \approx F_x\, \Delta x$$

130

(ΔW equals the area of the shaded rectangle in Figure 7.1.) The total work done by F_x as the particle is displaced from x_i to x_f is given approximately by the *sum* of such terms.

If we take such a sum, letting the widths of the displacements approach dx, the number of terms in the sum becomes very large and we get the actual work done:

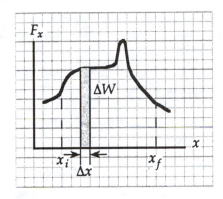

Figure 7.1

$$W = \lim_{\Delta x \to 0} \sum_{x_i}^{x_f} F_x \, \Delta x = \int_{x_i}^{x_f} F_x \, dx$$

The quantity on the right is a definite integral, which graphically represents the *area under the F_x versus x curve*, as in Figure 7.1. Appendix B6 of the text represents a brief review of integration operations with some examples.

REVIEW CHECKLIST

▷ Define the work done by a constant force, and realize that work is a scalar. Describe the work done by a force which *varies* with position. In the one-dimensional case, note that the work done equals the area under the F_x versus x curve.

▷ Take the scalar or dot product of any two vectors **A** and **B** using the definition $\mathbf{A \cdot B} \equiv A B \cos\theta$, or by writing **A** and **B** in unit vector form and using the multiplication table for unit vectors.

▷ Define the kinetic energy of an object of mass m moving with a speed v.

▷ Relate the work done by the net force on an object to the *change* in kinetic energy. The relation $W_{\text{net}} = \Delta K = K_f - K_i$ is called the work-energy theorem, and is valid whether or not the (resultant) force is constant. That is, if we know the net work done on a particle as it undergoes a displacement, we also know the *change* in its kinetic energy. This is the most important concept in this chapter, so you must understand it thoroughly.

▷ Define the concepts of average power and instantaneous power (the time rate of doing work).

SOLUTIONS TO SELECTED END-OF-CHAPTER PROBLEMS

1. If a person lifts a 20.0-kg bucket from a well and does 6.00 kJ of work, how deep is the well? Assume the speed of the bucket remains constant as it is lifted.

Solution Assuming the applied force is constant, and equal in magnitude to the weight of the bucket, we have:

$$W = Fd = mgh$$

$$h = \frac{W}{mg} = \frac{6.00 \times 10^3 \text{ J}}{(20.0 \text{ kg})\left(9.80 \text{ m/s}^2\right)} = 30.6 \text{ m} \quad \lozenge$$

3. A block of mass 2.5 kg is pushed 2.2 m along a frictionless horizontal table by a constant 16.0-N force directed 25° below the horizontal. Determine the work done by (a) the applied force, (b) the normal force exerted by the table, (c) the force of gravity, and (d) the net force on the block.

Solution $W = Fs \cos \theta$

(a) Work done by the applied force:

$$W_{appl} = (16.0 \text{ N}) (2.2 \text{ m}) \cos 25° = 31.9 \text{ J} \quad \lozenge$$

(b) Work done by normal force:

$$W_n = ns \cos \theta = ns \cos 90° = 0 \quad \lozenge$$

(c) Work done by force of gravity:

$$W_g = F_g s \cos \theta = mgs \cos 90° = 0 \quad \lozenge$$

(d) Net work done on the block:

$$W_{net} = W_{appl} + W_n + W_g = 31.9 \text{ J} \quad \lozenge$$

7. A horizontal force of 150 N is used to push a 40.0-kg box 6.00 m on a horizontal surface. If the box moves at constant speed, find (a) the work done by the 150-N force, (b) the energy lost due to friction, and (c) the coefficient of kinetic friction.

Solution

$s = 6.00$ m

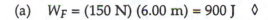

Since $a = 0$ (or v = const),

$$\Sigma F_x = 150 - f = 0$$

$$f = 150 \text{ N}$$

(a) $W_F = (150 \text{ N})(6.00 \text{ m}) = 900 \text{ J}$ ◊

(b) $W_f = fs \cos(180) = (150 \text{ N})(6.00 \text{ m})(-1) = -900 \text{ J}$ ◊

(c) $f = \mu n = \mu mg$

$$\mu = \frac{f}{mg} = \frac{150 \text{ N}}{(40.0 \text{ kg})(9.80 \text{ m/s}^2)} = 0.383 \text{ ◊}$$

10. Batman, whose mass is 80.0 kg, is holding on to the free end of a 12.0-m rope, the other end of which is fixed to a tree limb above. He is able to get the rope in motion as only Batman knows how, eventually getting it to swing enough that he can reach a ledge when the rope makes a 60° angle with the vertical. How much work was done against the force of gravity in this maneuver?

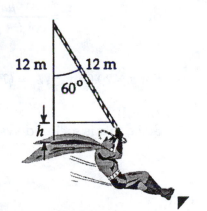

Solution

Batman's weight = mg = (80.0 kg)(9.80 m/s²) = 784 N

The work done is the weight of Batman times the vertical height his weight is raised.

This height, h, is $h = 12.0 \text{ m} - (12.0 \text{ m})\cos 60° = 6.00 \text{ m}$

Thus, $W = (784 \text{ N})(6.00 \text{ m}) = 4.70 \text{ kJ}$ ◊

11. A cart loaded with bricks has a total mass of 18.0 kg and is pulled at constant speed by a rope. The rope is inclined at 20.0° above the horizontal, and the cart moves 20.0 m on a horizontal surface. The coefficient of kinetic friction between ground and cart is 0.500. (a) What is the tension in the rope? (b) How much work is done on the cart by the rope? (c) What is the energy lost due to friction?

Solution

(a) Since v = constant, a = 0,

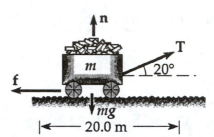

From $\sum F_x = 0$, (1) $T \cos \theta - f = 0$

From $\sum F_y = 0$, (2) $n + T \sin \theta - mg = 0$

By the friction equation,

(3) $f = \mu n = \mu(mg - T \sin \theta)$

Substituting (3) into (1) gives

$$T \cos \theta - \mu mg + \mu T \sin \theta = 0$$

or $$T = \frac{\mu mg}{\cos \theta + \mu \sin \theta} = \frac{(0.500)\,(18.0 \text{ kg})\,\left(9.80 \text{ m/s}^2\right)}{\cos 20.0° + (0.500)\, \sin 20.0°} = 79.4 \text{ N} \quad \Diamond$$

(b) For s = 20.0 m, $W_T = (T \cos \theta)s = (79.4 \text{ N})\,(\cos 20.0°)\,(20.0 \text{ m}) = 1.49 \text{ kJ} \quad \Diamond$

(c) Since this is a closed system,

$$W_{\text{net}} = W_T + W_f = 0,$$

and $$W_f = -W_T = -1.49 \text{ kJ} \quad \Diamond$$

17. A force $\mathbf{F} = (6\mathbf{i} - 2\mathbf{j})$ N acts on a particle that undergoes a displacement $\mathbf{s} = (3\mathbf{i} + \mathbf{j})$ m. Find (a) the work done by the force on the particle and (b) the angle between $\mathbf{F}$ and $\mathbf{s}$.

Solution We assume all numbers in this problem have three significant figures:

(a) $W = \mathbf{F}\cdot\mathbf{s} = (6\mathbf{i} - 2\mathbf{j})\cdot(3\mathbf{i} + \mathbf{j}) = (6.00 \text{ N})(3.00 \text{ m}) + (-2.00 \text{ N})(1.00 \text{ m})$

$W = 18.0 \text{ J} - 2.00 \text{ J} = 16.0 \text{ J} \lozenge$

(b) $|\mathbf{F}| = \sqrt{F_x^2 + F_y^2} = \sqrt{6.00^2 + (-2.00)^2} \text{ N} = 6.32 \text{ N}$

$|\mathbf{s}| = \sqrt{s_x^2 + s_y^2} = \sqrt{3.00^2 + 1.00^2} \text{ m} = 3.16 \text{ m}$

$W = Fs \cos \theta$

$\cos \theta = \dfrac{W}{Fs} = \dfrac{16.0 \text{ J}}{(6.32 \text{ N})(3.16 \text{ m})} = 0.8012$

$\theta = \cos^{-1}(0.8012) = 36.8° \ \lozenge$

---

20. Find the angle between $\mathbf{A} = -5\mathbf{i} - 3\mathbf{j} + 2\mathbf{k}$ and $\mathbf{B} = -2\mathbf{j} - 2\mathbf{k}$.

Solution   Their dot product is $\mathbf{A}\cdot\mathbf{B} = (-5\mathbf{i} - 3\mathbf{j} + 2\mathbf{k})\cdot(0\mathbf{i} - 2\mathbf{j} - 2\mathbf{k})$

Multiplying the components, $\mathbf{A}\cdot\mathbf{B} = -5(0) - 3(-2) + 2(-2) = 2$

We also have $|\mathbf{A}| = \sqrt{5^2 + 3^2 + 2^2} = 6.16$ and $|\mathbf{B}| = \sqrt{2^2 + 2^2} = 2.83$

So in $\mathbf{A}\cdot\mathbf{B} = |A|\,|B| \cos \theta$,

$$\theta = \cos^{-1}\left(\frac{\mathbf{A}\cdot\mathbf{B}}{AB}\right) = \cos^{-1}\left(\frac{2}{6.16 \times 2.83}\right) = 83.4° \ \lozenge$$

---

23. A body is subject to a force F_x that varies with position as in Figure P7.23. Find the work done by the force on the body as it moves (a) from $x = 0$ to $x = 5.0$ m, (b) from $x = 5.0$ m to $x = 10$ m, and (c) from $x = 10$ m to $x = 15$ m. (d) What is the total work done by the force over the distance $x = 0$ to $x = 15$ m?

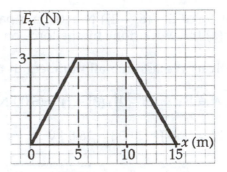

Figure P7.23

Solution

$W = \int F_x \, dx$ and W equals the area under the force-displacement curve.

(a) For the region $0 \leq x \leq 5.0$ m, $\qquad W = \dfrac{(3.0 \text{ N}) \, (5.0 \text{ m})}{2} = 7.50 \text{ J} \quad \lozenge$

(b) For the region $5.0 \text{ m} \leq x \leq 10$ m, $\qquad W = (3.0 \text{ N}) \, (5.0 \text{ m}) = 15.0 \text{ J} \quad \lozenge$

(c) For the region $10 \text{ m} \leq x \leq 15$ m, $\qquad W = \dfrac{(3.0 \text{ N}) \, (5.0 \text{ m})}{2} = 7.50 \text{ J} \quad \lozenge$

(d) For the region $0 \leq x \leq 15$ m, $\qquad W = (7.5 + 7.5 + 15) \text{ J} = 30.0 \text{ J} \quad \lozenge$

---

31. If it takes 4.00 J of work to stretch a Hooke's-law spring 10.0 cm from its unstressed length, determine the extra work required to stretch it an additional 10.0 cm.

Solution $\qquad W = \dfrac{1}{2}kx^2$

$$k = \frac{2W_{0-10}}{x_{0-10}^2} = \frac{2(4.00 \text{ J})}{(0.100 \text{ m})^2} = 800 \text{ N / m}$$

$$W_{0-20} = \frac{1}{2}kx_{0-20}^2$$

$$W_{0-20} = \left(\frac{1}{2}\right) (800 \text{ N/m}) (0.200 \text{ m})^2 = 16.0 \text{ J}$$

So, $\qquad\qquad\qquad \Delta W_{10-20} = W_{0-20} - W_{0-10} = 16.0 \text{ J} - 4.00 \text{ J} = 12.0 \text{ J} \quad \lozenge$

---

37. A 40-kg box initially at rest is pushed 5.0 m along a rough, horizontal floor with a constant applied horizontal force of 130 N. If the coefficient of friction between box and floor is 0.30, find (a) the work done by the applied force, (b) the energy lost due to friction, (c) the change in kinetic energy of the box, and (d) the final speed of the box.

Solution

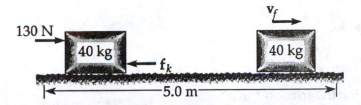

$\mu_k = 0.30$
$s = 5.0$ m
$v_0 = 0$

(a)   The applied force is *horizontal*.

Therefore, $W_F = \mathbf{F} \cdot \mathbf{s} = Fs = (130 \text{ N})(5.0 \text{ m}) = 650 \text{ J}$ ◊

(b) $f_k = \mu_k n = \mu_k mg = 0.30\,(40 \text{ kg})\left(9.80 \text{ m/s}^2\right) = 117.6 \text{ N}$

$W_{f_k} = \mathbf{f}_k \cdot \mathbf{s} = -f_k s = -(117.6 \text{ N})(5.0 \text{ m}) = -588 \text{ J}$ ◊

(c) $\Delta K = W_{net} = W_F + W_{f_k} = 650 \text{ J} + (-588 \text{ J}) = 62.0 \text{ J}$ ◊

(d) $\tfrac{1}{2}mv_f^2 - \tfrac{1}{2}mv_0^2 = W_{net}$

$v_f = \sqrt{\dfrac{2}{m}\left(W_{net} + \tfrac{1}{2}mv_0^2\right)}$

$v_f = \sqrt{\left(\dfrac{2}{40 \text{ kg}}\right)\left[62.0 \text{ J} + \tfrac{1}{2}(40 \text{ kg})(0)^2\right]}$

$v_f = 1.76 \text{ m/s}$ ◊

45. A sled of mass m is given a kick on a frozen pond, imparting to it an initial speed $v_i = 2$ m/s. The coefficient of kinetic friction between sled and ice is $\mu_k = 0.10$. Use energy considerations to find the distance the sled moves before stopping.

Solution

Step one: analyze the vertical forces.

$$\Sigma F_y = ma_y: \qquad +n - mg = 0,$$

$$or \qquad n = mg$$

Step two: Find an expression for the friction force.

$$f_k = \mu_k n = \mu_k\, mg$$

Step three: Apply the work-energy theorem. We do not consider the kick, but take the original point after the kick, with $v_o = 2$ m/s, and the final point where the sled stops moving. The weight and normal force, each at 90° to the direction of motion, do no work.

$$K_i + W_{net} = K_F, \qquad or \qquad \tfrac{1}{2}mv_o^2 - fs = 0$$

$$s = \frac{mv_o^2}{2f} = \frac{mv_o^2}{2\mu_k\, mg} = \frac{v_o^2}{2\mu_k\, g} = \frac{(2\text{ m/s})^2}{2(0.10)(9.80\text{ m/s}^2)}$$

$$s = 2.04 \text{ m} \quad \lozenge$$

---

47. A crate of mass 10.0 kg is pulled up a rough incline with an initial speed of 1.50 m/s. The pulling force is 100 N parallel to the incline, which makes an angle of 20.0° with the horizontal. The coefficient of kinetic friction is 0.400, and the crate is pulled 5.00 m. (a) How much work is done by gravity? (b) How much energy is lost due to friction? (c) How much work is done by the 100-N force? (d) What is the change in kinetic energy of the crate? (e) What is the speed of the crate after being pulled 5.00 m?

Solution

(a) The force of gravity is (10.0 kg) (9.80 m/s²) = 98.0 N straight down, at an angle of (90.0° + 20.0°) = 110.0° with the motion. The work done by gravity is

$$W_g = \mathbf{F} \cdot \mathbf{s} = (98.0 \text{ N})(5.00 \text{ m}) \cos 110.0° = -168 \text{ J} \quad \lozenge$$

(b) Take the y axis perpendicular to the incline.

Then from $\Sigma F_y = ma_y,$ $\quad +n - (98.0 \text{ N})\cos 20.0° = 0$

$$n = 92.1 \text{ N}$$

and $\quad f_k = \mu_k n = 0.400 \ (92.1 \text{ N}) = 36.8 \text{ N}$

So the energy lost due to friction is

$$\left| W_f \right| = \left| \mathbf{f}_k \cdot \mathbf{s} \right| = \left| (36.8 \text{ N})(5.00 \text{ m}) \cos 180° \right| = \left| -184 \text{ J} \right| = 184 \text{ J} \quad \lozenge$$

(c) $\quad W = \mathbf{F} \cdot \mathbf{s} = 100 \text{ N} \ (5.00 \text{ m}) \cos 0° = +500 \text{ J} \quad \lozenge$

(d) $\quad K_i + W_{\text{total}} = K_f$

$$\Delta K = K_f - K_i = W_g + W_f + W_{\text{applied force}} + W_n$$

$$= -168 \text{ J} - 184 \text{ J} + 500 \text{ J} + 0 = 148 \text{ J} \quad \lozenge$$

The normal force does zero work, because it is at 90° to the motion.

(e) $\quad K_i + W_{\text{total}} = K_f$

$$\tfrac{1}{2}(10.0 \text{ kg}) v_f^2 = \tfrac{1}{2}(10.0 \text{ kg})\left(1.50 \text{ m/s}^2\right) + 148 \text{ J} = 159 \text{ J}$$

$$v_f = \sqrt{\frac{2\left(159 \text{ kg} \cdot \text{m}^2 / \text{s}^2\right)}{10.0 \text{ kg}}} = 5.65 \text{ m/s} \quad \lozenge$$

49. A 4.00-kg block is given an initial speed of 8.00 m/s at the bottom of a 20.0° incline. The frictional force that retards its motion is 15.0 N. (a) If the block is directed up the incline, how far does it move before stopping? (b) Will it slide back down the incline?

Solution

$v_0 = 8\ m/s$

15.0 N

4 kg

20°

(a) If s is the displacement along the incline, the energy lost due to friction is $W_f = -fs$, while the work done by the force of gravity is

$$W_g = -mgs\ (\sin 20°)$$

Because the normal force and the component of weight $mg \cos 20°$ are both perpendicular to the displacement, they do no work. Hence, from the work-energy theorem,

$$W_{net} = \Delta K$$

$$W_f + W_g = \tfrac{1}{2}m\left(v_f^2 - v_o^2\right)$$

Noting that $v_f = 0$, $fs \cos(180°) - mgs \sin 20.0° = \tfrac{1}{2}m\left(0 - v_o^2\right)$

$$s = \frac{\tfrac{1}{2}m\left(v_o^2\right)}{f + mg\sin 20.0°} = \frac{\tfrac{1}{2}(4.00\ kg)(8.00\ m/s)^2}{15\ N + (4.00\ kg)\left(9.80\ m/s^2\right)(0.342)}$$

$$s = 4.51\ m \quad \lozenge$$

(b) On the incline the forces acting on the block along the surface are $mg \sin 20.0°$ and f; since $mg \sin 20.0° = 13.4\ N < 15\ N = f$, the block will *not* slide back down the incline.

53. A 700-N marine in basic training climbs a 10.0-m vertical rope at a constant speed in 8.00 s. What is his power output?

Solution

The marine must exert a 700 N upward force opposite the force of gravity to lift his body at constant speed. Then his muscles do work:

$$W = \mathbf{F} \cdot \mathbf{s} = 700\mathbf{j} \text{ N} \cdot 10.0\mathbf{j} \text{ m} = 7000 \text{ J}$$

The power he puts out is

$$P = \frac{W}{t} = \frac{7000 \text{ J}}{8.00 \text{ s}} = 875 \text{ W} \quad \Diamond$$

---

59. An outboard motor propels a boat through the water at 10.0 mi/h. The water resists the forward motion of the boat with a force of 15.0 lb. How much power is produced by the motor?

Solution $P = \mathbf{F} \cdot \mathbf{v}$

$P = Fv$ when $\mathbf{F}$ and $\mathbf{v}$ are parallel

$P = (15.0 \text{ lb})(10.0 \text{ mi/h})$

$$P = \left(150.0 \frac{\text{lb} \cdot \text{mi}}{\text{h}}\right)\left(\frac{5280 \text{ ft/mi}}{3600 \text{ s/h}}\right) = 220 \frac{\text{ft} \cdot \text{lb}}{\text{s}}$$

$$P = \left(220 \frac{\text{ft} \cdot \text{lb}}{\text{s}}\right)\left(\frac{1 \text{ hp}}{550 \text{ ft} \cdot \text{lb}}\right) = 0.400 \text{ hp} \quad \Diamond$$

---

65. A compact car of mass 900 kg has an overall motor efficiency of 15%. (That is, 15% of the energy supplied by the fuel is transformed into kinetic energy of the car.) (a) If burning one gallon of gasoline supplies 1.34×10^8 J of energy, find the amount of gasoline used in accelerating the car from rest to 55 mph. (b) How many such accelerations will one gallon provide? (c) The mileage claimed for the car is 38 mi/gal at 55 mph. What power is delivered to the wheels (to overcome frictional effects) when the car is driven at this speed?

Solution

(a) We first must convert $\left(\dfrac{55 \text{ mi}}{\text{h}}\right)\left(\dfrac{1609 \text{ m}}{\text{mi}}\right)\left(\dfrac{\text{h}}{3600 \text{ s}}\right) = 24.6 \text{ m/s}$

The engine must do work to provide kinetic energy to the chassis, according to

$$W_{net} = K_f - K_i = \tfrac{1}{2}mv_f^2 - \tfrac{1}{2}mv_i^2 .$$

$$W_{\text{engine output}} = \tfrac{1}{2}(900 \text{ kg})(24.6 \text{ m/s})^2 - 0 = 2.72 \times 10^5 \text{ J}$$

This output is only 15% of the chemical energy the engine takes in; the other 85% is lost as heat.

Since $\qquad \text{efficiency} = \dfrac{\text{useful energy output}}{\text{total energy input}} = \dfrac{2.72 \times 10^5 \text{ J}}{\text{total energy input}},$

$$\text{total energy input} = \dfrac{2.72 \times 10^5 \text{ J}}{0.15} = 1.82 \times 10^6 \text{ J}$$

and $\qquad \text{amount of gasoline} = \left(1.82 \times 10^6 \text{ J}\right)\left(\dfrac{1 \text{ gal}}{1.34 \times 10^8 \text{ J}}\right) = 0.0135 \text{ gal} \quad \Diamond$

(b) $\qquad\qquad$ 1 gal $= n$ (0.0135 gal)

$\qquad\qquad\qquad n = 73.8$ accelerations $\quad \Diamond$

(c) Consider driving 38 miles with energy input 1.34×10^8 J and output work $0.15\ (1.34 \times 10^8 \text{ J}) = 2.01 \times 10^7$ J. The time for this trip is 38 mi/(55 mi/h) = 0.691 h = 2490 s. So, the output power is

$$P = \frac{W}{t} = \frac{2.01 \times 10^7 \text{ J}}{2490 \text{ s}} = 8.08 \text{ kW} \quad \Diamond$$

69. A proton in a high-energy accelerator moves with a speed of $c/2$. Use the work-energy theorem to find the work required to increase its speed to (a) $0.75c$, (b) $0.995c$.

Solution

$K_i + W = K_f$:

$$\left(\frac{1}{\sqrt{1-v_i^2/c^2}}-1\right)mc^2 + W = \left(\frac{1}{\sqrt{1-v_f^2/c^2}}-1\right)mc^2$$

(a) $$\left(\frac{1}{\sqrt{1-0.5^2}}-1\right)(1.67\times10^{-27}\text{ kg})(3\times10^8\text{ m}/\text{s})^2 + W = \left(\frac{1}{\sqrt{1-0.75^2}}-1\right)(1.50\times10^{-10}\text{ J})$$

$$\left(\frac{1}{\sqrt{0.75}}-1\right)(1.50\times10^{-10}\text{ J}) + W = (0.512)(1.50\times10^{-10}\text{ J})$$

$$W = (0.512 - 0.155)(1.50\times10^{-10}\text{ J}) = 5.37\times10^{-11}\text{ J} \quad \Diamond$$

(b) $$(1.155-1)(1.50\times10^{-10}\text{ J}) + W = \left(\frac{1}{\sqrt{1-0.995^2}}-1\right)(1.50\times10^{-10}\text{ J})$$

$$W = (9.01 - 0.155)(1.50\times10^{-10}\text{ J}) = 1.33\times10^{-9}\text{ J} \quad \Diamond$$

This large energy input is required to boost the proton's speed close to that of light. Arbitrarily large energy inputs would fail to accelerate it to, or beyond, the speed of light.

75. A 4.0-kg particle moves along the x axis. Its position varies with time according to $x = t + 2.0t^3$, where x is in meters and t is in seconds. Find (a) the kinetic energy at any time t, (b) the acceleration of the particle and the force acting on it at time t, (c) the power being delivered to the particle at time t, and (d) the work done on the particle in the interval $t = 0$ to $t = 2.0$ s.

Solution

Given $m = 4.0$ kg and $x = t + 2.0t^3$, we find

(a) $\quad v = \dfrac{dx}{dt} = \dfrac{d}{dt}\left(t + 2.0t^3\right) = 1 + 6.0t^2$

$\quad\quad K = \tfrac{1}{2}mv^2 = \tfrac{1}{2}\left(4.0 \text{ kg}\right)\left(1 + 6.0t^2\right)^2 = \left(2.0 + 24t^2 + 72t^4\right) \text{J} \quad \Diamond$

(b) $\quad a = \dfrac{dv}{dt} = \dfrac{d}{dt}\left(1 + 6.0t^2\right) = 12t \text{ m/s}^2 \quad \Diamond$

$\quad\quad F = ma = 4.0(12t) = 48.0t \text{ N} \quad \Diamond$

(c) $\quad P = \dfrac{dW}{dt} = \dfrac{dK}{dt} = \dfrac{d}{dt}\left(2.0 + 24t^2 + 72t^4\right) = \left(48.0t + 288t^3\right) \text{W} \quad \Diamond$

$$[\text{or use} \quad P = Fv = 48.0t(1 + 6.0t^2)]$$

(d) $\quad W = K_f - K_i \quad$ where $\quad t_i = 0 \quad$ and $\quad t_f = 2.0$ s

$\quad\quad$ At $t_i = 0$, $K_i = 2.0$ J

$\quad\quad$ At $t_f = 2.0$ s, $K_f = [2.0 + 24(2.0)^2 + 72(2.0)^4] = 1250$ J

$\quad\quad$ Therefore, $\quad W = 1.25 \times 10^3$ J $\quad \Diamond$

$$\left[\text{or use} \quad W = \int_{t_i}^{t_f} P \, dt = \int_0^2 (48t + 288t^3) \, dt \text{ , etc.}\right]$$

77. A 2100-kg pile driver is used to drive a steel *I*-beam into the ground. The pile driver falls 5.0 m before contacting the beam, and it drives the beam 12 cm into the ground before coming to rest. Using energy considerations, calculate the average force the beam exerts on the pile driver while the pile driver is brought to rest.

Solution   Choose the initial point when the mass is elevated and the final point when it comes to rest again 5.12 m below. Two forces, gravity and the normal force exxerted by the beam on the pile driver, do work on the pile driver:

$$K_i + W_{net} = K_f$$

$$0 + mgs_w \cos 0° + ns_n \cos 180° = 0$$

where $s_w = 5.12$ m and $s_n = 0.12$ m.

In this situation, the weight vector is in the direction of motion and the beam exerts a force on the pile driver which is opposite the direction of motion.

$$(2100 \text{ kg}) \, (9.80 \text{ m/s}^2) \, (5.12 \text{ m}) + n \, (0.12 \text{ m}) \, (-1) = 0$$

Solve for *n*.
$$n = \frac{1.05 \times 10^5 \text{ J}}{0.12 \text{ m}} = 880 \text{ kN} \quad \Diamond$$

Additional Calculation: Show that the work done by gravity on an object can be respresented by mgh, where h is the vertical height that the object falls. Apply your results to the problem above.

$$F = mg$$

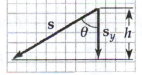

By the figure to the right, where **s** is the path of the object, and h is the height that the object falls,

$$h = |s_y| = s \cos \theta$$

So $mgh = Fs \cos \theta = \mathbf{F} \cdot \mathbf{s}$ $\Diamond$

In this problem, $mgh = n(s_n)$, or $(2100 \text{ kg})(9.80 \text{ m/s2})(5.12) = n(0.12 \text{ m})$

and $n = 880 \text{ kN}$ $\Diamond$

145

79. A 200-g block is pressed against a spring of force constant 1.40 kN/m until the block compresses the spring 10.0 cm. The spring rests at the bottom of a ramp inclined at 60.0° to the horizontal. Use energy considerations to determine how far up the incline the block moves before it stops (a) if there is no friction between block and ramp and (b) if the coefficient of kinetic friction is 0.400.

Solution   We have two forms of energy involved:

work done by gravity, $W_g = \mathbf{F \cdot s} = mg\,d\cos 150°,$

and work done by the spring: $W_s = \frac{1}{2}kx_m{}^2$

(a) Apply the work-energy theorem between the starting point and the point of maximum travel up the incline. The initial and final states of the block are stationary, so $\Delta K = 0$, and

$$W_{net} = W_s + W_g = 0$$

Substituting, $mgd\cos 150° + \frac{1}{2}kx_m{}^2 = 0$

and $d = \dfrac{-kx_m{}^2}{2mg\cos 150°} = \dfrac{(-1400 \text{ N/m})\,(0.100 \text{ m})^2}{2(0.200 \text{ kg})(9.80 \text{ m/s}^2)(-0.866)} = 4.12 \text{ m}$ ◊

(b)   In this case, we add a term representing the energy lost due to friction:

$$W_f = f_k d\cos 180° = -f_k d$$

The analysis of forces perpendicular to the incline reads

$$\Sigma F_y = ma_y \qquad +n - (1.96 \text{ N})\cos 60.0° = 0 \qquad n = 0.98 \text{ N}$$

Then the force of friction is $f_k = \mu_k n = 0.400 \times 0.\,980 \text{ N} = 0.392 \text{ N}$

Again applying our energy equation, with the additional term W_f, gives

$$0 + mgd\cos 150° + f_k d\cos 180° + \frac{1}{2}kx_m{}^2 = 0$$

$$(1.96 \text{ N})\,d\,(-0.866) + (0.392 \text{ N})\,d\,(-1.00) + 7.00 \text{ J} = 0$$

$$d = \frac{7.00 \text{ J}}{1.70 \text{ N} + 0.392 \text{ N}} = 3.35 \text{ m} \quad ◊$$

81. A 0.400-kg particle slides on a horizontal, circular track 1.50 m in radius. It is given an initial speed of 8.00 m/s. After one revolution, its speed drops to 6.00 m/s because of friction. (a) Find the energy lost due to friction in one revolution. (b) Calculate the coefficient of kinetic friction. (c) What is the total number of revolutions the particle makes before stopping?

Solution

(a) $W_f = \Delta K = \frac{1}{2}m(v_f^2 - v_i^2) = \frac{1}{2}(0.400 \text{ kg})\left[(6.00 \text{ m/s})^2 - (8.00 \text{ m/s})^2\right] = -5.60 \text{ J}$ ◊

(b) $W_f = -5.60 \text{ J} = -f\,\Delta s = -\mu mg\,(2\pi r)$

$$\mu = \frac{5.60 \text{ J}}{(0.400 \text{ kg})(9.80 \text{ m/s}^2)(2\pi)(1.50 \text{ m})} = 0.152 \quad ◊$$

(c) After n revolutions, $v = 0$ and all the initial kinetic energy K_i is lost due to friction.

$$K_i = \frac{1}{2}mv_i^2 = \frac{1}{2}(0.400 \text{ kg})(8.00 \text{ m/s})^2 = 12.8 \text{ J}$$

For one revolution, $W_f = -5.60 \text{ J}$, so for n revolutions

$$W_f = -5.60n \text{ J}$$

Since $W_f = \Delta K = -K_i$, we find

$$-5.60n = -12.8 \text{ J}$$

$$n = \frac{12.8}{5.60} = 2.29 \text{ revolutions} \quad ◊$$

89. The ball launcher in a pinball machine has a spring that has a force constant of 1.20 N/cm (Fig. P7.89). The surface on which the ball moves is inclined 10.0° with respect to the horizontal. If the spring is initially compressed 5.00 cm, find the launching speed of a 100-g ball when the plunger is released. Friction and the mass of the plunger are negligible.

Figure P7.89

Solution   Use the work-energy theorem.

$$W_{net} = \Delta K$$

$$W_s + W_g = \Delta K$$

$$\tfrac{1}{2}kx^2 - mgd\sin 10.0° = \tfrac{1}{2}m(v^2 - v_o{}^2)$$

$$v = \sqrt{\frac{2}{m}\left[\tfrac{1}{2}kd^2 - mgd\sin 10.0°\right]}$$

In this case,

$$x = d = 5.00\ \text{cm} = 5.00 \times 10^{-2}\ \text{m}, \quad k = 1.20\ \text{N/cm} = 120\ \text{N/m}, \quad \text{and} \quad v_o = 0$$

So,

$$v = \sqrt{\frac{2}{0.100\ \text{kg}}\left[\tfrac{1}{2}(120\ \text{N/m})(5.00 \times 10^{-2}\ \text{m})^2 - (0.100\ \text{kg})(9.80\ \text{m/s}^2)(5.00 \times 10^{-2}\ \text{m})(\sin 10.0°)\right]}$$

and

$$v = 1.68\ \text{m/s} \quad \Diamond$$

Chapter 8

Potential Energy and
Conservation of Energy

Chapter 8

POTENTIAL ENERGY
AND CONSERVATION OF ENERGY

INTRODUCTION

In Chapter 7 we introduced the concept of kinetic energy, which is the energy associated with the motion of an object. In this chapter we introduce another form of mechanical energy, *potential energy*, which is the energy associated with the position or configuration of an object. Potential energy can be thought of as stored energy that can either do work or be converted to kinetic energy.

The potential energy concept can be used only when dealing with a special class of forces called *conservative forces*. When only internal conservative forces, such as gravitational or spring forces, act within a system, the kinetic energy gained (or lost) by the system as its members change their relative positions is compensated by an equal energy loss (or gain) in potential energy. This is known as the *principle of conservation of mechanical energy*.

NOTES FROM SELECTED CHAPTER SECTIONS

8.1 Potential Energy

The work done on an object by the force of gravity is equal to the object's initial potential energy minus its final potential energy. The gravitational potential energy associated with an object depends only on the object's weight and its vertical height above the surface of the Earth. If the height above the surface increases, the potential energy will also increase; but the work done by the gravitational force will be negative. In working problems involving gravitational potential energy, it is necessary to choose an arbitrary reference level (or location) at which the potential energy is zero.

8.2 Conservative and Nonconservative Forces

A force is said to be *conservative* if the work done by that force on a body moving between any two points is independent of the path taken. In addition, the work done by a conservative force is zero when the body moves through any closed path and returns to its initial position. *Nonconservative* forces are those for which the work done on a particle moving between two points depends on the path. Furthermore, the work done by a nonconservative force (e.g. friction) around a closed path is not zero.

150

8.3 Conservative Forces and Potential Energy

It is possible to define a *potential energy function* associated with a conservative force such that the work done by the conservative force equals the negative of the change in the potential energy associated with the force.

8.4 Conservation of Energy

The law of *conservation of mechanical energy* states that the total mechanical energy of a system remains constant if the only force that does work on the system is a conservative force.

The *work done by the force of gravity* is equal to the initial value of the potential energy minus the final value of the potential energy.

8.5 Changes in Mechanical Energy When Nonconservative Forces Are Present

The *total work* done on a system by all forces (conservative and nonconservative) equals the *change in the kinetic energy* of the system. The work done by all *nonconservative* forces equals the change in the total *mechanical energy* of the system.

8.7 Energy Diagrams and the Equilibrium of a System

Positions of stable equilibrium correspond to points for which $U(x)$ has a minimum value; positions of unstable equilibrium correspond to those points for which $U(x)$ has a maximum value; finally, a position of neutral equilibrium corresponds to a region over which $U(x)$ remains constant.

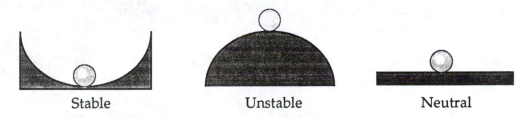

Stable   Unstable   Neutral

8.8 Conservation of Energy in General

The energy conservation principle can be generalized to include all forms of energies and energy transformations. Energy may be transformed from one form to another, but *the total energy of an isolated system remains constant*. That is, *energy cannot be created or destroyed*.

EQUATIONS AND CONCEPTS

The gravitational potential energy associated with an object at any point in space is the product of the object's weight and the vertical coordinate.

$$U_g \equiv mgy \tag{8.1}$$

The work done on any object by the gravitational force is equal to the change in the gravitational potential energy.

$$W_g = U_i - U_f \tag{8.2}$$

The units of energy (kinetic and potential) are the same as the units of work— joules.

In calculating the work done by the gravitational force, remember that the difference in potential energy between two points is independent of the location of the origin. Choose an origin which is convenient to calculate U_i and U_f for a particular situation.

The quantity $\frac{1}{2}kx^2$ is referred to as the *elastic potential energy* stored in a spring which has been deformed a distance x from the equilibrium position.

$$U_s \equiv \frac{1}{2}kx^2 \tag{8.4}$$

A potential energy function U can be defined for a conservative force.

$$\Delta U = U_f - U_i = -\int_{x_i}^{x_f} F_x \, dx \tag{8.6}$$

Total mechanical energy of a system is the sum of the kinetic and potential energies.

$$E = K + U \tag{8.8}$$

Note that both the gravitational force and the spring force satisfy the required properties of a conservative force. That is, the work done is path independent and is zero for any closed path.

The law of *conservation of mechanical energy* says that if only conservative forces act on a system, the sum of the kinetic and potential energies remains constant--or is conserved. According to this important conservation law, if the kinetic energy of the system increases by some amount, the potential energy must decrease by the same amount--and vice versa.

$$K_i + U_i = K_f + U_f \qquad (8.9)$$

If more than one conservative force acts on an object, then a potential energy term is associated with each force.

$$K_i + \Sigma U_i = K_f + \Sigma U_f \qquad (8.10)$$

If the only conservative force is the gravitational force, the equation for conservation of mechanical energy takes a special form.

$$\tfrac{1}{2}mv_i^2 + mgy_i = \tfrac{1}{2}mv_f^2 + mgy_f \qquad (8.11)$$

Conservation of mechanical energy for a mass-spring system is similar:

$$\tfrac{1}{2}mv_i^2 + \tfrac{1}{2}kx_i^2 = \tfrac{1}{2}mv_f^2 + \tfrac{1}{2}kx_f^2 \qquad (8.12)$$

When nonconservative forces (e.g. friction) are present, the change in kinetic energy due to the *net force* can be calculated, although the work done by the nonconservative forces cannot (in general) be calculated.

$$\int \mathbf{F}_{net} \cdot d\mathbf{x} = \Delta K \qquad (8.13)$$

There is an important relationship between a conservative force and the potential energy of the system on which the force acts. The conservative force equals the negative derivative of the system's potential energy with respect to x.

$$F_x = -\frac{dU}{dx} \qquad (8.15)$$

Mass and energy are conserved not separately, but as a single entity called *mass-energy*. The relationship between energy and mass is stated by the famous Einstein equation.

$$E = mc^2 \qquad (8.16)$$

Chapter 8

SUGGESTIONS, SKILLS, AND STRATEGIES

Choosing a Zero Level

In working problems involving gravitational potential energy, it is always necessary to choose a location at which the gravitational potential energy is zero. This choice is completely arbitrary because the important quantity is the *difference* in potential energy, and that difference is independent of the location of zero. It is often convenient, but not essential, to choose the surface of the Earth as the reference position for zero potential energy. In most cases, the statement of the problem suggests a convenient level to use.

Conservation of Energy

Take the following steps in applying the principle of conservation of energy:

- Define your system, which may consist of more than one object.

- Select a reference position for the zero point of gravitational potential energy.

- Determine whether or not nonconservative forces are present.

- If mechanical energy is conserved (that is, if only conservative forces are present), you can write the total initial energy at some point as the sum of the kinetic and potential energies at that point. Then, write an expression for the total final energy, $KE_f + PE_f$, at the final point of interest. Since mechanical energy is conserved, you can equate the two total energies and solve for the unknown.

- If nonconservative forces such as friction are present (and thus mechanical energy is not conserved), first write expressions for the total initial and total final energies. In this case, the difference between the two total energies is equal to the mechanical energy lost due to nonconservative force(s).

REVIEW CHECKLIST

▷ Recognize that the gravitational potential energy function, $U_g = mgy$, can be positive, negative, or zero, depending on the location of the reference level used to measure y. Be aware of the fact that although U depends on the origin of the coordinate system, the *change* in potential energy, $(U)_f - (U)_i$, is *independent* of the coordinate system used to define U.

154

▷ Understand that a force is said to be *conservative* if the work done by that force on a body moving between any two points is independent of the path taken. *Nonconservative* forces are those for which the work done on a particle moving between two points depends on the path. Account for nonconservative forces acting on a system using the work-energy theorem. In this case, the work done by all nonconservative forces equals the change in total mechanical energy of the system.

▷ Understand the distinction between kinetic energy (energy associated with motion), potential energy (energy associated with the position or coordinates of a system), and the total mechanical energy of a system. State the law of conservation of mechanical energy, noting that mechanical energy is conserved only when conservative forces act on a system. This extremely powerful concept is most important in all areas of physics.

SOLUTIONS TO SELECTED END-OF-CHAPTER PROBLEMS

1. A 4.00-kg particle moves from the origin to the position having coordinates $x = 5.00$ m and $y = 5.00$ m under the influence of gravity acting in the negative y direction (Fig. P8.1). Using Equation 7.2, calculate the work done by gravity in going from O to C along (a) OAC, (b) OBC, (c) OC. Your results should all be identical. Why?

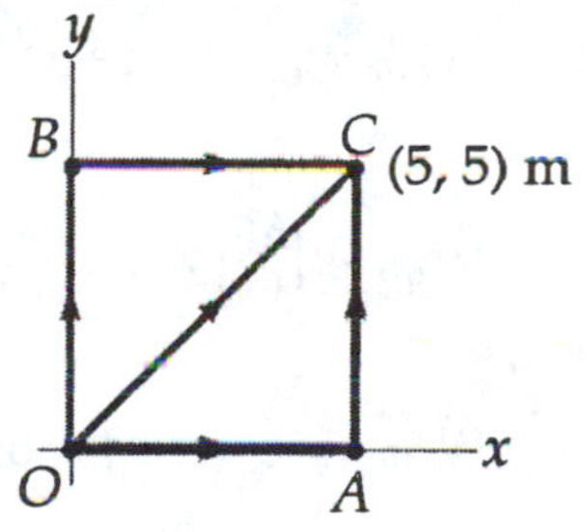

Solution $w = mg = (4.00 \text{ kg})(9.80 \text{ m/s}^2) = 39.2$ N

Figure P8.1

(a) $W_{OAC} = W_{OA} + W_{AC}$

$= wd_{OA} \cos 270° + wd_{AC} \cos 180°$

$= (39.2 \text{ N})(5.00 \text{ m})(0) + (39.2 \text{ N})(5.00 \text{ m})(-1) = -196 \text{ J}$ ◊

(b) $W_{OBC} = W_{OB} + W_{BC}$

$= (39.2 \text{ N})(5.00 \text{ m}) \cos 180° + (39.2 \text{ N})(5.00 \text{ m}) \cos 90° = -196 \text{ J}$ ◊

(c) $W_{OC} = wd_{OC} \cos 135° = (39.2 \text{ N})(5\sqrt{2} \text{ m})\left(-\frac{\sqrt{2}}{2}\right) = -196 \text{ J}$ ◊

The results should all be the same since gravitational forces are conservative.

5. A force acting on a particle moving in the xy plane is $F = (2y\mathbf{i} + x^2\mathbf{j})$ N, where x and y are in meters. The particle moves from the origin to a final position having coordinates $x = 5.0$ m and $y = 5.0$ m, as in Figure P8.1. Calculate the work done by F along (a) OAC, (b) OBC, (c) OC. (d) Is F conservative or nonconservative? Explain.

Solution

In evaluating the following integrals, remember $\mathbf{i} \cdot \mathbf{j} = \mathbf{j} \cdot \mathbf{j} = 1$ and $\mathbf{i} \cdot \mathbf{j} = 0$.

(a) $W_{OA} = \int_0^{5.0} (2y\mathbf{i} + x^2\mathbf{j}) \cdot (\mathbf{i}\,dx) = \int_0^{5.0} 2y\,dx = 2yx \Big]_{x=0,y=0}^{x=5,y=0} = 0$

$W_{AC} = \int_0^{5.0} (2y\mathbf{i} + x^2\mathbf{j}) \cdot (\mathbf{i}\,dy) = \int_0^{5.0} x^2\,dy = x^2y \Big]_{x=5.0,y=0}^{x=5.0,y=5.0} = 125$ J

$W_{OAC} = 0 + 125$ J $= 125$ J $\Diamond$

(b) $W_{OB} = \int_0^{5.0} (2y\mathbf{i} + x^2\mathbf{j}) \cdot (\mathbf{j}\,dy) = \int_0^{5.0} x^2\,dy = x^2y \Big]_{x=0...}^{x=0...} = 0$

$W_{BC} = \int_0^{5.0} (2y\mathbf{i} + x^2\mathbf{j}) \cdot (\mathbf{j}\,dx) = \int_0^{5.0} 2y\,dx = 2xy \Big]_{x=0,y=5.0}^{x=5.0,y=5.0} = 50$ J

$W_{OBC} = 0 + 50$ J $= 50$ J $\Diamond$

(c) $W_{OC} = \int (2y\mathbf{i} + x^2\mathbf{j}) \cdot (\mathbf{i}\,dx + \mathbf{j}\,dy) = \int (2y\,dx + x^2\,dy)$

Since $x = y$ along OC, $dx = dy$ and

$W_{OC} = \int_0^5 (2x + x^2)dx = 66.7$ J $\Diamond$

(d) **F** is non-conservative since the work done is path dependent. $\Diamond$

---

7. A single conservative force $F_x = (2.0x + 4.0)$ N acts on a 5.0-kg particle, where x is in meters. As the particle moves along the x axis from $x = 1.0$ m to $x = 5.0$ m, calculate (a) the work done by this force, (b) the change in the potential energy of the particle, and (c) its kinetic energy at $x = 5.0$ m if its speed at $x = 1.0$ m is 3.0 m/s.

Solution

(a) $\quad W_F = \int_{x_i}^{x_f} F_x\, dx$

where $\qquad F_x = (2.0x + 4.0)$ N, $\quad x_i = 1.0$ m, $\quad$ and $\quad x_f = 5.0$ m

therefore, $\qquad W_F = \int_{1.0\text{ m}}^{5.0\text{ m}} (2.0x + 4.0)dx\ \text{N} \cdot \text{m} = x^2 + 4.0x\big]_{1.0}^{5.0}\ \text{N} \cdot \text{m}$

and $\qquad W_F = 40\text{ J} \quad \Diamond$

(b) The change in potential energy equals the negative of the work done by the conservative force.

$\Delta U = -W_F = -40\text{ J} \quad \Diamond$

(c) When only conservative forces act,

$\Delta K + \Delta U = 0 \qquad$ (conservation of mechanical energy)

$K_f - K_i = -\Delta U \qquad \text{or} \qquad K_f = K_i - \Delta U$

$K_f = \tfrac{1}{2}mv_i^2 - \Delta U$

$K_f = \left(\tfrac{1}{2}\right)(5.0\text{ kg})(3.0\text{ m}/\text{s})^2 - (-40.0\text{ J})$

$K_f = 62.5\text{ J} \quad \Diamond$

9. A single constant force $\mathbf{F} = (3.0\mathbf{i} + 5.0\mathbf{j})$ N acts on a 4.0-kg particle. (a) Calculate the work done by this force if the particle moves from the origin to the point having the vector position $\mathbf{r} = (2.0\mathbf{i} - 3.0\mathbf{j})$ m. Does this result depend on the path? Explain. (b) What is the speed of the particle at $\mathbf{r}$ if its speed at the origin is 4.0 m/s? (c) What is the change in its potential energy?

Solution

(a) $W = \mathbf{F} \cdot \mathbf{s} = \mathbf{F} \cdot (\mathbf{r} - 0) = (3.0\mathbf{i}$ N $+ 5.0\mathbf{j}$ N $) \cdot (2.0\mathbf{i} - 3.0\mathbf{j})$ m

$W = (6.0 - 15.0)$ N·m $= -9.0$ J ◊

The work does not depend on the path because the force is constant.

Therefore, $$W = \int_i^f \mathbf{F} \cdot d\mathbf{s} = \mathbf{F} \cdot \int_i^f d\mathbf{s} = \mathbf{F} \cdot (\mathbf{r}_f - \mathbf{r}_i) \quad ◊$$

(b) $W = \Delta K = K_f - K_i$,   so $K_i + W = K_f$

$$\tfrac{1}{2}(4.0 \text{ kg})(4.0 \text{ m/s})^2 - 9.0 \text{ J} = \tfrac{1}{2}(4.0 \text{ kg})v_f^2$$

$$v_f = \sqrt{\frac{2 \times (32 - 9.0)\text{ J}}{4.0 \text{ kg}}} = 3.39 \text{ m/s} \quad ◊$$

(c) $U_f - U_i = -\int_i^f \mathbf{F} \cdot d\mathbf{s} = -W = +9.00$ J ◊

11. A bead slides without friction around a loop-the-loop (Fig. P8.11). If the bead is released from a height $h = 3.50R$, what is its speed at point A? How large is the normal force on it if its mass is 5.00 g?

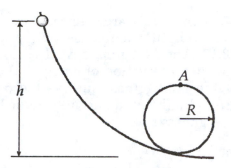

Solution It is convenient to choose the reference point of potential energy to be at the lowest point of the bead's motion.

Since $v_i = 0$,

Figure P8.11

$$E_i = K_i + U_i = 0 + mgh = mg(3.50R)$$

The total energy of the bead at point A can be written as

$$E_A = K_A + U_A = \tfrac{1}{2}mv_A{}^2 + mg(2R)$$

Since mechanical energy is conserved, $E_i = E_A$, and we get

$$\tfrac{1}{2}mv_A{}^2 + mg(2R) = mg(3.50R)$$

$$v_A{}^2 = 3.0gR \qquad \text{or} \qquad v_A = \sqrt{3.0gR} \quad \lozenge$$

To find the normal force at the top, it is useful to construct a free-body diagram as shown, where both **n** and m**g** are downward. Newton's second law gives

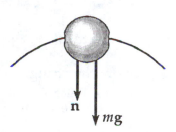

$$n + mg = \frac{mv_A{}^2}{R} = \frac{m(3.0gR)}{R} = 3.0mg$$

$$n = 3.0mg - mg = 2.0mg$$

$$n = 2\left(5.0 \times 10^{-3} \text{ kg}\right)\left(9.80 \text{ m}/\text{s}^2\right) = 0.0980 \text{ N} \quad \text{downward} \quad \lozenge$$

17. Two masses are connected by a light string passing over a light frictionless pulley as shown in Figure P8.17. The 5.0-kg mass is released from rest. Using the law of conservation of energy, (a) determine the speed of the 3.0-kg mass just as the 5.0-kg mass hits the ground. (b) Find the maximum height to which the 3.0-kg mass rises.

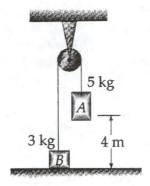

Figure P8.17

Solution

(a) Choose the initial point before release and the final point just before the larger mass hits the floor. The total energy of the two objects together remains constant, and the work-energy theorem for the two objects A and B is

$$(K_A + K_B + U_A + U_B)_i + \Delta K_{nc} = (K_A + K_B + U_A + U_B)_f$$

At the initial point, $K_{A,\,i}$, $K_{B,\,i}$, and $U_{B,\,i}$ go to zero; the change in kinetic energy ΔK_{nc} is zero, and $U_{A,f}$ is zero, so:

$$(5.0 \text{ kg})(9.80 \text{ m / s}^2)(4.0 \text{ m}) = \tfrac{1}{2}(5.0 \text{ kg})v_f^2 + \tfrac{1}{2}(3.0 \text{ kg})v_f^2 + (3.0 \text{ kg})(9.80 \text{ m / s}^2)(4.0 \text{ m})$$

$$(2.0 \text{ kg})(9.80 \text{ m / s}^2)(4.0 \text{ m}) = \tfrac{1}{2}(8.0 \text{ kg})v_f^2$$

$v_f = 4.43 \text{ m/s}$ ◊

(b) Now the string goes slack. The 5.0-kg mass loses all its mechanical energy, but the 3.0-kg mass becomes a projectile. Take the initial point at the previous final point, and the new final point at its maximum height:

$$(K + U_g)_i = (K + U_g)_f$$

$$\tfrac{1}{2}(3.0 \text{ kg})(4.43 \text{ m/s})^2 + (3.0 \text{ kg})(9.80 \text{ m/s}^2)(4.0 \text{ m}) = 0 + (3.0 \text{ kg})(9.80 \text{ m/s}^2)y_f$$

$y_f = 5.00 \text{ m}$ ◊

or 1.0 m higher than the height of the 5.0-kg mass when it was released.

---

19. A 5.0-kg block is set into motion up an inclined plane with an initial speed of 8.0 m/s (Fig. P8.19). The block comes to rest after traveling 3.0 m along the plane, which is inclined at an angle of 30° to the horizontal. Determine (a) the change in the block's kinetic energy, (b) the change in its potential energy, (c) the frictional force exerted on it (assumed to be constant). (d) What is the coefficient of kinetic friction?

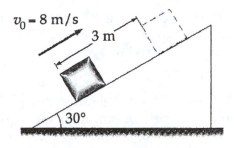

$v_0 = 8$ m/s

3 m

30°

Figure P8.19

Solution

(a) $\Delta K = K_f - K_i = 0 - \frac{1}{2}(5.0 \text{ kg})(8.0 \text{ m/s})^2 = -160 \text{ J}$ ◊

(b) $\Delta U = U_{gf} - U_{gi} = mgy_f - 0 = (5.0 \text{ kg})(9.80 \text{ m/s}^2)(3.0 \text{ m}) \sin 30°$

$\Delta U = 73.5 \text{ J}$ ◊

(c) $(K + U_g)_i + \Delta K_{nc} = (K + U_g)_f$

$\frac{1}{2}(5.0 \text{ kg})(8.0 \text{ m/s})^2 + 0 + f(3.0 \text{ m}) \cos 180° = 0 + (5.0 \text{ kg})(9.80 \text{ m/s}^2)(1.5 \text{ m})$

$160 \text{ J} - f(3.0 \text{ m}) = 73.5 \text{ J}$

$f = \dfrac{86.5 \text{ J}}{3.0 \text{ m}} = 28.8 \text{ N}$ ◊

(d) The forces perpendicular to the incline must add to zero:

$\Sigma F_y = 0$

$+n - mg \cos 30° = 0$

$n = (5.0 \text{ kg})(9.80 \text{ m/s}^2) \cos 30° = 42.4 \text{ N}$

Now, $f_k = \mu_k n$ gives $\mu_k = \dfrac{f_k}{n} = \dfrac{28.8 \text{ N}}{42.4 \text{ N}} = 0.679$ ◊

25. The coefficient of friction between the 3.0-kg mass and surface in Figure P8.25 is 0.40. The system starts from rest. What is the speed of the 5.0-kg mass when it has fallen 1.5 m?

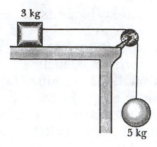

Figure P8.25

Solution We could solve this problem by using $\Sigma F = ma$ to give a pair of simultaneous equations in the unknown acceleration and tension; then we would have to solve a motion problem to find the final speed. It is easier to solve using the work-energy theorem.

For the 3.0-kg block, $\qquad w = mg = (3.0 \text{ kg})(9.80 \text{ m/s}^2) = 29.4 \text{ N}$

From $\Sigma F_y = ma_y$, $\qquad n - 29.4 \text{ N} = 0,$

and $\qquad n = 29.4 \text{ N}$

$$f_k = \mu_k n = 0.40 \times 29.4 \text{ N} = 11.8 \text{ N}$$

Now for two objects A and B, the work-energy theorem is

$$(K_A + K_B + U_A + U_B)_i + \Delta K_{nc} = (K_A + K_B + U_A + U_B)_f \qquad (1)$$

Choose the initial point before release and the final point after each block has moved 1.5 m. For the 5.0-kg block, the zero level of gravitational energy is at the final position. For the 3.0-kg block, choose $U_g = 0$ at the tabletop.

By substituting into Equation (1):

$$0 + 0 + 0 + m_B g y_{Bi} + f s \cos \theta = \tfrac{1}{2} m_A v_f^2 + \tfrac{1}{2} m_B v_f^2 + 0 + 0$$

$$(5.0 \text{ kg})(9.80 \text{ m/s}^2)(1.5 \text{ m}) + (11.8 \text{ N})(1.5 \text{ m}) \cos 180° = \tfrac{1}{2}(3.0 \text{ kg})v_f^2 + \tfrac{1}{2}(5.0 \text{ kg})v_f^2$$

$$73.5 \text{ J} - 17.6 \text{ J} = \tfrac{1}{2}(8.0 \text{ kg})v_f^2 \qquad \text{or} \qquad v_f = \sqrt{\frac{2 \times 55.9 \text{ J}}{8.0 \text{ kg}}} = 3.74 \text{ m/s} \quad \Diamond$$

33. A block of mass 0.250 kg is placed on top of a vertical spring of constant k = 5000 N/m and pushed downward, compressing the spring 0.100 m. After the block is released, it travels upward and then leaves the spring. To what maximum height above the point of release does it rise?

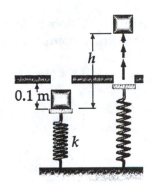

Solution   In both the initial and final states, the block is not moving. Therefore, the initial and final energies are:

$$E_i = K_i + U_i = 0 + (U_g + U_s)_i = 0 + \left(0 + \tfrac{1}{2}kx^2\right)$$

$$E_f = K_f + U_f = 0 + (U_g + U_s)_f = 0 + (mgh + 0)$$

Since $E_i = E_f$, $mgh = \tfrac{1}{2}kx^2$

and

$$h = \frac{kx^2}{2mg} = \frac{(5000 \text{ N}/\text{m})(0.100 \text{ m})^2}{2(0.250 \text{ kg})(9.80 \text{ m}/\text{s}^2)} = 10.2 \text{ m} \quad \Diamond$$

---

35. A 10.0-kg block is released from point A in Figure P8.35. The track is frictionless except for the portion BC, of length 6.00 m. The block travels down the track, hits a spring of force constant k = 2250 N/m, and compresses it 0.300 m from its equilibrium position before coming to rest momentarily. Determine the coefficient of kinetic friction between surface BC and block.

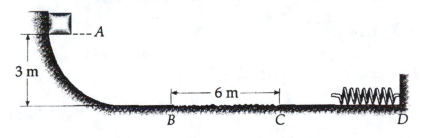

Figure P8.35

Solution   Choose the initial point when the block is at A and the final point when the spring is fully compressed.

$$(K+U_g+U_s)_A + \Delta K_{nc} = (K+U_g+U_s)_f$$

$$0+mgy_A+0+fs_{BC}\cos 180° = 0+0+\tfrac{1}{2}kx_f^2$$

$$(10.0 \text{ kg})(9.80 \text{ m/s}^2)(3.00 \text{ m}) - f(6.00 \text{ m}) = \tfrac{1}{2}(2250 \text{ N/m})(0.300 \text{ m})^2$$

$$294 \text{ J} - f(6.00 \text{ m}) = 101 \text{ J}$$

$$f = \frac{193 \text{ J}}{6.00 \text{ m}} = 32.1 \text{ N}$$

Now consider the vertical forces when the block is between B and C.

From $\Sigma F_y = 0$, $+n - mg = 0$

$$n = (10.0 \text{ kg})(9.80 \text{ m/s}^2) = 98.0 \text{ N}$$

Applying $f_k = \mu_k n$

$$\mu_k = \frac{f_k}{n} = \frac{32.1 \text{ N}}{98.0 \text{ N}} = 0.328 \quad \Diamond$$

Another way to look at the problem is to say that the original potential energy mgh goes into frictional energy loss, and into compressing the spring:

$$mgh = \mathbf{f} \cdot \mathbf{s} + \tfrac{1}{2}kx^2$$

or $\mathbf{f} \cdot \mathbf{s} = \mu mg \cdot s = mgh - \tfrac{1}{2}kx^2$

Substituting $m = 10.0 \text{ kg}$, $h = 3.00 \text{ m}$, $k = 2250 \text{ N/m}$, $x = 0.300 \text{ m}$, $s = 6.00 \text{ m}$

yields $$\mu = \frac{h}{s} - \frac{kx^2}{2mgs} = 0.328 \quad \Diamond$$

39. A potential energy function for a two-dimensional force is of the form $U = 3x^3y - 7x$. Find the force that acts at the point (x, y).

Solution

$$U = 3x^3y - 7x$$

$$F_x = -\frac{\partial U}{\partial x} = -\frac{\partial}{\partial x}\left(3x^3y - 7x\right) = -9x^2y + 7$$

$$F_y = -\frac{\partial U}{\partial y} = -\frac{\partial}{\partial y}\left(3x^3y - 7x\right) = -3x^3$$

Therefore, $\mathbf{F} = F_x\mathbf{i} + F_y\mathbf{j} = \left(7 - 9x^2y\right)\mathbf{i} - 3x^3\mathbf{j}$ ◊

---

43. A particle of mass m = 5.00 kg is released from point A on the frictionless track shown in Figure P8.43. Determine (a) the particle's speed at points B and C and (b) the net work done by the force of gravity in moving the particle from A to C.

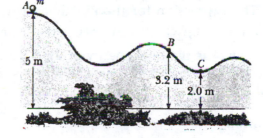

Figure P8.43

Solution

(a) There is no friction. The normal force, which is always perpendicular to the motion, does no work; so energy is conserved:

$$(K + U)_A + \Delta K_{nc} = (K + U)_B$$

$$0 + mgy_A + 0 = \tfrac{1}{2}mv_B^2 + mgy_B$$

The result does not depend on the mass of the particle:

$$v_B = \sqrt{2g(y_A - y_B)} = \sqrt{2(9.80 \text{ m/s}^2)(1.80 \text{ m})} = 5.94 \text{ m/s}$$ ◊

Again,

$$(K+U)_A = (K+U)_C$$

$$0 + mgy_A = \tfrac{1}{2}mv_C^2 + mgy_C$$

$$v_C = \sqrt{2g(y_A - y_C)} = \sqrt{2(9.80 \text{ m/s}^2)(3.0 \text{ m})} = 7.67 \text{ m/s} \quad \Diamond$$

(b) The work done by gravity is the negative of the change in gravitational potential energy:

$$W_g = -\Delta U = -(U_C - U_A) = U_A - U_C = mg(y_A - y_C)$$

$$W_g = -\Delta U = (5.0 \text{ kg})(9.80 \text{ m/s}^2)(3.0 \text{ m}) = 147 \text{ J} \quad \Diamond$$

---

45. The expression for the kinetic energy of a particle moving with speed v is given by Equation 7.20, which can be written as $K = \gamma mc^2 - mc^2$, where $\gamma = [1-(v/c)^2]^{-1/2}$. The term γmc^2 is the total energy of the particle, and the term mc^2 is its rest energy. A proton moves with a speed of 0.990c, where c is the speed of light. Find (a) its rest energy, (b) its total energy, and (c) its kinetic energy.

Solution

(a) $mc^2 = \left(1.67 \times 10^{-27} \text{ kg}\right)(3.00 \times 10^8 \text{ m/s})^2 = 1.50 \times 10^{-10} \text{ J} \quad \Diamond$

(b) $\gamma = \dfrac{1}{\sqrt{1 - v^2/c^2}} = \dfrac{1}{\sqrt{1 - 0.990^2}} = 7.09$

$\gamma mc^2 = 7.09 \ (1.50 \times 10^{-10} \text{ J}) = 1.07 \times 10^{-9} \text{ J} \quad \Diamond$

(c) $K = \gamma mc^2 - mc^2 = (7.09 - 1) \ 1.50 \times 10^{-10} \text{ J} = 9.15 \times 10^{-10} \text{ J} \quad \Diamond$

---

47. The particle described in Problem 46 (Fig. P8.46) is released from rest at A. The speed of the particle at B is 1.50 m/s. (a) What is its kinetic energy at B? (b) How much energy is lost due to friction as the particle moves from A to B? (c) Is it possible to determine μ from these results in any simple manner? Explain.

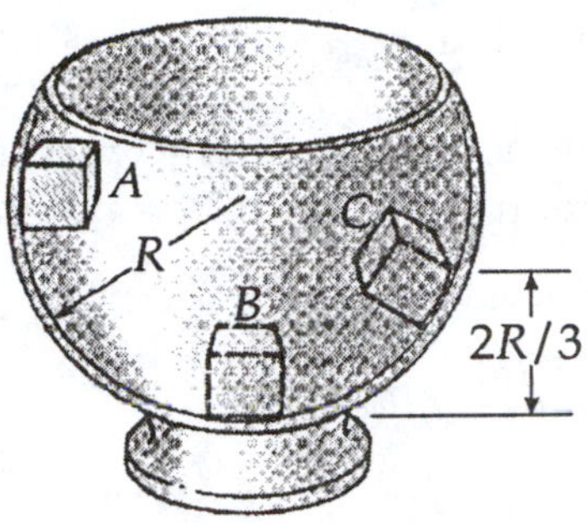

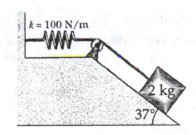

Figure P8.46

Solution   Let us take $U = 0$ at B. Since $v_i = 0$ at A,

$$K_A = 0 \quad \text{and} \quad U_A = mgR.$$

(a)   Since $v_B = 1.50$ m/s,

$$K_B = \tfrac{1}{2}mv_B{}^2 = \tfrac{1}{2}(0.200 \text{ kg})(1.50 \text{ m/s})^2 = 0.225 \text{ J} \quad \lozenge$$

(b)   At A, $E_i = K_A + U_A = 0 + mgR = (0.200 \text{ kg})(9.80 \text{ m/s}^2)(0.300 \text{ m}) = 0.588 \text{ J}$

At B, $E_f = K_B + U_B = 0.225 \text{ J}$

Hence, the energy lost $= E_i - E_f = 0.588 \text{ J} - 0.225 \text{ J} = 0.363 \text{ J} \quad \lozenge$

(c)   Even though the energy lost is known, both the normal force and the friction force change with position as the block slides on the inside of the bowl. Therefore, there is no easy way to find μ.

---

55. A 2.00-kg block situated on an incline is connected to a spring of negligible mass having a spring constant of 100 N/m (Fig. P8.55). The block is released from rest when the spring is unstretched, and the pulley is frictionless. The block moves 20.0 cm down the incline before coming to rest. Find the coefficient of kinetic friction between block and incline.

Figure P8.55

Solution

The nonconservative work (due to friction) must equal the change in the kinetic energy plus the change in the potential energy (gravitational and elastic). Therefore,

$$W_f = \Delta K + \Delta U = \Delta K + \Delta U_g + \Delta U_s$$

$$-\mu_k mgx \cos\theta = \Delta K + \tfrac{1}{2}kx^2 - mgx \sin\theta$$

but $\Delta K = 0$ because $v_i = v_f = 0$.

Thus,

$$-\mu_k(2.00 \text{ kg})(9.80 \text{ m}/\text{s}^2)(\cos 37°)(0.200 \text{ m})$$

$$= \frac{(100 \text{ N}/\text{m})(0.200 \text{ m})^2}{2} - 2(9.80 \text{ m}/\text{s}^2)(\sin 37°)(0.200 \text{ m})$$

and we find $\mu_k = 0.115$ ◊.

Note that in the above we had a *gain* in elastic potential energy for the spring and a *loss* in gravitational potential energy. The net loss in mechanical energy is equal to the energy lost due to friction.

---

57. A ball whirls around in a vertical circle at the end of a string. If the ball's total energy remains constant, show that the tension in the string at the bottom is greater than the tension at the top by six times the weight of the ball.

Solution   Applying Newton's second law at the bottom (*b*) and top (*t*) of the circular path gives

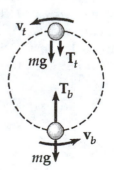

$$T_b - mg = \frac{mv_b^2}{R} \qquad (1)$$

$$-T_t - mg = -\frac{mv_t^2}{R} \qquad (2)$$

Subtracting (1) and (2) gives

$$T_b = T_t + 2mg + \frac{m(v_b^2 - v_t^2)}{R} \quad (3)$$

Also, energy must be conserved; that is, $\quad \Delta K + \Delta U = 0$

So,

$$\frac{1}{2}m(v_b^2 - v_t^2) + (0 - 2mgR) = 0$$

or

$$\frac{v_b^2 - v_t^2}{R} = 4g \quad (4)$$

Substituting (4) into (3) gives $\quad T_b = T_t + 6mg \quad \lozenge$

---

59. A 20.0-kg block is connected to a 30.0-kg block by a string that passes over a frictionless pulley. The 30.0-kg block is connected to a spring that has negligible mass and a force constant of 250 N/m, as in Figure P8.59. The spring is unstretched when the system is as shown in the figure, and the incline is frictionless. The 20.0-kg block is pulled 20.0 cm down the incline (so that the 30.0-kg block is 40.0 cm above the floor) and released from rest. Find the speed of each block when the 30.0-kg block is 20.0 cm above the floor (that is, when the spring is unstretched).

Solution Let x be the distance the spring is stretched from equilibrium ($x = 0.20$ m), which corresponds to the upward displacement of the 30.0-kg mass. Also let $U_g = 0$ be measured from the lowest position of the 20.0-kg mass when the system is released from rest. If v is the speed of each block as they pass through the original unstretched position, then

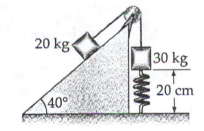

Figure P8.59

$$\Delta K + \Delta U_s + \Delta U_g = 0 \quad \text{gives:}$$

$$\left[\tfrac{1}{2}(m_1 + m_2)v^2 - 0\right] + \left(0 - \tfrac{1}{2}kx^2\right) + (m_2 \sin\theta - m_1)gx = 0$$

$$\tfrac{1}{2}(50.0 \text{ kg})v^2 - \tfrac{1}{2}(250 \text{ N/m})(0.20 \text{ m})^2 + \left[(20.0 \text{ kg}) \sin 40° - 30.0 \text{ kg}\right](9.80 \text{ m/s}^2)(0.20 \text{ m}) = 0$$

Solving for v gives $\quad v = 1.24$ m/s $\quad \lozenge$

65. A block of mass 0.500 kg is pushed against a horizontal spring of negligible mass, compressing the spring a distance of Δx (Fig. P8.65). The spring constant is 450 N/m. When released, the block travels along a frictionless, horizontal surface to point B, the bottom of a vertical circular track of radius $R = 1.00$ m, and continues to move up the track. The speed of the block at the bottom of the track is $v_B = 12$ m/s, and the

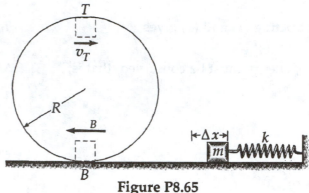

Figure P8.65

block experiences an average frictional force of 7.00 N while sliding up the track. (a) What is Δx? (b) What is the speed of the block at the top of the track? (c) Does the block reach the top of the track, or does it fall off before reaching the top?

Solution   Energy is conserved in the firing of the block. Therefore,

(a) $\frac{1}{2}kx^2 = \frac{1}{2}mv^2$, or $\frac{1}{2}(450 \text{ N/m})(\Delta x)^2 = \frac{1}{2}(0.500 \text{ kg})(12.0 \text{ m/s})^2$

   Thus, $\Delta x = 0.400$ m   ◊

(b)   To find speed of block at the top: $\Delta E = W_f$

$$\left(mgh_T + \tfrac{1}{2}mv_T{}^2\right) - \left(mgh_B + \tfrac{1}{2}mv_B{}^2\right) = -f(\pi R)$$

Substituting $mgh_T = (0.500 \text{ kg})(9.80 \text{ m/s}^2)(2.00 \text{ m}) = 9.80$ J,

We have $9.80 \text{ J} + \tfrac{1}{2}(0.500 \text{ kg})v_T{}^2 - \tfrac{1}{2}(0.500 \text{ kg})(12.0 \text{ m / s})^2 = -(7.00 \text{ N})(\pi)(1.00 \text{ m})$

and $0.250 v_T{}^2 = 4.21$

Thus, $v_T = 4.10$ m/s   ◊

(c)   Block falls if $a_c < g$

$$a_c = \frac{v_T{}^2}{R} = \frac{(4.10 \text{ m / s})^2}{1.00 \text{ m}} = 16.8 \text{ m / s}^2$$

Therefore, $a_c > g$. Some downward normal force is required along with the block's weight to constitute the central force, and the block stays on the track.   ◊

Chapter 9

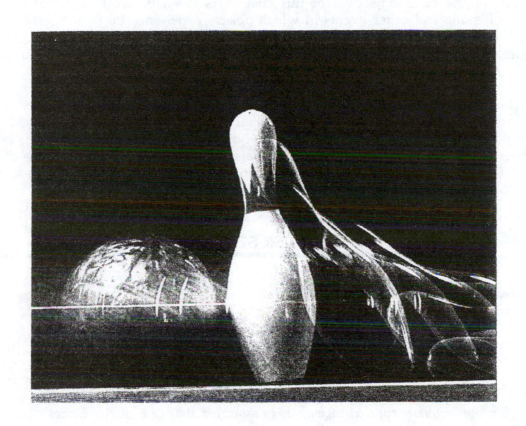

Linear Momentum and Collisions

LINEAR MOMENTUM AND COLLISIONS

INTRODUCTION

One of the main objectives of this chapter is to enable you to understand and analyze collisions and other events in which objects experience large accelerations as a result of very large forces which act for a short time interval. As a first step, we shall introduce the concept of *momentum*, a term that is used in describing objects in motion. Momentum is defined as the product of mass and velocity.

The concept of momentum leads us to a second conservation law, that of conservation of momentum. This law is especially useful for treating problems that involve collisions between objects and for analyzing rocket propulsion. The concept of the center of mass of a system of particles is also introduced, and we shall see that the motion of a system of particles can be represented by the motion of one representative particle located at the center of mass.

NOTES FROM SELECTED CHAPTER SECTIONS

9.1 Linear Momentum and Its Conservation
9.2 Impulse and Momentum

If two particles of masses m_1 and m_2 form an *isolated system*, then the total momentum of the system remains constant. The *time rate of change of the momentum* of a particle is equal to the resultant force on the particle. The *impulse* of a force equals the change in momentum of the particle on which the force acts. Under the *impulse approximation*, it is assumed that one of the forces acting on a particle is of short time duration but of much greater magnitude than any of the other forces.

9.3 Collisions
9.4 Elastic and Inelastic Collisions in One Dimension

For *any type of collision*, the total momentum before the collision equals the total momentum just after the collision.

In an *inelastic collision*, the total momentum is conserved; however, the total kinetic energy is not conserved.

In a *perfectly inelastic collision*, the two colliding objects stick together following the collision.

In an *elastic collision*, both momentum and kinetic energy are conserved.

9.5   Two-Dimensional Collisions

The law of conservation of momentum is not restricted to one-dimensional collisions. If two masses undergo a *two-dimensional* (glancing) *collision* and there are no external forces acting, the total momentum in each of the x, y, and z directions is conserved.

9.6   The Center of Mass

The position of the center of mass of a system can be described as the average position of the system's mass. If **g** is constant over a mass distribution, then the center of gravity will coincide with the center of mass.

9.7   Motion of a System of Particles

The center of mass of a system of particles moves like an imaginary particle of mass M (equal to the total mass of the system) under the influence of the resultant external force on the system.

9.8   Rocket Propulsion

The operation of a rocket depends on the law of conservation of linear momentum applied to a system of particles (the rocket plus its ejected fuel). The thrust of the rocket is the force exerted on it by the exhaust gases.

EQUATIONS AND CONCEPTS

The *linear momentum* **p** of a particle is defined as the product of its mass m with its velocity **v**. This equation is equivalent to three component scalar equations, one along each of the coordinate axes.

$$\mathbf{p} \equiv m\mathbf{v} \tag{9.1}$$

The time rate of change of the linear momentum of a particle is equal to the resultant force acting on the particle. This is Newton's second law for a particle.

$$\mathbf{F} = \frac{d\mathbf{p}}{dt} \tag{9.3}$$

When two particles interact with each other (but are otherwise isolated from their surroundings), Newton's third law *tells us that the force of particle 1 on particle 2 is equal and opposite to the force of particle 2 on particle 1.* Since force is the time rate of change of momentum (Newton's second law), one finds that the total momentum of the isolated pair of particles is conserved.

$$\mathbf{p}_{tot} = \mathbf{p}_1 + \mathbf{p}_2 = \text{constant} \tag{9.4}$$

In general, when the *external force* acting on a system of particles is zero, the *total linear momentum of the system is conserved.* This important statement is known as the *law of conservation of momentum.* It is especially useful in treating problems involving collisions between two bodies.

$$\mathbf{p}_{1i} + \mathbf{p}_{2i} = \mathbf{p}_{1f} + \mathbf{p}_{2f} \tag{9.5}$$

The *impulse* of a force $\mathbf{F}$ acting on a particle *equals the change in momentum of the particle.* This is known as the impulse-momentum theorem.

$$\mathbf{I} \equiv \int_{t_i}^{t_f} \mathbf{F}\, dt = \Delta\mathbf{p} \tag{9.9}$$

To calculate impulse, we usually define a *time-averaged force* $\mathbf{F}$ which would give the same impulse to the particle as the actual time-varying force over the time interval Δt.

$$\overline{\mathbf{F}} \equiv \frac{1}{\Delta t} \int_{t_i}^{t_f} \mathbf{F}\, dt \tag{9.10}$$

The average force can be thought of as the constant force that would give the same change in momentum over the time interval Δt as the applied impulse.

$$I = \Delta p = \overline{F} \Delta t \qquad (9.11)$$

In the impulse approximation, the force F appearing in Equation 9.9 acts for a short time and is much larger than any other force present. This approximation is usually assumed in problems involving collisions, where the force is the contact force between the particles during the collision.

It is useful to consider two particular types of collisions that can occur between two bodies. *An elastic collision is one in which both linear momentum and kinetic energy are conserved.* An *inelastic collision is one in which only linear momentum is conserved.* A perfectly inelastic collision is an inelastic collision in which the two bodies stick together after the collision. Note that momentum is conserved in any type of collision. Furthermore, note that when we say that the momentum is conserved, we are speaking about the momentum of the *entire system.* That is, the momentum of each particle may change as the result of the collision, but the momentum of the system remains unchanged.

$$p_1 + p_2 = \text{constant}$$
$$\text{(Elastic)}$$
$$K_1 + K_2 = \text{constant}$$

$$p_1 + p_2 = \text{constant} \qquad \text{(Inelastic)}$$

When two particles moving along a straight line collide and stick together (perfectly inelastic collision), the common velocity after the collision can be calculated in terms of the two mass values and the two initial velocities.

$$v_f = \frac{m_1 v_{1i} + m_2 v_{2i}}{m_1 + m_2} \qquad (9.14)$$

When two particles undergo a perfectly elastic collision, both momentum and kinetic energy are conserved. When such a collision occurs, *the relative velocity before the collision equals the negative of the relative velocity of the two particles following the collision.*

$$v_{1i} - v_{2i} = -\left(v_{1f} - v_{2f}\right) \tag{9.19}$$

When the masses and initial speeds of both particles are known, the final speeds can be calculated.

$$v_{1f} = \left(\frac{m_1 - m_2}{m_1 + m_2}\right)v_{1i} + \left(\frac{2m_2}{m_1 + m_2}\right)v_{2i} \tag{9.20}$$

$$v_{2f} = \left(\frac{2m_1}{m_1 + m_2}\right)v_{1i} + \left(\frac{m_2 - m_1}{m_1 + m_2}\right)v_{2i} \tag{9.21}$$

An important special use occurs *when the second particle (m_2, the "target") is initially at rest.* Remember the appropriate algebraic signs (designating direction) must be included for v_{1i} and v_{2i}.

$$v_{1f} = \left(\frac{m_1 - m_2}{m_1 + m_2}\right)v_{1i} \tag{9.22}$$

$$v_{2f} = \left(\frac{2m_1}{m_1 + m_2}\right)v_{1i} \tag{9.23}$$

The *x coordinate of the center of mass of n particles* whose individual coordinates are $x_1, x_2, x_3, \ldots$ and whose masses are $m_1, m_2, m_3, \ldots$ is given by Equation 9.28. The y and z coordinates of the center of mass are defined by similar expressions. The center of mass of a homogeneous, symmetric body must lie on an axis of symmetry.

$$x_{CM} \equiv \frac{\Sigma m_i x_i}{\Sigma m_i} \tag{9.28}$$

The center of mass for a collection of particles can be located by its position vector.

$$\mathbf{r}_{CM} \equiv \frac{\Sigma m_i \mathbf{r}_i}{M} \tag{9.30}$$

For an extended object, the center of mass can be calculated by integrating over the total length, area, or volume which includes the total mass M.

$$\mathbf{r}_{CM} = \frac{1}{M}\int \mathbf{r}\, dm \tag{9.33}$$

176

In this expression for the *velocity of the center of mass of a system of particles*, $\mathbf{v}_i$ is the velocity of the ith particle and M is the total mass of the system.

$$\mathbf{v}_{CM} = \frac{\Sigma m_i \mathbf{v}_i}{M} \qquad (9.34)$$

The *total momentum of a system of particles* is equal to the total mass M multiplied by the velocity of the center of mass.

$$\mathbf{p}_{tot} = M\mathbf{v}_{CM} = \text{constant} \qquad (9.39)$$

$$(\text{when } \Sigma\mathbf{F}_{ext} = 0)$$

The *acceleration of the center of mass of a system of particles* depends on the value of the acceleration for each of the individual particles.

$$\mathbf{a}_{CM} = \frac{1}{M}\sum m_i \mathbf{a}_i \qquad (9.36)$$

Newton's second law applied to a system of particles says that the *resultant external force* acting on the system *equals the time rate of change of the total momentum.* This form of Newton's second law must be used when the mass of the system changes. If the net external force on the system is zero, then the total momentum of the system remains constant.

$$\Sigma\mathbf{F}_{ext} = \frac{d\mathbf{p}_{tot}}{dt} \qquad (9.38)$$

The principle behind the operation of a rocket is the law of conservation of momentum as applied to the rocket and its ejected fuel. If a rocket moves in the absence of gravity and ejects fuel with an exhaust velocity v_e, *its change in velocity is proportional to the exhaust velocity,* where M_i and M_f refer to its initial and final mass values for the rocket.

$$\Delta v = v_f - v_i = v_e \ln\left(\frac{M_i}{M_f}\right) \qquad (9.41)$$

The thrust on a rocket increases as the exhaust speed increases and as the burn rate increases.

$$\text{Thrust} = \left|v_e \frac{dM}{dt}\right| \qquad (9.42)$$

SUGGESTIONS, SKILLS, AND STRATEGIES

The following procedure is recommended when dealing with problems involving collisions between two objects:

- Set up a coordinate system and define your velocities with respect to that system. That is, objects moving in the direction selected as the positive direction of the x axis are considered as having a positive velocity and negative if moving in the negative x direction. It is convenient to have the x axis coincide with one of the initial velocities.

- In your sketch of the coordinate system, draw all velocity vectors with labels and include all the given information.

- Write expressions for the momentum of each object before and after the collision. (In two-dimensional collision problems, write expressions for the x and y components of momentum before and after the collision.) Remember to include the appropriate signs for the velocity vectors.

- Now write expressions for the *total* momentum *before* and *after* the collision and equate the two. (For two-dimensional collisions, this expression should be written for the momentum in both the x and y directions.) It is important to emphasize that it is the momentum of the *system* (the two colliding objects) that is conserved, not the momentum of the individual objects.

- If the collision is *inelastic*, you should then proceed to solve the momentum equations for the unknown quantities.

- If the collision is *elastic*, kinetic energy is also conserved, so you can equate the total kinetic energy before the collision to the total kinetic energy after the collision. This gives an additional relationship between the various velocities. The conservation of kinetic energy for elastic collisions leads to the expression $v_{1i} - v_{2i} = -(v_{1f} - v_{2f})$, which is often easier to use in solving elastic collision problems than is an expression for conservation of kinetic energy.

REVIEW CHECKLIST

▷ The impulse of a force acting on a particle during some time interval equals the *change* in momentum of the particle, and the impulse equals the area under the force-time graph.

▷ The momentum of any isolated system (one for which the net external force is zero) is conserved, regardless of the nature of the forces between the masses which comprise the system.

▷ There are two types of collisions that can occur between two particles, namely elastic and inelastic collisions. Recognize that a *perfectly* inelastic collision is an inelastic collision in which the colliding particles stick together after the collision, and hence move as a composite particle.

▷ The conservation of linear momentum applies not only to head-on collisions (one-dimensional), but also to glancing collisions (two- or three-dimensional). For example, in a two-dimensional collision, the total momentum in the x direction is conserved and the total momentum in the y direction is conserved.

▷ The equations for momentum and kinetic energy can be used to calculate the final velocities in a two-body head-on elastic collision; and to calculate the final velocity and the change of kinetic energy in a two-body system for a completely inelastic collision.

SOLUTIONS TO SELECTED END-OF-CHAPTER PROBLEMS

7. An estimated force-time curve for a baseball struck by a bat is shown in Figure P9.7. From this curve, determine (a) the impulse delivered to the ball, (b) the average force exerted on the ball, and (c) the peak force exerted on the ball.

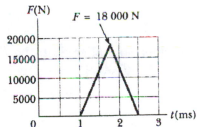

Figure P9.7

Solution

(a) $I = \int F\,dt =$ area under the F-t graph

$$= \left(\frac{0 + 18000\ \text{N}}{2}\right)\left(2.5 \times 10^{-3}\ \text{s} - 1.0 \times 10^{-3}\ \text{s}\right) = 13.5\ \text{N·s} \quad \Diamond$$

(b) $\bar{F} = \dfrac{\int F\,dt}{\Delta t} = \dfrac{13.5\ \text{N·s}}{(2.5 - 1.0)10^{-3}\ \text{s}} = 9000\ \text{N} \quad \Diamond$

(c) From the graph, $F_{max} = 18000\ \text{N}$ $\Diamond$

13. A 0.15-kg baseball is thrown with a speed of 40 m/s. It is hit straight back at the pitcher with a speed of 50 m/s. (a) What is the impulse delivered to the baseball? (b) Find the average force exerted by the bat on the ball if the two are in contact for 2.0×10^{-3} s. Compare this with the weight of the ball and determine whether or not the impulse approximation is valid in the situation.

Solution

(a) Take the x axis to be directed from the pitcher toward the batter. Take initial and final points just before and after the bat contacts the ball. Then,

$$I = \Delta p = p_f - p_i$$

$$I = (0.15 \text{ kg})(-50 \text{ m}/\text{s})i - (0.15 \text{ kg})(40 \text{ m}/\text{s})i$$

$$= -13.5i \text{ kg} \cdot \text{m}/\text{s} = 13.5 \text{ kg} \cdot \text{m}/\text{s} \quad \text{toward the pitcher} \quad \Diamond$$

(b) $\bar{F} = \dfrac{I}{\Delta t} = \dfrac{-13.5i \text{ kg} \cdot \text{m}/\text{s}}{2.0 \times 10^{-3} \text{ s}} = -6.75i \text{ kN} \quad \Diamond$

The weight of the ball is

$$w = mg = (0.15 \text{ kg})(9.80 \text{ m}/\text{s}^2) = -1.5 \text{ N} \quad \text{(down)}$$

This is smaller than the force of the bat by 5000 times, so the impulse approximation is good. The vertical velocity of the ball changes negligibly in 2 ms. $\quad \Diamond$

---

15. A 3.0-kg steel ball strikes a wall with a speed of 10 m/s at an angle of 60° with the surface. It bounces off with the same speed and angle (Fig. P9.15). If the ball is in contact with the wall for 0.20 s, what is the average force exerted on the ball by the wall?

Solution

$\Delta p = F \Delta t$

$\Delta p_y = m(v_{fy} - v_{iy}) = m(v \cos 60°) - mv \cos 60° = 0$

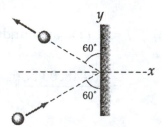

Figure P9.15

$$\Delta p_x = m\left(v_{fx} - v_{ix}\right) = m(-v\sin 60° - v\sin 60°) = -2mv\sin 60°$$

$$\Delta p_x = -2(3.0 \text{ kg})(10 \text{ m}/\text{s})(0.866) = -52 \text{ kg}\cdot\text{m}/\text{s}$$

$$\overline{\mathbf{F}} = \frac{\Delta \mathbf{p}}{\Delta t} = \frac{\Delta p_x \mathbf{i}}{\Delta t} = \frac{-52\mathbf{i} \text{ kg}\cdot\text{m}/\text{s}}{0.20 \text{ s}} = -260\mathbf{i} \text{ N} \quad \lozenge$$

21. A 45.0-kg girl is standing on a plank that has a mass of 150 kg. The plank, originally at rest, is free to slide on a frozen lake, which is a flat, frictionless supporting surface. The girl begins to walk along the plank at a constant speed of 1.5 m/s relative to the plank. (a) What is her speed relative to the ice surface? (b) What is the speed of the plank relative to the ice surface?

Solution   Let $\mathbf{v}_g$ = velocity of the girl relative to the ice
$\mathbf{v}_{gp}$ = velocity of the girl relative to the plank
$\mathbf{v}_p$ = velocity of the plank relative to the ice

The girl and the plank exert forces on each other, but the ice isolates them from outside horizontal forces. Therefore, the net momentum is zero for the combined girl plus plank system.

$$0 = m_g\mathbf{v}_g + m_p\mathbf{v}_p$$

Further, the relation among relative speeds can be written:

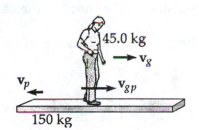

$$\mathbf{v}_g = \mathbf{v}_{gp} + \mathbf{v}_p$$

$$\mathbf{v}_g = \left(1.5\mathbf{i} + \mathbf{v}_p\right) \text{ m}/\text{s}$$

We substitute:

$$0 = (45.0 \text{ kg})(1.5\mathbf{i} \text{ m}/\text{s} + \mathbf{v}_p) + (150 \text{ kg})\,\mathbf{v}_p$$

$$(195 \text{ kg})\mathbf{v}_p = (-45.0 \text{ kg})(1.5\mathbf{i} \text{ m}/\text{s})$$

(b) $\mathbf{v}_p = -0.346\mathbf{i} \text{ m}/\text{s} \quad \lozenge$

(a) $\mathbf{v}_g = 1.5\mathbf{i} - 0.346\mathbf{i} \text{ m}/\text{s} = 1.15\mathbf{i} \text{ m}/\text{s} \quad \lozenge$

25. A 10.0-g bullet is stopped in a block of wood (m = 5.00 kg). The speed of the bullet-plus-wood combination immediately after the collision is 0.600 m/s. What was the original speed of the bullet?

Solution We suppose the block of wood was originally stationary. The total momentum of the combined bullet plus block system is constant over the short time during which relative motion stops:

$$\mathbf{p}_{1i} + \mathbf{p}_{2i} = \mathbf{p}_{1f} + \mathbf{p}_{2f}$$

$$(0.010 \text{ kg})\mathbf{v}_{1i} + 0 = (0.010 \text{ kg})(0.600 \text{ m/s})\mathbf{i} + (5.00 \text{ kg})(0.600 \text{ m/s})\mathbf{i}$$

$$\mathbf{v}_{1i} = \frac{(5.01 \text{ kg})(0.600 \text{ m/s})\mathbf{i}}{0.010 \text{ kg}} = 301\mathbf{i} \text{ m/s} \quad \lozenge$$

---

29. A neutron in a reactor makes an elastic head-on collision with the nucleus of a carbon atom initially at rest. (a) What fraction of the neutron's kinetic energy is transferred to the carbon nucleus? (b) If the initial kinetic energy of the neutron is 1.6×10^{-13} J, find its final kinetic energy and the kinetic energy of the carbon nucleus after the collision. (The mass of the carbon nucleus is about 12 times the mass of the neutron.)

Solution (a) This a perfectly elastic head-on collision, so we use the equation:

$$\mathbf{v}_{1i} - \mathbf{v}_{2i} = -\left(\mathbf{v}_{1f} - \mathbf{v}_{2f}\right)$$

Let object 1 be the neutron, and object 2 be the carbon nucleus, with $m_2 = 12m_1$.

Since $\mathbf{v}_{2i} = 0$, $\qquad\qquad \mathbf{v}_{2f} = \mathbf{v}_{1i} + \mathbf{v}_{1f}$

Now, by conservation of momentum, $\qquad m_1\mathbf{v}_{1i} + m_2\mathbf{v}_{2i} = m_1\mathbf{v}_{1f} + m_2\mathbf{v}_{2f}$

or $\qquad\qquad m_1\mathbf{v}_{1i} = m_1\mathbf{v}_{1f} + 12m_1\mathbf{v}_{2f}$

Substituting our velocity equation, $\qquad \mathbf{v}_{1i} = \mathbf{v}_{1f} + 12(\mathbf{v}_{1i} + \mathbf{v}_{1f})$

We solve $-11\mathbf{v}_{1i} = 13 \mathbf{v}_{1f}$: $\qquad \mathbf{v}_{1f} = -\left(\frac{11}{13}\right)\mathbf{v}_{1i}$, and $\qquad \mathbf{v}_{2f} = \mathbf{v}_{1i} - \left(\frac{11}{13}\right)\mathbf{v}_{1i} = \left(\frac{2}{13}\right)\mathbf{v}_{1i}$

The neutron's original kinetic energy is $\frac{1}{2}m_1v_{1i}{}^2$

The carbon's final kinetic energy is

$$\frac{1}{2}m_2v_{2f}{}^2 = \frac{1}{2}(12m_1)\left(\frac{2}{13}\right)^2 v_{1i}{}^2 = \left(\frac{48}{169}\right)\left(\frac{1}{2}\right)m_1v_{1i}{}^2$$

So, $\left(\frac{48}{169}\right) = 0.284,$ or   28% of the total energy is transferred. ◊

(b) For the carbon nucleus, $K_{2f} = (0.284)(1.60 \times 10^{-13} \text{ J}) = 4.54 \times 10^{-14} \text{ J}$ ◊

The collision is perfectly elastic, so the neutron retains the rest of the energy,

$$K_{1f} = (1.60 - 0.454) \times 10^{-13} \text{ J} = 1.15 \times 10^{-13} \text{ J} \quad ◊$$

---

35. A 12-g bullet is fired into a 100-g wooden block initially at rest on a horizontal surface. After impact, the block slides 7.5 m before coming to rest. If the coefficient of friction between block and surface is 0.65, what was the speed of the bullet immediately before impact?

Solution   Since the collision is *totally inelastic*, and momentum is conserved,

$$m_1v_1 = (m_1 + m_2)v_2$$

After impact, the change in kinetic energy is equal to the energy lost due to friction:

$$\frac{1}{2}(m_1 + m_2)v_2{}^2 = f_f L = \mu(m_1 + m_2)Lg$$

Solving for v_2: $v_2 = \sqrt{2\mu Lg} = \sqrt{2(0.65)(7.5 \text{ m})(9.80 \text{ m/s}^2)} = 9.77 \text{ m/s}$

Now using the momentum conservation equation:

$$v_1 = \left(\frac{m_1 + m_2}{m_1}\right)v_2$$

$$v_1 = \left(\frac{0.112 \text{ kg}}{0.012 \text{ kg}}\right)9.77 \text{ m/s} = 91.2 \text{ m/s} \quad ◊$$

---

37. Consider a frictionless track ABC as shown in Figure P9.37. A block of mass m_1 = 5.00 kg is released from A. It makes a head-on elastic collision with a block of mass m_2 = 10.0 kg at B, initially at rest. Calculate the maximum height to which m_1 rises after the collision.

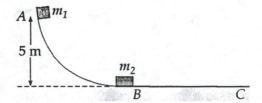

Figure P9.37

Solution

First, let us find the velocity of m_1 at B just *before* the collision. From conservation of energy, and the fact that $v_A = 0$, we get

$$K_A + U_A = K_B + U_B$$

$$0 + mgh = \tfrac{1}{2}mv_B^2 + 0$$

$$v_B = \sqrt{2gh} = \sqrt{2(9.80 \text{ m/s}^2)(5.00 \text{ m})} = 9.90 \text{ m/s}$$

Now use Equation 9.22 to get the velocity of m_1 just *after* the collision:

$$v_{1f} = \left(\frac{m_1 - m_2}{m_1 + m_2}\right)v_{1i} = \left(\frac{5.00 - 10.0}{5.00 + 10.0}\right)v_{1i} = -\tfrac{1}{3}(9.90 \text{ m/s}) = -3.30 \text{ m/s}$$

Thus, the 5-kg mass (m_1) moves to the *left* after the collision, while the 10-kg mass (m_2) moves to the *right*. To find the maximum height to which m_1 rises, we again apply conservation of energy to m_1 and find

$$m_1gh' = \tfrac{1}{2}m_1v_{1f}^2$$

$$h' = \frac{v_{1f}^2}{2g} = \frac{(-3.30 \text{ m/s})^2}{2(9.80 \text{ m/s}^2)} = 0.556 \text{ m} \quad \Diamond$$

41. A 3.00-kg mass with an initial velocity of 5.00i m/s collides with and sticks to a 2.00-kg mass with an initial velocity of –3.00j m/s. Find the final velocity of the composite mass.

Solution   Momentum is conserved, with both masses having the same final velocity:

$$m_1\mathbf{v}_{1i} + m_2\mathbf{v}_{2i} = m_1\mathbf{v}_{1f} + m_2\mathbf{v}_{2f}$$

$$(3.00\ kg)(5.00i\ m/s) + (2.00\ kg)(-3.00j\ m/s) = (3.00\ kg + 2.00\ kg)\mathbf{v}_f$$

$$\mathbf{v}_f = \frac{15.0i - 6.00j}{5.00}\ m/s = (3.00i - 1.20j)\ m/s \quad \lozenge$$

Related Calculation: Compute the kinetic energy both before and after the collision; show that kinetic energy is not conserved.

$$K_{1i} + K_{2i} = \frac{1}{2}(3.00\ kg)(5.00\ m/s)^2 + \frac{1}{2}(2.00\ kg)(3.00\ m/s)^2 = 46.5\ J$$

$$K_{1f} + K_{2f} = \frac{1}{2}(5.00\ kg)\left[(3.00\ m/s)^2 + (1.20\ m/s)^2\right] = 26.1\ J \quad \lozenge$$

---

43. An unstable nucleus of mass 17×10^{-27} kg initially at rest disintegrates into three particles. One of the particles, of mass 5.0×10^{-27} kg, moves along the y axis with a speed of 6.0×10^6 m/s. Another particle, of mass 8.4×10^{-27} kg, moves along the x axis with a speed of 4.0×10^6 m/s. Find (a) the velocity of the third particle and (b) the total energy given off in the process.

Solution

(a)   With three particles, the total final momentum is $m_1\mathbf{v}_{1f} + m_2\mathbf{v}_{2f} + m_3\mathbf{v}_{3f}$, and it must be zero to equal the original momentum.

The mass of the third particle is

$$m_3 = (17 - 5.0 - 8.4)(10^{-27}\ kg) = 3.6 \times 10^{-27}\ kg$$

Because the total momentum is zero, we get

$$0 = \left(5.0 \times 10^{-27} \text{ kg}\right)\left(6.0 \times 10^{6} \text{ m/s}\right)\mathbf{j} + \left(8.4 \times 10^{-27} \text{ kg}\right)\left(4.0 \times 10^{6} \text{ m/s}\right)\mathbf{i} + \left(3.6 \times 10^{-27} \text{ kg}\right)\mathbf{v}_{3f}$$

So $\qquad \mathbf{v}_{3f} = \dfrac{(-3.00\mathbf{j} - 3.36\mathbf{i})\left(10^{-20} \text{ kg} \cdot \text{m/s}\right)}{3.6 \times 10^{-27} \text{ kg}} = \left(-9.33 \times 10^{6}\mathbf{i} - 8.33 \times 10^{6}\mathbf{j}\right) \text{ m/s} \quad \Diamond$

(b) The original kinetic energy is zero. The final kinetic energy is $K = K_{1f} + K_{2f} + K_{3f}$.

$$K = \tfrac{1}{2}\left(5.0 \times 10^{-27} \text{ kg}\right)\left(6.0 \times 10^{6} \text{ m/s}\right)^{2} + \tfrac{1}{2}\left(8.4 \times 10^{-27} \text{ kg}\right)\left(4.0 \times 10^{6} \text{ m/s}\right)^{2}$$

$$+ \tfrac{1}{2}\left(3.6 \times 10^{-27} \text{ kg}\right)\left(9.33^{2} + 8.33^{2}\right)\left(10^{12} \text{ m}^{2}/\text{s}^{2}\right)$$

and

$$K = \left(9.0 \times 10^{-14} \text{ J}\right) + \left(6.72 \times 10^{-14} \text{ J}\right) + \left(28.2 \times 10^{-14} \text{ J}\right) = 4.39 \times 10^{-13} \text{ J} \quad \Diamond$$

---

47. A billiard ball moving at 5.00 m/s strikes a stationary ball of the same mass. After the collision, the first ball moves at 4.33 m/s at an angle of 30.0° with respect to the original line of motion. Assuming an elastic collision (and ignoring friction and rotational motion), find the struck ball's final velocity.

Solution   Since all the balls have the same mass, call each mass m. Take the x axis in the direction of the original motion. Call $\mathbf{v}_{2f}$ the velocity of the second ball after the collision, as in the figure, with an angle ϕ.

The x component of momentum is conserved:

$$m(5.00 \text{ m/s}) = m(4.33 \text{ m/s}) \cos 30° + mv_{2fx}$$

and $v_{2fx} = 1.25$ m/s

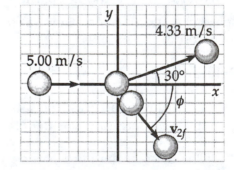

The y component of momentum is conserved:

$$0 = m(4.33 \text{ m/s}) \sin 30° + mv_{2fy}$$

and $v_{2fy} = -2.17$ m/s

$$\mathbf{v}_{2f} = 1.25\mathbf{i} - 2.17\mathbf{j} \quad \text{(or 2.50 m/s at } -60.0°) \quad \Diamond$$

---

53. The mass of the Sun is 329,390 Earth masses, and the mean distance from the center of the Sun to the center of the Earth is 1.496×10^8 km. Treating the Earth and Sun as particles, with each mass concentrated at its respective geometric center, how far from the center of the Sun is the center of mass of the Earth-Sun system? Compare this distance with the mean radius of the Sun (6.960×10^5 km).

Solution
$$r_{CM} = \frac{(329,390 M_e)(0) + M_e \left(1.496 \times 10^8 \text{ km} \right)}{329,390 M_e + M_e} = 454 \text{ km} \quad \Diamond$$

The center of mass of the Sun-Earth system is less than $1/100$ of the Sun's radius from the center of the Sun. $\Diamond$

---

57. A uniform piece of sheet steel is shaped as in Figure P9.57. Compute the x and y coordinates of the center of mass of the piece.

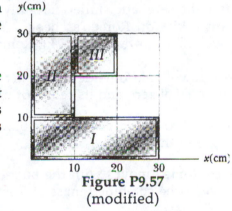
Figure P9.57
(modified)

Solution Think of the sheet as composed of three sections, and consider the mass of each section to be at the geometric center of that section. Define the mass per unit area to be σ, and number the rectangles as shown.

$m_I = (30 \text{ cm})(10 \text{ cm})\sigma \quad CM_I = (15 \text{ cm, } 5 \text{ cm})$
$m_{II} = (10 \text{ cm})(20 \text{ cm})\sigma \quad CM_{II} = (5 \text{ cm, } 20 \text{ cm})$
$m_{III} = (10 \text{ cm})(10 \text{ cm})\sigma \quad CM_{III} = (15 \text{ cm, } 25 \text{ cm})$

The overall CM is at:

$$\mathbf{r}_{CM} = \frac{(300\sigma \text{ cm}^3)(15\mathbf{i} + 5\mathbf{j}) + (200\sigma \text{ cm}^3)(5\mathbf{i} + 20\mathbf{j}) + (100\sigma \text{ cm}^3)(15\mathbf{i} + 25\mathbf{j})}{(300 + 200 + 100)\sigma \text{ cm}^2}$$

$$= \frac{(45\mathbf{i} + 15\mathbf{j} + 10\mathbf{i} + 40\mathbf{j} + 15\mathbf{i} + 25\mathbf{j})}{6} \text{ cm}$$

$$= (11.7\mathbf{i} + 13.3\mathbf{j}) \text{ cm} \quad \Diamond$$

If we chose any other division of the original shape, the answer would be the same.

---

59. A 2.0-kg particle has a velocity (2.0i – 3.0j) m/s, and a 3.0-kg particle has a velocity (1.0i + 6.0j) m/s. Find (a) the velocity of the center of mass and (b) the total momentum of the system.

Solution Use $\mathbf{v}_{CM} = \dfrac{m_1\mathbf{v}_1 + m_2\mathbf{v}_2}{m_1 + m_2}$ and $\mathbf{p}_{CM} = (m_1 + m_2)\mathbf{v}_{CM}$:

(a) $\mathbf{v}_{CM} = \dfrac{(2.0\text{ kg})\left[(2.0i - 3.0j)\text{ m}/\text{s}\right] + (3.0\text{ kg})\left[(1.0i + 6.0j)\text{ m}/\text{s}\right]}{(2.0\text{ kg} + 3.0\text{ kg})} = (1.40i + 2.40j)\text{ m}/\text{s}$ ◊

(b) $\mathbf{p}_{CM} = (2.0\text{ kg} + 3.0\text{ kg})\left[(1.4i + 2.4j)\text{ m}/\text{s}\right] = (7.0i + 12.0j)\text{ kg}\cdot\text{m}/\text{s}$ ◊

---

62. Romeo entertains Juliet by playing his guitar from the rear of their boat in still water. After the serenade, Juliet carefully moves to the rear of the boat (away from shore) to plant a kiss on Romeo's cheek. If the 80-kg boat is facing shore and the 55-kg Juliet moves 2.7 m toward the 77-kg Romeo, how far does the boat move toward shore?

Solution No outside forces act on the boat-plus-lovers system, so its momentum is conserved at zero and its center of mass stays fixed: $x_{CM,\,i} = x_{CM,\,f}$.

Define K to be the point where they kiss, and Δx_J and Δx_b as shown in the figure.

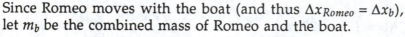

Since Romeo moves with the boat (and thus $\Delta x_{Romeo} = \Delta x_b$), let m_b be the combined mass of Romeo and the boat.

Then, $\qquad m_J \Delta x_J + m_b \Delta x_b = 0$

Choosing the x axis to point away from the shore,

$$(55\text{ kg})\Delta x_J + (77\text{ kg} + 80\text{ kg})\Delta x_b = 0$$

and $\qquad\qquad\qquad \Delta x_J = -2.85\,\Delta x_b$

As Juliet moves away from shore, the boat and Romeo glide toward the shore until the original 2.7 m gap between them is closed:

$$\Delta x_J - \Delta x_b = 2.7\text{ m}$$

Substituting, we find $\qquad \Delta x_b = -0.700\text{ m, or }\ 0.700\text{ m}\ \ \text{towards the shore}$ ◊

---

64. The first stage of a Saturn V space vehicle consumes fuel at the rate of 1.5×10^4 kg/s, with an exhaust speed of 2.6×10^3 m/s. (a) Calculate the thrust produced by these engines. (b) Find the initial acceleration of the vehicle on the launch pad if its initial mass is 3.0×10^6 kg. [*Hint:* You must include the force of gravity to solve (b).]

Solution

(a) The impulse due to the thrust, F, is equal to the change in momentum as fuel is exhausted from the rocket.

$$F = \frac{dp}{dt} = \frac{d}{dt}(mv_e)$$

Since v_e is a constant exhaust velocity, $\qquad F = v_e\left(\frac{dm}{dt}\right)$

The term $\dfrac{dm}{dt}$ is nothing more than the fuel consumption rate.

So, $\qquad F = \left(2.6 \times 10^3 \text{ m/s}\right)\left(1.5 \times 10^4 \text{ kg/s}\right) = 3.90 \times 10^7 \text{ N} \quad \lozenge$

(b) Applying the equation $\Sigma F = ma$, and substituting, we have:

$$\left(3.9 \times 10^7 \text{ N}\right) - \left(3.0 \times 10^6 \text{ kg}\right)\left(9.80 \text{ m/s}^2\right) = \left(3.0 \times 10^6 \text{ kg}\right)a$$

$$a = \frac{\left(3.9 \times 10^7 \text{ N}\right) - \left(29.4 \times 10^6 \text{ N}\right)}{3.0 \times 10^6 \text{ kg}} = 3.20 \text{ m/s}^2 \quad \lozenge$$

67. Fuel aboard a rocket has a density of 1.4×10^3 kg/m^3 and is ejected with a speed of 3.0×10^3 m/s. If the engine is to provide a thrust of 2.5×10^6 N, what volume of fuel must be burned per second?

Solution

$$\text{Thrust} = v_e\left|\frac{dM}{dt}\right| = 2.5\times10^6 \text{ N}$$

$$\left|\frac{dM}{dt}\right| = \frac{2.5\times10^6 \text{ N}}{3.0\times10^3 \text{ m/s}} = 833 \text{ kg/s}$$

Since

$$\rho = 1.4\times10^3 \text{ kg/m}^3, \quad \text{and} \quad \rho = M/V,$$

it follows that

$$\frac{dV}{dt} = \frac{1}{\rho}\left|\frac{dM}{dt}\right| = \frac{833 \text{ kg/s}}{1.4\times10^3 \text{ kg/m}^3} = 0.595 \text{ m}^3/\text{s} \quad \lozenge$$

---

69. A golf ball (m = 46 g) is struck a blow that makes an angle of 45° with the horizontal. The drive lands 200 m away on a flat fairway. If the golf club and ball are in contact for 7.0 ms, what is the average force of impact? (Neglect air resistance.)

Solution Use the equation for the range of a projectile:

$$R = \left(\frac{v_0^2 \sin 2\theta_0}{g}\right)$$

or

$$v_0 = \sqrt{\frac{Rg}{\sin 2\theta_0}} = \sqrt{\frac{(200 \text{ m})(9.80 \text{ m/s}^2)}{\sin[(2)(45°)]}} = 44.3 \text{ m/s}$$

$$I = \overline{F}\Delta t = p_f - p_i$$

so

$$\overline{F} = \frac{p_f - p_i}{\Delta t} = \frac{(0.046 \text{ kg})(44.3 \text{ m/s})}{7.0\times10^{-3} \text{ s}} = 291 \text{ N} \quad \lozenge$$

---

73. An attack helicopter is equipped with a 20-mm cannon that fires 130-g shells in the forward direction with a muzzle speed of 800 m/s. The fully loaded helicopter has a mass of 4000 kg. A burst of 160 shells is fired in a 4.0-s interval. What is the resulting average force on the helicopter and by what amount is its forward speed reduced?

Solution The impulse imparted to the shell equals the change in momentum:

$$\overline{F}(\Delta t) = \Delta(mv)$$

$$\overline{F} = v\left(\frac{\Delta m}{\Delta t}\right) = (800 \text{ m/s})\frac{(160 \text{ shells})(0.130 \text{ kg/shell})}{4.0 \text{ s}} = 4160 \text{ N} \quad \Diamond$$

Note that, since the velocity of the shells is much greater than the velocity of the helicopter, it is not necessary to use a relative velocity for this calculation. To find the amount its forward speed is reduced, use $M \Delta v = mv$, or

$$\Delta v = \frac{m}{M}v = \frac{(160)(0.130 \text{ kg})}{4000 \text{ kg}}(800 \text{ m/s}) = 4.16 \text{ m/s} \quad \Diamond$$

---

79. An 80.0-kg astronaut is working on the engines of his ship, which is drifting through space with a constant velocity. The astronaut, wishing to get a better view of the Universe, pushes against the ship and later finds himself 30.0 m behind the ship. Without a thruster, the only way to return to the ship is to throw his 0.500-kg wrench directly away from the ship. If he throws the wrench with a speed of 20.0 m/s, how long does it take the astronaut to reach the ship?

Solution No external force acts on the system (astronaut plus wrench), so the total momentum is constant. Since the final momentum (wrench plus astronaut) must be zero, we have final momentum = initial momentum = 0, or

$$m_{\text{wrench}}v_{\text{wrench}} + m_{\text{astronaut}}v_{\text{astronaut}} = 0$$

Thus $\qquad v_{\text{astronaut}} = -\frac{m_{\text{wrench}}v_{\text{wrench}}}{m_{\text{astronaut}}} = -\frac{(0.500 \text{ kg})(20.0 \text{ m/s})}{80.0 \text{ kg}} = -0.125 \text{ m/s},$

At this speed, the time to travel to the ship is

$$t = \frac{30.0 \text{ m}}{0.125 \text{ m/s}} = 240 \text{ s} = 4.00 \text{ minutes} \quad \Diamond$$

81. A chain of length L and total mass M is released from rest with its lower end just touching the top of a table, as in Figure P9.81a. Find the force exerted by the table on the chain after the chain has fallen through a distance x, as in Figure P9.81b. (Assume each link comes to rest the instant it reaches the table.)

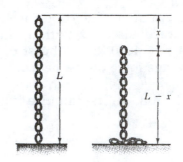

Figure P9.81

Solution   The force exerted by the table is equal to the change in momentum of each of the links in the chain.

By calculus' chain rule of derivatives,

$$F_1 = \frac{dp}{dt} = \frac{d(mv)}{dt} = v\frac{dm}{dt} + m\frac{dv}{dt}$$

We choose to account for the change in momentum of each link by having it pass from our area of interest just before it hits the table, so that

$$v\frac{dm}{dt} \neq 0 \quad \text{and} \quad m\frac{dv}{dt} = 0$$

[Alternatively, we could account for the change in momentum of each link by decelerating it from a velocity v to 0, in a distance of $dL/2$. In that case, the reverse would be true, but the final solution would be the same.]

Since the mass per unit length is uniform, we can express an each link of length dx as having a mass dm :

$$dm = \left(\frac{M}{L}\right)dx$$

The magnitude of the force due to the falling chain is the force that will be necessary to stop each of the elements dm.

$$F_1 = v\frac{dm}{dt} = v\left(\frac{M}{L}\right)\frac{dx}{dt} = \left(\frac{M}{L}\right)v^2$$

After falling a distance x, the square of the velocity of each link $v^2 = 2gx$ (from kinematics), hence

$$F_1 = \frac{2Mgx}{L}$$

The links already on the table have a total length x , and their weight exerts a force F_2 on the table:

$$F_2 = \frac{Mgx}{L}$$

Hence, the *total* force on the table is

$$F_{total} = F_1 + F_2 = \frac{3Mgx}{L} \quad \lozenge$$

That is, *the total force is three times the weight of the chain on the table at that instant.*

---

85. A 5.00-g bullet moving with an initial speed of 400 m/s is fired into, and passes through, a 1.00-kg block, as in Figure P9.85. The block, initially at rest on a frictionless, horizontal surface, is connected to a spring of force constant 900 N/m. If the block moves 5.00 cm to the right after impact, find (a) the speed at which the bullet emerges from the block and (b) the energy lost in the collision.

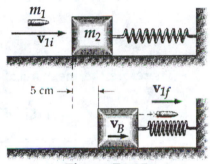

Figure P9.85

Solution   First find the initial velocity of the block, using conservation of energy during compression of the spring. Note that conservation of momentum does not apply here.

$$\tfrac{1}{2}m_2v_B^2 = \tfrac{1}{2}kx^2$$

$$\tfrac{1}{2}[1.00 \text{ kg}]v_B^2 = \tfrac{1}{2}[900 \text{ N}/\text{m}][0.0500 \text{ m}]^2$$

$$v_B = \sqrt{2.25} \text{ m}/\text{s} = 1.50 \text{ m}/\text{s}$$

(a)   When the bullet collides with the block, it is the momentum that is conserved:

$$m_1 v_{1i} + m_2 v_{2i} = m_1 v_{1f} + m_2 v_B$$

$$v_{1f} = \frac{m_1 v_{1i} - m_2 v_B}{m_1} = \frac{\left(5.00 \times 10^{-3}\ \text{kg}\right)(400\ \text{m/s}) - (1.00\ \text{kg})(1.50\ \text{m/s})}{5.00 \times 10^{-3}\ \text{kg}} = 100\ \text{m/s} \quad \Diamond$$

(b)   We use the work-energy theorem to find the energy lost in the collision. Before the collision, the block is motionless, and the bullet's energy is:

$$K_1 = \tfrac{1}{2} m v_{1i}^2 = \tfrac{1}{2}[0.00500\ \text{kg}][400\ \text{m/s}]^2 = 400\ \text{J}$$

After the collision, the energy is:

$$K_2 = \tfrac{1}{2} m_1 v_{1f}^2 + \tfrac{1}{2} m_2 v_B^2 = \tfrac{1}{2}[0.00500\ \text{kg}][100\ \text{m/s}]^2 + \tfrac{1}{2}[1.00\ \text{kg}][1.50\ \text{m/s}]^2 = 26.1\ \text{J}$$

Therefore the mechanical energy lost in the collision is:

$$|\Delta K| = |K_2 - K_1| = |26.1\ \text{J} - 400\ \text{J}| = 374\ \text{J} \quad \Diamond$$

Chapter 10

Rotation of a Rigid Object
About a Fixed Axis

ROTATION OF A RIGID OBJECT
ABOUT A FIXED AXIS

INTRODUCTION

When an extended object, such as a wheel, rotates about its axis, the motion cannot be analyzed by treating the object as a particle, since at any given time different parts of the object have different velocities and accelerations. For this reason, it is convenient to consider an extended object as a large number of particles, each with its own velocity and acceleration.

In dealing with the rotation of an object, analysis is greatly simplified by assuming the object to be rigid. A rigid object is defined as one that is nondeformable or, to say the same thing another way, one in which the distances between all pairs of particles remain constant. In this chapter, we treat the rotation of a rigid object about a fixed axis, commonly referred to as *pure rotational motion*.

NOTES FROM SELECTED CHAPTER SECTIONS

10.1 Angular Velocity and Angular Acceleration

Pure rotational motion refers to the motion of a rigid body about a fixed axis.

One *radian* (rad) is the angle subtended by an arc length equal to the radius of the arc.

In the case of *rotation about a fixed axis*, every particle on the rigid body has the same angular velocity and the same angular acceleration.

The angular displacement (θ), angular velocity (ω), and angular acceleration (α) are analogous to linear displacement (x), linear velocity (v), and linear acceleration (a), respectively. The variables, θ, ω, and α, differ dimensionally from the variables x, v, and a, only by a length factor.

10.2 Rotational Kinematics: Rotational Motion with Constant Angular Acceleration

The *kinematic expressions* for rotational motion under constant angular acceleration are of the *same form* as those for linear motion under constant linear acceleration with the substitutions $x \rightarrow \theta$, $v \rightarrow \omega$, and $a \rightarrow \alpha$.

10.3 Relationships Between Angular and Linear Quantities

When a rigid body rotates about a fixed axis, every part of the body has the same angular velocity and the same angular acceleration. However, different parts of the body, in general, have different linear velocities and different linear accelerations.

10.4 Rotational Energy

From the definition of moment of inertia, we see that it has dimensions of ML^2 (kg·m^2 in SI units). It plays the role of mass in *all* rotational equations. Although we shall commonly refer to the quantity $\frac{1}{2}I\omega^2$ as the rotational kinetic energy, it is not a new form of energy. It is ordinary kinetic energy. It is important to recognize the analogy between kinetic energy associated with linear motion, $\frac{1}{2}mv^2$, and rotational kinetic energy, $\frac{1}{2}I\omega^2$. The quantities I and ω in rotational motion are analogous to m and v in linear motion, respectively.

The *total kinetic energy* of a body in rolling motion is the sum of the rotational kinetic energy about the center of mass and the translational kinetic energy of the center of mass.

10.6 Torque

Torque is a physical quantity which is the measure of the tendency of a force to cause rotation of a body about a specified axis. It is important to remember that torque must be defined with respect to a *specific axis* of rotation. Torque, which has the *SI units* of N·m, must not be confused with force.

10.8 Work, Power, and Energy in Rotational Motion

The *work-energy theorem in rotational motion* states that the net work done by external forces in rotating a rigid body about a fixed axis equals the change in the body's rotational kinetic energy.

EQUATIONS AND CONCEPTS

When a particle moves along a circular path of radius r, the distance traveled by the particle is called the arc length, s. The radial line from the center of the path to the particle sweeps out an angle, θ.

$$\theta = \frac{s}{r}$$
(10.1b)

The angle θ is the ratio of two lengths (arc length to radius) and hence is a dimensionless quantity. However, it is common practice to refer to the angle as being in units of radians. In calculations, the relationship between radians and degrees is the following:

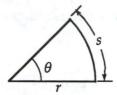

$$\theta^{rad} = \left(\frac{\pi}{180°}\right)\theta^{deg}$$

The *average angular speed* $\overline{\omega}$ of a particle or body rotating about a fixed axis equals the ratio of the angular displacement $\Delta\theta$ to the time interval Δt, where θ is measured in radians.

$$\overline{\omega} = \frac{\Delta\theta}{\Delta t}$$
(10.2)

The *instantaneous angular speed* ω is defined as the limit of the average angular velocity as Δt approaches zero.

$$\omega = \frac{d\theta}{dt}$$
(10.3)

The *average angular acceleration* $\overline{\alpha}$ of a rotating body is defined as the ratio of the change in angular velocity to the time interval Δt.

$$\overline{\alpha} = \frac{\Delta\omega}{\Delta t}$$
(10.4)

The *instantaneous angular acceleration* equals the limit of the average angular acceleration as Δt approaches zero.

$$\alpha = \frac{d\omega}{dt}$$
(10.5)

If a particle or body rotates about a fixed axis with *constant* angular acceleration, we can apply the *equations of rotational kinematics*.

$$\omega = \omega_0 + \alpha t \qquad (10.6)$$

$$\theta = \theta_0 + \omega_0 t + \tfrac{1}{2}\alpha t^2 \qquad (10.7)$$

$$\omega^2 = \omega_0{}^2 + 2\alpha(\theta - \theta_0) \qquad (10.8)$$

If a rigid body rotates about a fixed axis, the linear speed of any point on the body a distance r from the axis of rotation is related to the angular speed through the relation $v = r\omega$. Similarly, the tangential acceleration of any point on the body is related to the angular acceleration through the relation $a_t = r\alpha$. Note that *every point on the body has the same ω and α, but not every point has the same v and a_t*

$$v = r\omega \qquad (10.9)$$

$$a_t = r\alpha \qquad (10.10)$$

The *moment of inertia of a system of particles* is defined by Equation 10.14, where m_i is the mass of the ith particle and r_i is its distance from a specified axis. Note that I has SI units of kg·m².

$$I = \Sigma m_i r_i{}^2 \qquad (10.14)$$

The *kinetic energy* of a rigid body rotating with an angular speed ω about some axis is proportional to the square of the angular speed. Note that it does not represent a new form of energy. It is simply a convenient form for representing rotational kinetic energy.

$$K_R = \tfrac{1}{2}I\omega^2 \qquad (10.15)$$

The *torque τ* due to an applied force has a magnitude given by the product of the force and its moment arm d, where d equals the *perpendicular distance* from the rotation axis to the line of action of **F**. Torque is a measure of the ability of a force to rotate a body about a specified axis. Note that the torque depends on the axis of rotation, which must be specified when τ is evaluated.

$$\tau \equiv rF\sin\phi = Fd \qquad (10.18)$$

The *net torque* acting on a rigid body about some axis is equal to the product of the moment of inertia and angular acceleration, where I is the moment of inertia about the axis of rotation. This is only true for a plane laminar body or for the case when the axis of rotation is a principal axis.

$$\tau_{net} = I\alpha \qquad (10.20)$$

If a net torque τ acts on a rigid body, the *power supplied to the body* at any instant is proportional to the angular speed.

$$P = \tau\omega \qquad (10.22)$$

The *work-energy theorem* says that the net work done by external forces in rotating a rigid body about a fixed axis equals the change in the body's rotational kinetic energy.

$$W = \tfrac{1}{2}I\omega^2 - \tfrac{1}{2}I\omega_0^2 \qquad (10.23)$$

SUGGESTIONS, SKILLS, AND STRATEGIES

You should know how to calculate the moment of inertia of a system of particles about a specified axis. The technique is straightforward, and consists of applying $I = \Sigma m_i r_i^2$, where m_i is the mass of the ith particle and r_i is the distance from the axis of rotation to the particle.

Once the moment of inertia about an axis through the center of mass I_{CM} is known, you can easily evaluate the moment of inertia about any axis parallel to the axis through the center of mass using the *parallel axis theorem* :

$$I = I_{CM} + Md^2$$

Figure 10.1

where d is the distance between the two axes.

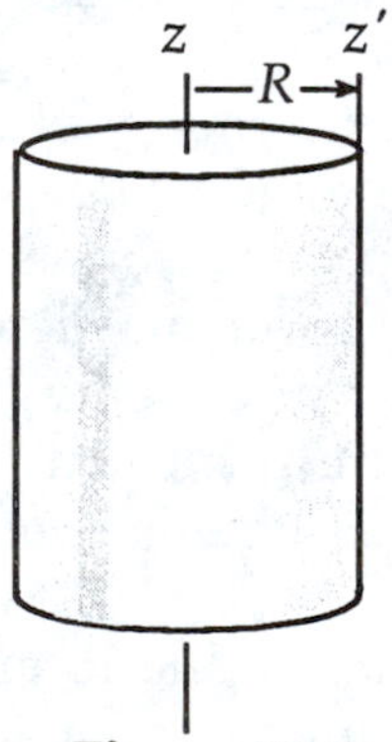

For example, the moment of inertia of a solid cylinder about an axis through its center (the z axis in Figure 10.1) is given by $I_z = \tfrac{1}{2}MR^2$. Hence, the moment of inertia about the z' axis located a distance $d = R$ from the z axis is

$$I_{z'} = I_z + MR^2 = \tfrac{1}{2}MR^2 + MR^2 = \tfrac{3}{2}MR^2$$

REVIEW CHECKLIST

▷ Quantitatively, the angular displacement, speed, and acceleration for a rigid body system in rotational motion are related to the distance traveled, tangential speed, and tangential acceleration. The linear quantity is calculated by multiplying the angular quantity by the radius arm for an object or point in that system.

▷ If a body rotates about a fixed axis, every particle on the body has the same angular speed and angular acceleration. For this reason, rotational motion can be simply described using these quantities. The formulas which describe angular motion are analogous to the corresponding set of formulas pertaining to linear motion.

▷ Calculate the moment of inertia I of a system of particles or a rigid body about a specific axis. Note that the value of I depends on (a) the mass distribution and (b) the axis about which the rotation occurs. The parallel-axis theorem is useful for calculating I about an axis parallel to one that goes through the center of mass.

▷ Understand the concept of torque associated with a force, noting that the torque associated with a force has a magnitude equal to the force times the moment arm. Furthermore, note that the value of the torque depends on the origin about which it is evaluated.

▷ Recognize that the work-energy theorem can be applied to a rotating rigid body. That is, the net work done on a rigid body rotating about a fixed axis equals the change in its rotational kinetic energy.

SOLUTIONS TO SELECTED END-OF-CHAPTER PROBLEMS

1. A wheel starts from rest and rotates with constant angular acceleration to an angular speed of 12.0 rad/s in 3.00 s. Find (a) the magnitude of the angular acceleration of the wheel and (b) the angle in radians through which it rotates in this time.

Solution

(a) $\alpha = \dfrac{\omega - \omega_0}{t} = \dfrac{(12.0 - 0)\ \text{rad/s}}{3.00\ \text{s}} = 4.00\ \text{rad/s}^2$ ◊

(b) $\theta = \omega_0 t + \frac{1}{2}\alpha t^2 = \frac{1}{2}\left(4.00\ \text{rad/s}^2\right)(3.00\ \text{s})^2 = 18.0\ \text{rad}$ ◊

5. An electric motor rotating a grinding wheel at 100 rev/min is switched off. Assuming constant negative angular acceleration of magnitude 2.00 rad/s², (a) how long does it take the wheel to stop? (b) Through how many radians does it turn during the time found in (a)?

Solution
$$\omega_0 = 100\,\frac{rev}{min}\left(2\pi\,\frac{rad}{rev}\right)\left(\frac{1\,min}{60.0\,s}\right) = 10.47\ rad/s$$

$$\omega = 0 \qquad \alpha = -2.00\ rad/s^2$$

(a) $\omega = \omega_0 + \alpha t$

$$t = \frac{\omega - \omega_0}{\alpha} = \frac{0 - (10.5\ rad/s)}{-2.00\ rad/s^2} = 5.24\ s \quad \Diamond$$

(b) $\omega^2 - \omega_0^2 = 2\alpha(\theta - \theta_0)$

$$\theta - \theta_0 = \frac{\omega^2 - \omega_0^2}{2\alpha} = \frac{0 - (10.5\ rad/s)^2}{2(-2.00\ rad/s^2)} = 27.4\ rad \quad \Diamond$$

Note also in part (b) that since a constant acceleration is acting for time t,

$$\theta = \overline{\omega}t = \left(\frac{10.5 + 0\ rad/s}{2}\right)(5.24\ s) = 27.4\ rad \quad \Diamond$$

---

9. A racing car travels on a circular track of radius 250 m. If the car moves with a constant linear speed of 45.0 m/s, find (a) its angular speed and (b) the magnitude and direction of its acceleration.

Solution (a) $\omega = \dfrac{v}{r} = \dfrac{45.0\ m/s}{250\ m} = 0.180\ rad/s \quad \Diamond$

(b) With no change in speed, the car has no tangential acceleration. The acceleration is centripetal and acts toward the center.

$$a_c = \frac{v^2}{r} = \frac{(45.0\ m/s)^2}{250\ m} = 8.10\ m/s^2 \quad \Diamond$$

---

11. A wheel 2.00 m in diameter rotates with a constant angular acceleration of 4.00 rad/s^2. The wheel starts at rest at $t = 0$, and the radius vector at point P on the rim makes an angle of 57.3° with the horizontal at this time. At $t = 2.00$ s, find (a) the angular speed of the wheel, (b) the linear speed and acceleration of the point P, and (c) the position of the point P.

Solution   Given $r = 1.00$ m, $\alpha = 4.00$ rad/s^2, $\omega_0 = 0$, and $\theta_0 = 57.3° = 1$ rad

(a) $\omega = \omega_0 + \alpha t = 0 + \alpha t$

At $t = 2.00$ s, $\omega = \left(4.00 \text{ rad/s}^2\right)(2.00 \text{ s}) = 8.00 \text{ rad/s}$ ◊

(b) $v = r\omega = (1.00 \text{ m})(8.00 \text{ rad/s}) = 8.00 \text{ m/s}$ ◊

$a_c = r\omega^2 = (1.00 \text{ m})(8.00 \text{ rad/s})^2 = 64.0 \text{ m/s}^2$ ◊

$a_t = r\alpha = (1.00 \text{ m})\left(4.00 \text{ rad/s}^2\right) = 4.00 \text{ m/s}^2$ ◊

The magnitude of the total acceleration is:

$$a = \sqrt{a_c^2 + a_t^2} = \sqrt{\left(64.0 \text{ m/s}^2\right)^2 + \left(4.00 \text{ m/s}^2\right)^2} = 64.1 \text{ m/s}^2 \text{ ◊}$$

The direction of the total acceleration vector makes an angle ϕ with respect to the radius to point P:

$$\phi = \tan^{-1}\left(\frac{a_t}{a_c}\right) = \tan^{-1}\left(\frac{4.00}{64.0}\right) = 3.58° \text{ ◊}$$

(c) $\theta = \theta_0 + \omega_0 t + \frac{1}{2}\alpha t^2 = (1.00 \text{ rad}) + \frac{1}{2}\left(4.00 \text{ rad/s}^2\right)(2 \text{ s})^2 = 9.00 \text{ rad}$

But θ is the total angle through which point P has passed, and is greater than one revolution. The position of point P is found by subtracting one revolution.

P is at $9.00 \text{ rad} - 2\pi \text{ rad} = 2.72 \text{ rad}$ ◊

13. A disk 8.00 cm in radius rotates at a constant rate of 1200 rev/min about its central axis. Determine (a) its angular speed, (b) the linear speed at a point 3.00 cm from its center, (c) the radial acceleration of a point on the rim, and (d) the total distance a point on the rim moves in 2.00 s.

Solution (a) $\omega = 2\pi f = (2\pi \text{ rad/rev})\left(\dfrac{1200 \text{ rev/min}}{60 \text{ s/min}}\right) = 125.7 \text{ rad/s} = 126 \text{ rad/s}$ ◊

(b) $v = \omega R = (125.7 \text{ rad/s})(0.0300 \text{ m}) = 3.77 \text{ m/s}$ ◊

(c) $a_c = \omega^2 R = (125.7 \text{ rad/s})^2(0.0800 \text{ m}) = 1.26 \times 10^3 \text{ m/s}^2$ ◊

(d) $s = R\theta = R\omega t = (8.00 \times 10^{-2} \text{ m})(125.7 \text{ rad/s})(2.00 \text{ s}) = 20.1 \text{ m}$ ◊

---

17. The four particles in Figure P10.17 are connected by rigid rods of negligible mass. The origin is at the center of the rectangle. If the system rotates in the xy plane about the z axis with an angular speed of 6.00 rad/s, calculate (a) the moment of inertia of the system about the z axis and (b) the rotational energy of the system.

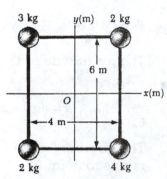

Figure P10.17

Solution (a) All four particles are at the same distance r from the z axis:

$$r^2 = (3.00 \text{ m})^2 + (2.00 \text{ m})^2 = 13.00 \text{ m}^2$$

Therefore,

$$I_z = \sum m_i r_i^2$$

$$= (3.00 \text{ kg})(13.00 \text{ m}^2) + (2.00 \text{ kg})(13.00 \text{ m}^2) + (4.00 \text{ kg})(13.00 \text{ m}^2) + (2.00 \text{ kg})(13.00 \text{ m}^2)$$

$$= 143 \text{ kg} \cdot \text{m}^2 \quad ◊$$

(b) $K_R = \frac{1}{2} I_z \omega^2 = \frac{1}{2}(143 \text{ kg} \cdot \text{m}^2)(6 \text{ rad/s})^2 = 2.57 \text{ kJ}$ ◊

---

19. Three particles are connected by rigid rods of negligible mass lying along the y axis (Fig. P10.19). If the system rotates about the x axis with an angular speed of 2.00 rad/s, find (a) the moment of inertia about the x axis and the total rotational energy evaluated from $\frac{1}{2}I\omega^2$ and (b) the linear speed of each particle and the total energy evaluated from $\sum\frac{1}{2}m_i v_i^2$.

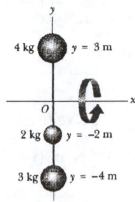

Figure P10.19

Solution

(a) $I = \Sigma mr^2$

$I = (4.00 \text{ kg})(3.00 \text{ m})^2 + (2.00 \text{ kg})(-2.00 \text{ m})^2 + (3.00 \text{ kg})(-4.00 \text{ m})^2$

$I = 92.0 \text{ kg} \cdot \text{m}^2$ ◊

$K_R = \frac{1}{2}I\omega^2 = \frac{1}{2}(92.0 \text{ kg} \cdot \text{m}^2)(2.00 \text{ rad}/\text{s})^2 = 184 \text{ J}$ ◊

(b) The 4.00-kg mass moves at $v = r\omega = (3.00 \text{ m})(2.00 \text{ rad}/\text{s}) = 6.00 \text{ m}/\text{s}$
For the 2.00-kg mass, $v = r\omega = (2.00 \text{ m})(2.00 \text{ rad}/\text{s}) = 4.00 \text{ m}/\text{s}$
For the 3.00-kg mass, $v = r\omega = (4.00 \text{ m})(2.00 \text{ rad}/\text{s}) = 8.00 \text{ m}/\text{s}$

$K = \sum\frac{1}{2}mv^2$

$K = \frac{1}{2}(4.00 \text{ kg})(6.00 \text{ m}/\text{s})^2 + \frac{1}{2}(2.00 \text{ kg})(4.00 \text{ m}/\text{s})^2 + \frac{1}{2}(3.00 \text{ kg})(8.00 \text{ m}/\text{s})^2$

$K = 184 \text{ J}$ ◊

Because we are evaluating the energy at a moment in time, the energy can be calculated as rotational or translational.

25. Find the net torque on the wheel in Figure P10.25 about the axle through O if $a = 10$ cm and $b = 25$ cm.

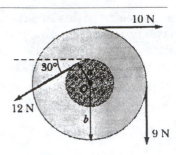

Figure P10.25

Solution

$\Sigma\tau = \Sigma Fd$

$= (12.0 \text{ N})(0.10 \text{ m}) - (10.0 \text{ N})(0.25 \text{ m}) - (9.0 \text{ N})(0.25 \text{ m})$

$= -3.55 \text{ N} \cdot \text{m}$ (away) ◊

Note that the 30° angle is not required for the solution. Note also that the 10.0-N and 9.0-N forces both produce clockwise, negative torques.

---

30. A model airplane whose mass is 0.75 kg is tethered by a wire so that it flies in a circle 30 m in radius. The airplane engine provides a net thrust of 0.80 N perpendicular to the tethering wire. (a) Find the torque the net thrust produces about the center of the circle. (b) Find the angular acceleration of the airplane when it is in level flight. (c) Find the linear acceleration of the airplane tangent to its flight path.

Solution

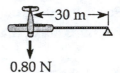

(a)　$\tau = Fd = (0.80 \text{ N})(30 \text{ m}) = 24 \text{ N} \cdot \text{m}$ ◊

(b)　$I = mr^2 = (0.75 \text{ kg})(30 \text{ m})^2 = 675 \text{ kg} \cdot \text{m}^2$

$\Sigma\tau = I\alpha$

$\alpha = \dfrac{\Sigma\tau}{I} = \dfrac{24 \text{ N} \cdot \text{m}}{675 \text{ kg} \cdot \text{m}^2} = 0.0356 \text{ rad}/\text{s}^2 = 0.0356 \text{ rad}/\text{s}^2$ ◊

(c)　$a = r\alpha = (30 \text{ m})(0.0356/\text{s}^2) = 1.07 \text{ m}/\text{s}^2$ ◊

We could also find this linear acceleration from $\Sigma F = ma$:

$$a = \frac{\Sigma F}{m} = \frac{0.80 \text{ N}}{0.75 \text{ kg}} = 1.07 \text{ m}/\text{s}^2$$

---

33. (a) A uniform solid disk of radius R and mass M is free to rotate on a frictionless pivot through a point on its rim (Fig. P10.33). If the disk is released from rest in the position shown by the green circle, what is the speed of its center of mass when the disk reaches the position indicated by the dashed circle? (b) What is the speed of the lowest point on the disk in the dashed position? (c) Repeat part (a) for a uniform hoop.

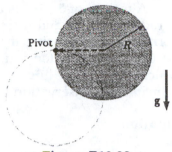

Figure P10.33

Solution   We cannot use the equation $\omega^2 - \omega_0{}^2 = 2\alpha\left(\dfrac{\pi}{2}\right)$

to find ω, because α is not constant. Instead, we use conservation of energy. To identify the change in gravitational energy, think of the height through which the center of mass fails. To identify the final kinetic energy, think of the motion as rolling on a one-point track, with center-of-mass translational kinetic energy plus energy of rotation about the center of mass:

$$(K_{\text{trans}} + K_{\text{rot}} + U_g)_i + \Delta K_{\text{nc}} = (K_{\text{trans}} + K_{\text{rot}} + U_g)_f$$

$$0 + 0 + MgR + 0 = \tfrac{1}{2}Mv_{\text{CM}}{}^2 + \tfrac{1}{2}I\omega^2 + 0$$

Since $\omega = \dfrac{v_{\text{CM}}}{R}$ and $I = \tfrac{1}{2}mR^2$ (for a disk),

$$MgR = \tfrac{1}{2}Mv_{\text{CM}}{}^2 + \frac{\tfrac{1}{2}\left(\tfrac{1}{2}MR^2\right)v_{\text{CM}}{}^2}{R^2} = \tfrac{3}{4}Mv_{\text{CM}}{}^2$$

(a) $v_{\text{CM}} = \sqrt{\dfrac{4gR}{3}}$ ◊

(b)   Twice as far from the axis, the bottom of the disk moves twice as fast through space.

$$\omega = \frac{v}{r} = \frac{v_{\text{CM}}}{R} = \frac{1}{R}\sqrt{\frac{4gR}{3}}, \qquad \text{so} \qquad v_b = \omega r = 2\omega R = 2\sqrt{\frac{4gR}{3}}$$ ◊

(c)   The hoop has a larger moment of inertia, so it turns more slowly. In this case $I = 2MR^2$, and conservation of energy gives:

$$MgR = \tfrac{1}{2}Mv_{\text{CM}}{}^2 + \frac{\tfrac{1}{2}\left(MR^2\right)v_{\text{CM}}{}^2}{R^2} = Mv_{\text{CM}}{}^2, \qquad \text{so} \qquad v_{\text{CM}} = \sqrt{gR}$$ ◊

35.   A weight of 50.0 N is attached to the free end of a light string wrapped around a pulley of radius 0.250 m and mass 3.00 kg. The pulley is free to rotate in a vertical plane about the horizontal axis passing through its center. The weight is released 6.00 m above the floor. (a) Determine the tension in the string, the acceleration of the mass, and the speed with which the weight hits the floor. (b) Find the speed calculated in part (a) by using the principle of conservation of energy.

Solution

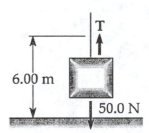

(a)   If it is a uniform disk, the pulley has moment of inertia

$$I = \tfrac{1}{2}MR^2 = \tfrac{1}{2}3.00 \text{ kg}(0.250 \text{ m})^2 = 0.0938 \text{ kg} \cdot \text{m}^2$$

The forces on it are shown, including a normal force exerted by its axle.

$\Sigma\tau = I\alpha$ becomes

$$n(0) + w(0) + T(0.250 \text{ m}) = \left(0.0938 \text{ kg} \cdot \text{m}^2\right)(a\,/\,0.250 \text{ m}) \qquad (1)$$

where we have applied $a_t = r\alpha$ to the point of contact between string and pulley.

The counterweight has mass

$$m = \frac{w}{g} = \frac{50.0 \text{ N}}{9.80 \text{ m}\,/\,\text{s}^2} = 5.10 \text{ kg}$$

For this mass, $\Sigma F_y = ma_y$ becomes $\qquad\qquad +T - 50.0 \text{ N} = (5.10 \text{ kg})(-a) \qquad (2)$

Note carefully the minus sign expressing that its acceleration is downward. We now have our two equations in the unknowns T and a for the two linked objects.

We substitute: $\quad \left[50.0 \text{ N} - (5.10 \text{ kg})a\right](0.250 \text{ m}) = 0.0938 \text{ kg} \cdot \text{m}^2(a\,/\,0.250 \text{ m})$

$$12.5 \text{ N} \cdot \text{m} - (1.28 \text{ kg} \cdot \text{m})a = (0.375 \text{ kg} \cdot \text{m})a$$

$$12.5 \text{ N} \cdot \text{m} = a(1.65 \text{ kg} \cdot \text{m}), \text{ or} \qquad\qquad a = 7.57 \text{ m}\,/\,\text{s}^2$$

and $\qquad\qquad T = 50.0 \text{ N} - 5.10 \text{ kg}(7.57 \text{ m/s}^2) = 11.4 \text{ N} \quad \Diamond$

For the motion of the weight,

$$v^2 = v_0{}^2 + 2a(x - x_0) = 0^2 + 2(7.57 \text{ m/s}^2)(6.00 \text{ m})$$

$$v = 9.53 \text{ m/s (down)} \quad \lozenge$$

(b) The work-energy theorem can take account of multiple objects more easily than Newton's second law. Like your bratty cousins, the work-energy theorem grows between visits; now it reads:

$$\left(K_1 + K_{2,rot} + U_{g1} + U_{g2}\right)_0 = \left(K_1 + K_{2,rot} + U_{g1} + U_{g2}\right)_f$$

$$0 + 0 + m_1 g y_{10} + 0 + 0 = \tfrac{1}{2} m_1 v_{1f}{}^2 + \tfrac{1}{2} I_2 \omega_{2f}{}^2 + 0 + 0$$

Now note that $v = \omega r$ as the string unwinds from the pulley.

$$50.0 \text{ N}(6.00 \text{ m}) = \tfrac{1}{2}(5.10 \text{ kg})v_f{}^2 + \tfrac{1}{2}\left(0.0938 \text{ kg} \cdot \text{m}^2\right)\left(\frac{v_f}{0.250 \text{ m}}\right)^2$$

$$300 \text{ N} \cdot \text{m} = \tfrac{1}{2}(5.10 \text{ kg})v_f{}^2 + \tfrac{1}{2}(1.50 \text{ kg})v_f{}^2$$

$$v_f = \sqrt{\frac{2(300 \text{ N} \cdot \text{m})}{6.60 \text{ kg}}} = 9.53 \text{ m/s} \quad \lozenge$$

---

41. A 4.00-m length of light nylon cord is wound around a uniform cylindrical spool of radius 0.500 m and mass 1.00 kg. The spool is mounted on a frictionless axle and is initially at rest. The cord is pulled from the spool with a constant acceleration of magnitude 2.50 m/s². (a) How much work has been done on the spool, when it reaches an angular speed of 8.00 rad/s? (b) Assuming there is enough cord on the spool, how long does it take the spool to reach this angular speed? (c) Is there enough cord on the spool?

Solution

(a) $W = \Delta K_R = \frac{1}{2}I\omega^2 - \frac{1}{2}I\omega_0^2$

$$= \frac{1}{2}I\left(\omega^2 - \omega_0^2\right) \qquad \text{where} \qquad I = \frac{1}{2}mR^2$$

$$= \left(\frac{1}{2}\right)\left(\frac{1}{2}\right)(1.00 \text{ kg})(0.500 \text{ m})^2\left[\left(8.00 \frac{\text{rad}}{\text{s}}\right)^2 - 0\right] = 4.00 \text{ J} \quad \Diamond$$

(b) $\omega = \omega_0 + \alpha t \qquad \text{where} \qquad \alpha = \dfrac{a}{r} = \dfrac{2.50 \text{ m/s}^2}{0.500 \text{ m}} = 5.00 \text{ rad/s}^2$

$$t = \frac{\omega - \omega_0}{\alpha} = \frac{(8.00 \text{ rad/s} - 0)}{5.00 \text{ rad/s}^2} = 1.60 \text{ s} \quad \Diamond$$

(c) $\theta = \theta_0 + \omega_0 t + \frac{1}{2}\alpha t^2$

$$\theta = 0 + 0 + \frac{1}{2}\left(5.00 \text{ rad/s}^2\right)(1.60 \text{ s})^2 = 6.40 \text{ rad}$$

The length of string pulled from the spool, $\qquad s = r\theta = (0.500 \text{ m})(6.40 \text{ rad}) = 3.20 \text{ m}$

When the spool reaches an angular velocity of 8 rad/s, 1.60 s will have elapsed and 3.20 m of cord will have been removed from the spool. Our answer is *yes*. $\quad \Diamond$

---

43. A long uniform rod of length L and mass M is pivoted about a horizontal, frictionless pin through one end. The rod is released from rest in a vertical position as in Figure P10.43. At the instant the rod is horizontal, find (a) its angular speed, (b) the magnitude of its angular acceleration, (c) the x and y components of the acceleration of its center of mass, and (d) the components of the reaction force at the pivot.

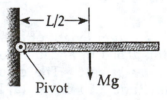

Figure P10.43

Solution

Since only conservative forces are acting on the bar, use conservation of energy:

(a)
$$\Delta K + \Delta U = 0$$

$$K_f - K_i + U_f - U_i = 0$$

Take the zero level of potential energy at the level of the pivot, and assume the mass of the bar to be located at the center of mass. Under these conditions $U_f = 0$ and $U_i = mgL/2$. Using the equation above,

$$\left(\frac{1}{2}I\omega^2 - 0\right) + \left(0 - mg\frac{L}{2}\right) = 0 \qquad \text{and} \qquad \omega = \sqrt{\frac{mgL}{I}}$$

For a bar rotating about an axis through one end, $I = mL^2/3$.

Therefore,
$$\omega = \sqrt{\frac{mgL}{\frac{1}{3}mL^2}} = \sqrt{\frac{3g}{L}} \quad \Diamond$$

(b) $\sum \tau = I\alpha \qquad$ or $\qquad mg\left(\frac{L}{2}\right) = \left(\frac{mL^2}{3}\right)\alpha, \qquad$ and $\qquad \alpha = \frac{3g}{2L} \quad \Diamond$

(c) $a_x = a_c = r\omega^2 = \left(\frac{L}{2}\right)\left(\frac{3g}{L}\right) = \frac{3g}{2} \quad \Diamond$

Since this is *centripetal* acceleration, it is directed along the *negative* horizontal:

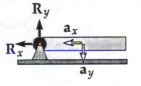

$$a_y = a_t = r\alpha = \frac{L}{2}\alpha = \frac{3g}{4} \quad \Diamond$$

(d) Using $\sum \mathbf{F} = m\mathbf{a}$, we have

$$R_x = ma_x = \frac{3mg}{2} \quad \text{in the } \textit{negative} \text{ direction} \quad \Diamond$$

$$R_y - mg = -ma_y$$

$$R_y = m(g - a_y) = m\left(g - \frac{3g}{4}\right) = \frac{mg}{4} \quad \Diamond$$

51. The blocks shown in Figure P10.51 are connected by a string of negligible mass passing over a pulley of radius $R = 0.250$ m and moment of inertia I. The block on the incline is moving up with a constant acceleration of magnitude $a = 2.00$ m/s^2. (a) Determine T_1 and T_2, the tensions in the two parts of the string, and (b) find the moment of inertia of the pulley.

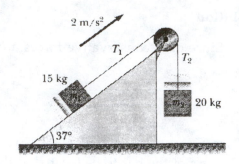

Figure P10.51

Solution

(a)   The 15.0-kg block weighs

$$w = mg = (15.0 \text{ kg})(9.80 \text{ m/s}^2) = 147 \text{ N}$$

We assume the surface of the incline to be frictionless. Taking the x axis as directed up the incline, $\Sigma F_x = ma_x$ yields:

$$(-147 \text{ N})\sin 37° + T_1 = (15.0 \text{ kg})(2.00 \text{ m/s}^2)$$

$$T_1 = 118 \text{ N} \quad \lozenge$$

For the 20.0-kg block, we have $\Sigma F_y = ma_y$,   or

$$T_2 - (20.0 \text{ kg})(9.80 \text{ m/s}^2) = (20.0 \text{ kg})(-2.00 \text{ m/s}^2)$$

So, $T_2 = 156 \text{ N} \quad \lozenge$

(b)   Now for the pulley,

$$\alpha = \frac{a}{r} = \frac{-2.00 \text{ m/s}^2}{0.250 \text{ m}} = -8.00 \text{ rad/s}^2, \quad \text{taking positive to be counterclockwise.}$$

$$\Sigma \tau = I\alpha, \quad \text{or} \quad (+118 \text{ N})(0.250 \text{ m}) - (156 \text{ N})(0.250 \text{ m}) = I(-8.00 \text{ rad/s}^2)$$

$$I = \frac{9.50 \text{ N} \cdot \text{m}}{8.00 \text{ rad/s}^2} = 1.19 \text{ kg} \cdot \text{m}^2 \quad \lozenge$$

53. As a result of friction, the angular speed of a wheel changes with time according to

$$\omega = \frac{d\theta}{dt} = \omega_0 e^{-\sigma t}$$

where ω_0 and σ are constants. The angular speed changes from 3.50 rad/s at $t = 0$ to 2.00 rad/s at $t = 9.30$ s. Use this information to determine σ and ω_0. Then determine (a) the magnitude of the angular acceleration at $t = 3.00$ s, (b) the number of revolutions the wheel makes in the first 2.50 s, and (c) the number of revolutions it makes before coming to rest.

Solution $\omega = \omega_0 e^{-\sigma t}$

When $t = 0$, $\omega = 3.50$ rad/s, and $e^0 = 1$,

so $\omega_{t=0} = \omega_0 e^{-\sigma(t=0)} = 3.50$ rad/s gives $\omega_0 = 3.50$ rad/s ◊

We now calculate σ: $2.00 \text{ rad/s} = (3.50 \text{ rad/s}) e^{-\sigma(9.30 \text{ s})}$

$0.571 = e^{-\sigma(9.30 \text{ s})}$

$\left[\ln(0.571) = -5.60\right] = \left[\ln\left(e^{-9.30\sigma}\right) = -9.30\sigma\right]$

and $\sigma = 0.0602 \text{ s}^{-1}$ ◊

(a) At all times, $\alpha = \frac{d\omega}{dt} = \frac{d}{dt}\left[\omega_0 e^{-\sigma t}\right] = -\sigma\omega_0 e^{-\sigma t}$

At $t = 3.00$ s, $\alpha = -\left(0.0602 \text{ s}^{-1}\right)(3.50 \text{ rad/s}) e^{-1.806} = -0.176 \text{ rad/s}^2$

(b) From the given equation, $d\theta = \omega_0 e^{-\sigma t} dt$,

and

$$\theta = \int_{0 \text{ s}}^{2.50 \text{ s}} \omega_0 e^{-\sigma t} dt = \left. \frac{\omega_0}{-\sigma} e^{-\sigma t} \right|_{0 \text{ s}}^{2.50 \text{ s}} = \frac{\omega_0}{-\sigma} \left(e^{-2.50\sigma} - 1 \right)$$

Substituting and solving, $\theta = -58.2(0.860 - 1) \text{ rad} = 8.12 \text{ rad}$

or

$$\theta = (8.12 \text{ rad}) \left(\frac{1 \text{ rev}}{2\pi \text{ rad}} \right) = 1.29 \text{ rev} \quad \Diamond$$

(c) The motion continues to a finite limit, as ω approaches zero and t goes to infinity. From part (b), the total angular displacement is

$$\theta = \left. \frac{\omega_0}{-\sigma} e^{-\sigma t} \right|_0^\infty = \frac{\omega_0}{-\sigma} \left(e^{-\infty} - e^0 \right) = \frac{\omega_0}{-\sigma} (0 - 1) = \frac{\omega_0}{-\sigma}$$

Substituting, we can calculate:

$$\theta = 58.2 \text{ rad},$$

or

$$\left(\frac{1 \text{ rev}}{2\pi \text{ rad}} \right) (58.2 \text{ rad}) = 9.26 \text{ rev} \quad \Diamond$$

Chapter 11

Rolling Motion, Angular Momentum, and Torque

Chapter 11

ROLLING MOTION, ANGULAR MOMENTUM, AND TORQUE

INTRODUCTION

In the previous chapter we learned how to treat the rotation of a rigid body about a fixed axis. This chapter deals in part with the more general case, where the axis of rotation is not fixed in space. We begin by describing the rolling motion of an object. Next, we define a vector product, a convenient mathematical tool for expressing such quantities as torque and angular momentum. The central point of this chapter is to develop the concept of the angular momentum of a system of particles, a quantity that plays a key role in rotational dynamics. In analogy to the conservation of linear momentum, we find that angular momentum is always conserved. This conservation law is a special case of the result that the time rate of change of the total angular momentum of any system equals the resultant external torque acting on the system.

NOTES FROM SELECTED CHAPTER SECTIONS

11.1 Rolling Motion of a Rigid Body

The *total kinetic energy* of a body undergoing rolling motion is the sum of the rotational kinetic energy about the center of mass and the translational kinetic energy of the center of mass.

11.3 Angular Momentum of a Particle

The *torque* acting on a particle is equal to the time rate of change of its angular momentum.

11.4 Rotation of a Rigid Body About a Fixed Axis

Although the points on a rigid body rotating about a fixed axis may not experience the same force, linear acceleration, or linear velocity, every point on the body has the same angular acceleration and angular velocity at any instant. Therefore, at any instant the rigid body as a whole is characterized by specific values for angular acceleration, net torque, and angular velocity.

The *work-energy theorem in rotational motion* states that the net work done by external forces in rotating a symmetric rigid body about a fixed axis equals the change in the body's rotational kinetic energy.

The total kinetic energy of an object undergoing rolling motion is the sum of a rotational kinetic energy about the center of mass and the translational kinetic energy of the center of mass.

11.5 Conservation of Angular Momentum

The *total angular momentum* of a system is constant if the resultant external torque acting on the system is zero. The resultant torque acting about the center of mass of a body equals the time rate of change of angular momentum, regardless of the motion of the center of mass.

EQUATIONS AND CONCEPTS

If a uniform body of circular cross section rolls on a rough surface without slipping, the *speed and acceleration of the center of mass* are simply related to the angular speed and angular acceleration.

$$v_{CM} = R\omega \tag{11.1}$$

$$a_{CM} = R\alpha \tag{11.2}$$

The total kinetic energy of a rigid body rolling on a rough surface can be expressed as the sum of the rotational kinetic energy about the center of mass and the translational kinetic energy of the center of mass.

$$K = \tfrac{1}{2}I_{CM}\omega^2 + \tfrac{1}{2}Mv_{CM}^2 \tag{11.4}$$

If a body rolls down an incline *without slipping*, one can use conservation of energy to find the *velocity of the center of mass* as the body falls through a vertical distance h, starting from rest. From this expression, one can also find the acceleration of the center of mass.

$$v_{CM} = \sqrt{\frac{2gh}{1 + \dfrac{I_{CM}}{MR^2}}} \tag{11.6}$$

The *torque* acting on a particle whose vector position is **r** can be expressed as **r** × **F**, where **F** is the external force acting on the particle. Torque also depends on the choice of the origin and has the SI unit of N·m.

$$\tau \equiv \mathbf{r} \times \mathbf{F} \qquad (11.7)$$

The *cross product* of any two vectors **A** and **B** is a vector **C** whose magnitude is given by $AB \sin \theta$ and whose direction is perpendicular to the plane formed by **A** and **B**. The sense of **C** can be determined from the right-hand rule.

$$\mathbf{C} = \mathbf{A} \times \mathbf{B} \qquad (11.8)$$

$$|\mathbf{C}| \equiv AB \sin \theta \qquad (11.9)$$

The *angular momentum* of a particle whose linear momentum is **p** and whose vector position is **r** is defined as $\mathbf{L} = \mathbf{r} \times \mathbf{p}$. The SI unit of angular momentum is kg·m/s². Note that both the magnitude and direction of **L** depend on the choice of origin.

$$\mathbf{L} \equiv \mathbf{r} \times \mathbf{p} \qquad (11.15)$$

If the same origin is used to define **L** and τ, then the *torque* on the particle *equals the time rate of change of its angular momentum.* This expression is the rotational analog of Newton's second law, $\mathbf{F} = \dfrac{d\mathbf{p}}{dt}$, and is the basic equation for treating rotating rigid bodies and rotating particles.

$$\tau = \frac{d\mathbf{L}}{dt} \qquad (11.19)$$

The angular momentum of a system of particles is obtained by taking the vector sum of the individual angular momenta about some point in an inertial frame. The individual momenta may change with time, which can change the total angular momentum. However, the total angular momentum of the system will only change if a net *external* torque acts on the system. In fact, *the net torque acting on a system of particles equals the time rate of change of the total angular momentum.*

$$\Sigma \tau_{ext} = \frac{d\mathbf{L}}{dt} \qquad (11.20)$$

The *magnitude of the angular momentum of a rigid body* in the form of a plane lamina rotating in the *x-y* plane about a *fixed axis* (the *z* axis) is given by the product $I\omega$, where I is the moment of inertia about the axis of rotation and ω is the angular speed.

$$L_z = I\omega \qquad (11.21)$$

The *law of conservation of angular momentum* states that if the resultant external torque acting on a system is zero, the total angular momentum is constant. This follows from Equation 11.20.

$$\text{If} \quad \Sigma \tau_{ext} = \frac{d\mathbf{L}}{dt} = 0 \qquad (11.24)$$

$$\mathbf{L} = \text{constant} \qquad (11.25)$$

If a zero net torque acts on a body rotating about a fixed axis, and the moment of inertia changes from I_i to I_f, then the conservation of angular momentum can be used to find the final angular speed in terms of the initial angular speed.

$$I_i\omega_i = I_f\omega_f = \text{constant} \qquad (11.27)$$

SUGGESTIONS, SKILLS, AND STRATEGIES

The operation of the vector or cross product is used for the first time in this chapter. (Recall that the angular momentum **L** of a particle is defined as $\mathbf{L} = \mathbf{r} \times \mathbf{p}$, while torque is defined by the expression $\tau = \mathbf{r} \times \mathbf{F}$.) Let us briefly review the cross-product operation and some of its properties.

If **A** and **B** are any two vectors, their cross product, written as $\mathbf{A} \times \mathbf{B}$, is also a vector **C**. That is,

$$\mathbf{C} = \mathbf{A} \times \mathbf{B}$$

where the magnitude of **C** is given by

$$C = |\mathbf{C}| = AB \sin\theta$$

and θ is the angle between **A** and **B** as in Figure 11.1.

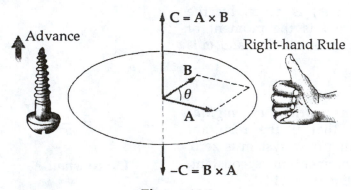

Figure 11.1

The direction of **C** is perpendicular to the plane formed by **A** and **B**, and its sense is determined by the right-hand rule. You should practice this rule for various choices of vector pairs. Note that $\mathbf{B} \times \mathbf{A}$ is directed *opposite* to $\mathbf{A} \times \mathbf{B}$. That is, $\mathbf{A} \times \mathbf{B} = -\mathbf{B} \times \mathbf{A}$.

This follows from the right-hand rule. You should not confuse the cross product of two vectors, which is a vector quantity, with the dot product of two vectors, which is a scalar quantity. (Recall that the dot product is defined as $\mathbf{A} \cdot \mathbf{B} = AB \cos\theta$.

Note that the cross product of any vector with itself is zero. That is, $\mathbf{A} \times \mathbf{A} = 0$ since $\theta = 0°$, and $\sin(0) = 0$.

Very often, vectors will be expressed in unit vector form, and it is convenient to make use of the multiplication table for unit vectors. Note that $\mathbf{i}$, $\mathbf{j}$, and $\mathbf{k}$ represent a set of mutually orthogonal vectors as shown in Figure 11.3.

$$\mathbf{i} \times \mathbf{i} = \mathbf{j} \times \mathbf{j} = \mathbf{k} \times \mathbf{k} = 0$$

$$\mathbf{i} \times \mathbf{j} = -\mathbf{j} \times \mathbf{i} = \mathbf{k}$$

$$\mathbf{j} \times \mathbf{k} = -\mathbf{k} \times \mathbf{j} = \mathbf{i}$$

$$\mathbf{k} \times \mathbf{i} = -\mathbf{i} \times \mathbf{k} = \mathbf{j}$$

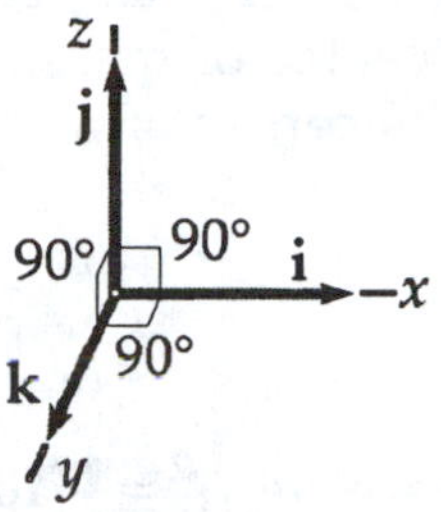

Figure 11.3

For example, if $\mathbf{A} = 3.0\mathbf{i} + 5.0\mathbf{j}$ and $\mathbf{B} = 4.0\mathbf{j}$, then we can take the cross-product:

$$\mathbf{A} \times \mathbf{B} = (3.0\mathbf{i} + 5.0\mathbf{j}) \times (4.0\mathbf{j}) = 3.0\mathbf{i} \times 4.0\mathbf{j} + 5.0\mathbf{j} \times 4.0\mathbf{j} = 12.0\mathbf{k}$$

REVIEW CHECKLIST

▷ Define the cross product (magnitude and direction) of any two vectors, $\mathbf{A}$ and $\mathbf{B}$, and state the various properties of the cross product.

▷ Define the angular momentum $\mathbf{L}$ of a particle moving with a velocity $\mathbf{v}$ relative to a specified point, and the torque τ acting on the particle relative to that point. Note that both $\mathbf{L}$ and τ are quantities which depend on the choice of the origin since they involve the vector position $\mathbf{r}$ of the particle. (That is, $\mathbf{L} = \mathbf{r} \times \mathbf{p}$ and $\tau = \mathbf{r} \times \mathbf{F}$.)

▷ Apply the conservation of angular momentum principle to a body rotating about a fixed axis, in which the moment of inertia changes due to a change in the mass distribution.

▷ Describe the center of mass motion of a rigid body which undergoes both rotation about some axis and translation in space. Note that for pure rolling motion of an object such as a sphere or cylinder, the total kinetic energy can be expressed as the sum of a rotational kinetic energy about the center of mass plus the translational energy of the center of mass.

Chapter 11

SOLUTIONS TO SELECTED END-OF-CHAPTER PROBLEMS

1. A cylinder of mass 10.0 kg rolls without slipping on a horizontal surface. At the instant its center of mass has a speed of 10.0 m/s, determine (a) the translational kinetic energy of its center of mass, (b) the rotational energy about its center of mass, and (c) its total energy.

Solution

$V_{CM} = 10 m/s$
$m = 10 kg$

(a) $K_{trans} = \frac{1}{2}mv_{CM}^2 = \frac{1}{2}(10.0 \text{ kg})(10.0 \text{ m/s})^2 = 500 \text{ J}$ ◊

(b) Call the radius of the cylinder R. An observer at the center sees the rough surface and the circumference of the cylinder moving at 10.0 m/s, so the angular speed of the cylinder is:

$$\omega = \frac{v_{CM}}{R} = \frac{10.0 \text{ m/s}}{R}$$

The moment of inertia $\qquad I_{CM} = \frac{1}{2}mR^2,$

so $\qquad K_{rot} = \frac{1}{2}I_{CM}\omega^2 = \left(\frac{1}{2}\right)\left[\frac{1}{2}(10.0 \text{ kg})R^2\right]\left(\frac{10.0 \text{ m/s}}{R}\right)^2 = 250 \text{ J}$ ◊

(c) $K_{tot} = 500 \text{ J} + 250 \text{ J} = 750 \text{ J}$ ◊

---

3. (a) Determine the acceleration of the center of mass of a uniform solid disk rolling down an incline and compare this acceleration with that of a uniform hoop. (b) What is the minimum coefficient of friction required to maintain pure rolling motion for the disk?

Solution $\qquad \Sigma F_x = mg\sin\theta - f = ma_{CM}$ $\qquad\qquad$ (1)

$$\Sigma F_y = n - mg\cos\theta = 0 \qquad\qquad (2)$$

$$\tau = fr = I_{CM}\alpha = \frac{I_{CM}a_{CM}}{r} \qquad\qquad (3)$$

(a) For a disk, $I_{CM} = \frac{1}{2}mr^2$. From (3) we find $f = \frac{\left[\frac{1}{2}mr^2\right]a_{CM}}{r^2} = \frac{1}{2}ma_{CM}$.

Substituting this into (1) gives

$$mg\sin\theta - \frac{1}{2}ma_{CM} = ma_{CM} \quad \text{and} \quad a_{CM} = \frac{2}{3}g\sin\theta \quad \Diamond$$

For a hoop, $I_{CM} = mr^2$

From (3), $\qquad f = \frac{mr^2 a_{CM}}{r^2} = ma_{CM}$

Substituting this into (1) gives

$$mg\sin\theta - ma_{CM} = ma_{CM} \quad \text{and} \quad a_{CM} = \frac{1}{2}g\sin\theta \quad \Diamond$$

Therefore, $\qquad \dfrac{\left(a_{CM}\right)_{sphere}}{\left(a_{CM}\right)_{hoop}} = \dfrac{\frac{2}{3}g\sin\theta}{\frac{1}{2}g\sin\theta} = \dfrac{4}{3} \quad \Diamond$

(b) From (2) we find $n = mg\cos\theta$, and $\qquad f = \mu n = \mu mg\cos\theta$

Likewise, from equation (1), $\qquad f = mg\sin\theta - ma_{CM}$

Setting these two equations equal,

$$\mu mg\cos\theta = mg\sin\theta - \frac{2}{3}mg\sin\theta = \frac{mg\sin\theta}{3}$$

so $\qquad \mu = \frac{1}{3}\left(\frac{\sin\theta}{\cos\theta}\right) = \frac{1}{3}\tan\theta \quad \Diamond$

7. Two vectors are given by $\mathbf{A} = -3\mathbf{i} + 4\mathbf{j}$ and $\mathbf{B} = 2\mathbf{i} + 3\mathbf{j}$. Find (a) $\mathbf{A} \times \mathbf{B}$ and (b) the angle between $\mathbf{A}$ and $\mathbf{B}$.

Solution

(a) $\mathbf{A} \times \mathbf{B} = (-3\mathbf{i} + 4\mathbf{j}) \times (2\mathbf{i} + 3\mathbf{j})$

$$= (-6\mathbf{i} \times \mathbf{i}) - (9\mathbf{i} \times \mathbf{j}) + (8\mathbf{j} \times \mathbf{i}) + (12\mathbf{j} \times \mathbf{j})$$

$$= 0 - 9\mathbf{k} + 8(-\mathbf{k}) + 0 = -17\mathbf{k} \quad \Diamond$$

(b) Since $|\mathbf{A} \times \mathbf{B}| = |\mathbf{A}||\mathbf{B}||\sin\theta|$,

$$\theta = \sin^{-1}\frac{|\mathbf{A} \times \mathbf{B}|}{|\mathbf{A}||\mathbf{B}|} = \sin^{-1}\left[\frac{17}{\sqrt{3^2 + 4^2}\sqrt{2^2 + 3^2}}\right] = 71° \quad \Diamond$$

Related Calculation: To solidify your understanding, review taking the dot product of these same two vectors, and again find the angle between them. For these next calculations, assume all given values to be known to 3 significant figures.

$$\mathbf{A} \cdot \mathbf{B} = (-3.00\mathbf{i} + 4.00\mathbf{j}) \cdot (2.00\mathbf{i} + 3.00\mathbf{j}) = -6.00 + 12.0 = 6.0 \quad \Diamond$$

$$\mathbf{A} \cdot \mathbf{B} = |\mathbf{A}||\mathbf{B}|\cos\theta$$

$$\theta = \text{Arccos}\frac{|\mathbf{A} \cdot \mathbf{B}|}{|\mathbf{A}||\mathbf{B}|} = \text{Arccos}\frac{6.0}{\sqrt{3.00^2 + 4.00^2}\sqrt{2.00^2 + 3.00^2}} = 71° \quad \Diamond$$

11. If $|\mathbf{A} \times \mathbf{B}| = \mathbf{A} \cdot \mathbf{B}$, what is the angle between $\mathbf{A}$ and $\mathbf{B}$?

Solution

We are given the condition $|\mathbf{A} \times \mathbf{B}| = \mathbf{A} \cdot \mathbf{B}$. This says that

$$AB \sin\theta = AB \cos\theta \quad \text{and} \quad \tan\theta = 1.$$

$$\theta = 45° \quad \text{and} \quad \theta = 225° \quad \text{both satisfy this condition.} \quad \Diamond$$

15. A light rigid rod 1.00 m in length rotates in the xy plane about a pivot through the rod's center. Two particles of mass 4.00 kg and 3.00 kg are connected to its ends (Fig. P11.15). Determine the angular momentum of the system about the origin at the instant the speed of each particle is 5.00 m/s.

Solution The moment of inertia of the compound object about its center is:

$$I = \sum mr^2 = (4.00 \text{ kg})(0.500 \text{ m})^2 + (3.00 \text{ kg})(0.500 \text{ m})^2$$

$$I = 1.75 \text{ kg} \cdot \text{m}^2$$

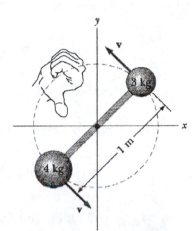

Its angular speed is $\quad \omega = \dfrac{v}{r} = \dfrac{(5.00 \text{ m/s})}{\frac{1}{2}(1.00 \text{ m})} = 10.0 \text{ rad/s}$

By the right-hand rule, we find that this is directed out of the plane.

Figure P11.15

So its angular momentum is

$$\mathbf{L} = I\omega = (1.75 \text{ kg} \cdot \text{m}^2)(10.0 \text{ rad/s})$$

$$= 17.5 \text{ kg} \cdot \text{m}^2/\text{s} \quad \text{out of the plane} \quad \Diamond$$

We could also compute this from:

$$\mathbf{L} = \sum m\mathbf{r} \times \mathbf{v}$$

$$\mathbf{L} = (4.00 \text{ kg})(0.500 \text{ m})(5.00 \text{ m/s}) \quad \text{out of the plane}$$

$$+ (3.00 \text{ kg})(0.500 \text{ m})(5.00 \text{ m/s}) \quad \text{out of the plane}$$

$$= 17.5 \text{ kg} \cdot \text{m}^2/\text{s} \quad \text{out of the plane} \quad \Diamond$$

17. The position vector of a particle of mass 2.0 kg is given as a function of time by $r = (6.0i + 5.0tj)$ m. Determine the angular momentum of the particle as a function of time.

Solution The velocity of the particle is

$$v = \frac{dr}{dt} = \frac{d}{dt}(6.0i \text{ m} + 5.0tj \text{ m}) = 5.0j \text{ m/s}$$

The angular momentum is

$$L = r \times p = mr \times v = (2.0 \text{ kg})(6.0i \text{ m} + 5.0tj \text{ m}) \times 5.0j \text{ m/s}$$

$$= (60.0 \text{ kg} \cdot \text{m}^2/\text{s})i \times j + (50.0t \text{ kg} \cdot \text{m}^2/\text{s})j \times j$$

$$= 60.0k \text{ kg} \cdot \text{m}^2/\text{s}, \text{ constant in time} \quad \Diamond$$

---

23. A particle of mass m is shot with an initial velocity v_0 making an angle θ with the horizontal as shown in Figure P11.23. The particle moves in the gravitational field of the Earth. Find the angular momentum of the particle about the origin when the particle is (a) at the origin, (b) at the highest point of its trajectory, and (c) just before it hits the ground. (d) What torque causes its angular momentum to change?

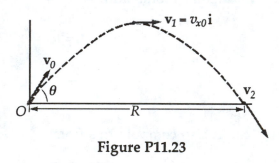

Figure P11.23

Solution $L = r \times mv$

(a) Since $r = 0$, $L_0 = 0$

(b) At the highest point of the trajectory,

$$L_1 = \left(\frac{v_0^2 \sin 2\theta}{2g}i + \frac{(v_0 \sin \theta)^2}{2g}j\right) \times mv_{x0}i$$

$$= \frac{-m(v_0 \sin \theta)^2 v_0 \cos \theta}{2g}k \quad \Diamond$$

(c) $\quad L_2 = R\mathbf{i} \times m\mathbf{v}_2,$ where $R = \dfrac{v_0^2 \sin 2\theta}{g}$

$$= m\left[R\mathbf{i} \times \left(v_0\mathbf{i}\cos\theta - v_0\mathbf{j}\sin\theta\right)\right]$$

$$= -mRv_0\mathbf{k}\sin\theta = \dfrac{-mv_0^3 \sin 2\theta \sin\theta}{g}\mathbf{k} \quad \Diamond$$

(d) The downward force of gravity exerts a torque in the –z direction. $\quad \Diamond$

---

25. A particle of mass 0.400 kg is attached to the 100-cm mark of a meter stick of mass 0.100 kg. The meter stick rotates on a horizontal, frictionless table with an angular speed of 4.00 rad/s. Calculate the angular momentum of the system when the stick is pivoted about an axis (a) perpendicular to the table through the 50.0-cm mark and (b) perpendicular to the table through the 0-cm mark.

Solution $\quad M = 0.100$ kg $\quad m = 0.400$ kg $\quad D = 1.00$ m $\quad \omega = 4.00$ rad/s

(a) For the meter stick, $I_{a,m} = \dfrac{MD^2}{12}$

For the weight, the parallel axis theorem gives us $I_{a,w} = I_{a,w} + md^2 = 0 + \dfrac{mD^2}{4}$

Thus, $\quad I_A = 0.108$ kg·m^2

and $\quad L_A = I_A\omega = \left(0.108 \text{ kg·m}^2\right)\left(4.00 \text{ rad/s}\right) = 0.433 \text{ kg·m}^2/\text{s} \quad \Diamond$

(b) Again using the parallel axis theorem,

$$I_{B,m} = I_{0,m} + \dfrac{1}{2}md^2 = \dfrac{MD^2}{12} + \dfrac{MD^2}{4} = \dfrac{MD^2}{3}$$

Likewise, $\quad I_{B,w} = I_{0,w} + md^2 = 0 + mD^2$

$$I_B = 0.433 \text{ kg·m}^2$$

and $\quad L_B = I_B\omega = \left(0.433 \text{ kg·m}^2\right)\left(4.00 \text{ rad/s}\right) = 1.73 \text{ kg·m}^2/\text{s} \quad \Diamond$

---

29. A 60-kg woman stands at the rim of a horizontal turntable having a moment of inertia of 500 kg·m² and a radius of 2.00 m. The turntable is initially at rest, and is free to rotate about a frictionless, vertical axle through its center. The woman then starts walking around the rim clockwise (as viewed from above the system) at a constant speed of 1.50 m/s relative to the Earth. (a) In what direction and with what angular speed does the turntable rotate? (b) How much work does the woman do to set the turntable into motion?

Solution (a) The table rotates in a direction opposite to that in which the woman walks. There are no external torques acting on the system; therefore, from conservation of angular momentum, we have

$$L_f = L_i = 0 \qquad \text{so} \qquad L_f = I_w\omega_w + I_t\omega_t = 0$$

and

$$\omega_t = -\frac{I_w}{I_t}\omega_w = -\left(\frac{m_w r^2}{I_t}\right)\left(\frac{v_w}{r}\right)$$

$$\omega_t = -\frac{(60.0\ \text{kg})(2.00\ \text{m})(1.50\ \text{m}/\text{s})}{500\ \text{kg}\cdot\text{m}^2} = -0.360\ \text{rad}/\text{s} = 0.360\ \text{rad/s CCW} \quad \Diamond$$

(b) Work done = ΔK

$$W = K_f - 0 = \tfrac{1}{2}m_{\text{woman}}v_{\text{woman}}^2 + \tfrac{1}{2}I_{\text{table}}\omega_{\text{table}}^2$$

$$W = \tfrac{1}{2}(60.0\ \text{kg})(1.50\ \text{m/s})^2 + \tfrac{1}{2}\left(500\ \text{kg}\cdot\text{m}^2\right)(0.360\ \text{rad/s})^2$$

$$W = 99.9\ \text{J} \quad \Diamond$$

Related Questions: (a) Why is angular momentum conserved? (b) Why is mechanical energy not conserved? (c) Is linear momentum conserved?

(a) Because the axle exerts no torque on the woman-plus-turntable system; only torques from outside the system can change the total angular momentum.

(b) The internal forces of the woman pushing backward on the turntable does positive work, converting chemical into kinetic energy.

(c) No. If the woman starts walking north, she pushes south on the turntable. Its axle holds it still against linear motion by pushing north on it, and this outside force delivers northward linear momentum into the system.

33. A wooden block of mass M resting on a frictionless horizontal surface is attached to a rigid rod of length l and of negligible mass (Fig. P11.33). The rod is pivoted at the other end. A bullet of mass m traveling parallel to the horizontal surface and normal to the rod with speed v hits the block and gets embedded in it. (a) What is the angular momentum of the bullet- block system? (b) What fraction of the original kinetic energy is lost in the collision?

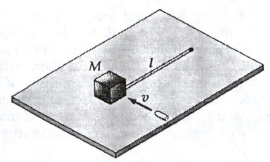

Figure P11.33

Solution   The initial angular momentum is due only to that of the bullet of mass m. Taking the origin at the pivot point,

(a) $\mathbf{L} = \mathbf{r} \times m\mathbf{v}$. Here, $\mathbf{r}$ is perpendicular to $\mathbf{v}$. So $\sin\theta = 1$, and $L = lmv$ ◊

Since there are no external torques acting on the system, the angular momentum remains constant.

(b)   The kinetic energy after the collision is

$$K_f = \tfrac{1}{2}(m+M)v_f^2$$

where M is the mass of the block, and v_f is the speed of the bullet and the block together.

From conservation of angular momentum, $L_i = L_f$,   or $lmv = l(m+M)v_f$.

So $v_f = \left(\dfrac{m}{m+M}\right)v$

and $K_f = \tfrac{1}{2}(m+M)v_f^2 = \tfrac{1}{2}(m+M)\left(\dfrac{m}{m+M}\right)^2 v^2 = \tfrac{1}{2}\left(\dfrac{m^2}{m+M}\right)v^2$

The initial kinetic energy $K_i = \tfrac{1}{2}mv^2$, so the fraction of the kinetic energy that is "lost" will be:

$$\text{Fraction} = \frac{\Delta K}{K_i} = \frac{K_i - K_f}{K_i} = \frac{\tfrac{1}{2}mv^2 - \tfrac{1}{2}\left(\dfrac{m^2}{m+M}\right)v^2}{\tfrac{1}{2}mv^2} = \frac{M}{m+M}\quad ◊$$

39. A string is wound around a uniform disk of radius R and mass M. The disk is released from rest with the string vertical and its top end tied to a fixed support (Fig. P11.39). As the disk descends, show that (a) the tension in the string is one third of the weight of the disk, (b) the magnitude of the acceleration of the center of mass is $2g/3$, and (c) the speed of the center of mass is $(4gh/3)^{1/2}$. Verify your answer to (c) using the energy approach.

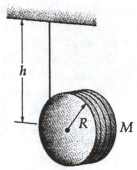

Figure P11.39

Solution

$\Sigma F = ma$ yields $\qquad \Sigma F = T - Mg = -Ma$

$\Sigma \tau = m\alpha$ becomes $\qquad \Sigma \tau = TR = I\alpha = \frac{1}{2}MR^2\left(\frac{a}{R}\right)$

(a) Solving the above two equations for a, we find

$$a = \frac{Mg - T}{M} \quad \text{and} \quad a = \frac{2T}{M};$$

Setting these two equations equal, $\qquad T = \frac{Mg}{3}$ ◊

(b) $\quad a = \frac{2T}{M} = \left(\frac{2}{M}\right)\left(\frac{Mg}{3}\right) \quad \text{or} \quad a = \frac{2}{3}g$ ◊

(c) Requiring conservation of mechanical energy, we have

$$\Delta U + \Delta K_{rot} + \Delta K_{trans} = 0$$

$$mg\Delta h + \frac{1}{2}I\omega^2 + \frac{1}{2}mv^2 = 0$$

$$(0 - mgh) + \frac{1}{2}\left(\frac{1}{2}MR^2\right)\omega^2 - 0 + \left(\frac{1}{2}Mv^2 - 0\right) = 0$$

When there is no slipping, $\quad \omega = \frac{v}{R} \quad \text{and} \quad v = \sqrt{\frac{4gh}{3}}$ ◊

43. This problem describes a method of determining the moment of inertia of an irregularly shaped object such as the payload for a satellite. Figure P11.43 shows one method of determining I experimentally. A mass m is suspended by a cord wound around the inner shaft (radius r) of a turntable supporting the object. When the mass is released from rest, it descends uniformly a distance h, acquiring a speed v. Show that the moment of inertia I of the equipment (including the turntable) is $mr^2(2gh/v^2 - 1)$.

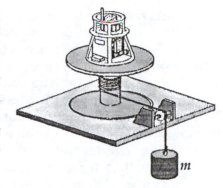

Figure P11.43

Solution

If the friction is negligible, energy is conserved as the counterweight unwinds. Each point on the cord moves at a linear speed of $v = \omega r$, where r is the radius of the drum.

The energy conservation equation gives us:

$$\left(K_1 + K_2 + U_{g1} + U_{g2}\right)_o + W_{nc} = \left(K_1 + K_2 + U_{g1} + U_{g2}\right)_f$$

Solving, we have

$$0 + 0 + mgh + 0 + 0 = \tfrac{1}{2}mv^2 + \tfrac{1}{2}I\omega^2 + 0 + 0$$

$$mgh = \tfrac{1}{2}mv^2 + \tfrac{1}{2}\frac{Iv^2}{r^2}$$

$$2mgh - mv^2 = I\frac{v^2}{r^2}$$

and finally,

$$I = mr^2\left(\frac{2gh}{v^2} - 1\right) \quad \Diamond$$

49. A mass m is attached to a cord passing through a small hole in a frictionless, horizontal surface (Fig. P11.49). The mass is initially orbiting with speed v_0 in a circle of radius r_0. The cord is then slowly pulled from below, decreasing the radius of the circle to r. (a) What is the speed of the mass when the radius is r? (b) Find the tension in the cord as a function of r. (c) How much work W is done in moving m from r_0 to r? (*Note*: The tension depends on r.) (d) Obtain numerical values for v, T, and W when $r = 0.100$ m, $m = 50.0$ g, $r_0 = 0.300$ m, and $v_0 = 1.50$ m/s.

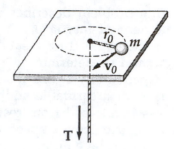

Figure P11.49

Solution

(a) Although an external force (tension of rope) acts on the mass, no external torques act. Therefore L = constant, and

$$mv_0r_0 = mvr \quad \text{and} \quad v = \frac{v_0 r_0}{r} \quad \lozenge$$

(b) $T = mv^2/r$. Substituting for v from (a), we find $\quad T = m\dfrac{v_0^2 r_0^2}{r^3} \quad \lozenge$

(c) $\quad W = \Delta K = \frac{1}{2}m\left(v^2 - v_0^2\right) = \dfrac{mv_0^2}{2}\left(\dfrac{r_0^2}{r^2} - 1\right) \quad \lozenge$

(d) Substituting the given values into the previous equations, we find

$$v = 4.50 \text{ m/s} \quad \lozenge$$

$$T = 10.1 \text{ N} \quad \lozenge$$

$$W = 0.450 \text{ J} \quad \lozenge$$

51. A trailer with loaded weight **w** is being pulled by a vehicle with a force **F**, as in Figure P11.51. The trailer is loaded such that its center of mass is located as shown. Neglect the force of rolling friction and assume the trailer has an acceleration of magnitude a. (a) Find the vertical component of **F** in terms of the given parameters. (b) If $a = 2.00$ m/s^2 and $h = 1.50$ m, what must be the value of d in order that $F_y = 0$ (no vertical load on the vehicle)? (c) Find F_x and F_y given that $w = 1500$ N, $d = 0.800$ m, $L = 3.00$ m, $h = 1.50$ m, and $a = -2.00$ m/s^2.

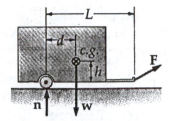

Figure P11.51

Solution

(a) $\Sigma F = ma$ gives us $\qquad F_x = ma \qquad$ and $\qquad F_y - w + n = 0$

When taken about the origin, $\Sigma\tau = ma = 0$ gives

$$F_y (L - d) + F_x h - nd = 0$$

Solving, $\qquad\qquad\qquad F_y L - F_y d + mah - nd = 0$

$$F_y L - wd + nd + mah - nd = 0$$

and $\qquad\qquad\qquad F_y = \dfrac{w}{L}\left(d - \dfrac{ah}{g}\right) \qquad \lozenge$

(b) If $F_y = 0$, then $\qquad\qquad d = \dfrac{ah}{g} = \dfrac{\left(2.00 \text{ m/s}^2\right)\left(1.50 \text{ m}\right)}{9.80 \text{ m/s}^2} = 0.306 \text{ m} \qquad \lozenge$

(c) Using the given data,

$$F_x = -306 \text{ N} \qquad \text{and} \qquad F_y = 553 \text{ N} \quad \lozenge$$

53. Two astronauts (Fig. P11.53), each having a mass of 75 kg, are connected by a 10-m rope of negligible mass. They are isolated in space, orbiting their center of mass at speeds of 5.0 m/s. Calculate (a) the magnitude of the angular momentum of the system by treating the astronauts as particles and (b) the rotational energy of the system. By pulling on the rope, the astronauts shorten the distance between them to 5.0 m. (c) What is the new angular momentum of the system? (d) What are their new speeds? (e) What is the new rotational energy of the system? (f) How much work is done by the astronauts in shortening the rope?

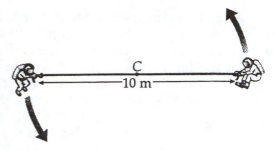

Figure P11.53

Solution

(a) $L = |m\mathbf{r} \times \mathbf{v}|$

In this case, $\mathbf{r}$ and $\mathbf{v}$ are perpendicular, so the magnitude of $\mathbf{L}$ about C is

$$L = \sum mrv = 2(75 \text{ kg})(5.0 \text{ m})(5.0 \text{ m/s}) = 3.75 \times 10^3 \text{ kg} \cdot \text{m}^2 / \text{s} \quad \lozenge$$

(b) $K = \frac{1}{2}mv^2 + \frac{1}{2}mv^2 = \frac{1}{2}(75 \text{ kg})(5.0 \text{ m/s})^2(2) = 1.88 \times 10^3 \text{ J} \quad \lozenge$

(c)   With a lever arm of zero, the rope tension generates no torque about C. Thus, the angular momentum is unchanged: $L = 3.75 \times 10^3 \text{ kg} \cdot \text{m}^2 / \text{s} \quad \lozenge$

(d)   Again, $L = 2mrv$

$$3.75 \times 10^3 \text{ kg} \cdot \text{m}^2 / \text{s} = 2(75 \text{ kg})(2.5 \text{ m})(v \sin 90°)$$

$$v = 10.0 \text{ m/s} \quad \lozenge$$

(e) $K = 2\left(\frac{1}{2}mv^2\right) = 2\left(\frac{1}{2}75 \text{ kg}\right)(10 \text{ m/s})^2 = 7.50 \times 10^3 \text{ J} \quad \lozenge$

(f) $W_{nc} = K_f - K_0 = 7.50 \times 10^3 \text{ J} - 1.88 \times 10^3 \text{ J}$

$$W_{nc} = 5.62 \times 10^3 \text{ J} \quad \lozenge$$

55. Toppling chimneys often break apart in mid-fall because the mortar between the bricks cannot withstand much tension force. As the chimney falls, this tension supplies the centripetal forces on the topmost segments that they need to keep them traveling in an arc. For simplicity, let us model the chimney as a uniform rod of length l pivoted at the lower end. The rod starts at rest in a vertical position (with the pivot at the bottom) and falls over under the influence of gravity. What fraction of the length of the rod has a tangential acceleration greater than $g \sin \theta$, where θ is the angle the chimney makes with the vertical axis?

Solution As the rod falls, the lever arm of its weight increases, so the total torque on it and its angular acceleration increase.

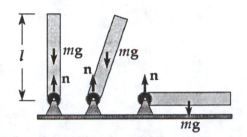

Now, consider how it is accelerating in the diagram to the right, just before it hits the ground.

Here, $\Sigma\tau = I\alpha$ becomes: $\dfrac{mgl}{2} = \dfrac{1}{3}ml^2\alpha$

so $\alpha = \dfrac{3g}{2l}$

Consider a point at a distance r along the rod from the pivot. Its tangential acceleration is

$$a_t = r\alpha = \frac{3gr}{2l}$$

For this to be greater than g, we require:

$$\frac{3gr}{2l} > g \quad \text{and} \quad r > \frac{2l}{3}$$

The outermost one third of the rod will have a tangential acceleration greater than g. ◊

---

57. A spool of wire of mass M and radius R is unwound under a constant force F (Fig. P11.57). Assuming the spool is a uniform solid cylinder that doesn't slip, show that (a) the acceleration of the center of mass is $4F/3M$ and (b) the force of friction is to the *right* and equal in magnitude to $F/3$. (c) If the cylinder starts from rest and rolls without slipping, what is the speed of its center of mass after it has rolled through a distance d?

Solution

To keep the spool from slipping, there must be static friction acting at its contact point with the floor. Assume friction is directed to the left. Then the full set of Newton's second-law equations are:

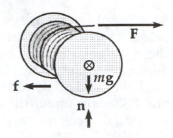

Figure P11.57
(modified)

$$\sum F_x = ma_x: \qquad F - f = Ma$$

$$\sum F_y = ma_y: \qquad -Mg + n = 0$$

$$\sum \tau_{CM} = I_{CM}\alpha_{CM}: \qquad -FR + Mg(0) + n(0) - fR = \tfrac{1}{2}MR^2\left(-\frac{a}{R}\right)$$

We have written $\alpha_{CM} = \dfrac{-a}{R}$ to describe the clockwise rotation. Regarding F, M, and R as known, the first and third of these equations allow us to solve for f and a:

$$f = F - Ma \qquad \text{and} \qquad -F - f = -\frac{1}{2}Ma$$

(a) $\quad F + F - Ma = \dfrac{1}{2}Ma \qquad 2F = \dfrac{3}{2}Ma \qquad a = \dfrac{4F}{3M} \qquad \Diamond$

(b) $\quad f = F - M\left(\dfrac{4F}{3M}\right) = F - \dfrac{4F}{3} = -\dfrac{F}{3} \qquad \Diamond$

The negative sign means that the force of friction is opposite the direction we assumed. That is, **f** is to the right, to oppose the possibility of the wheel spinning like a car stuck in the snow.

(c) Since a is constant, we can use

$$v^2 = v_0^2 + 2a(x - x_0) = 0 + 2\left(\frac{4F}{3M}\right)d$$

$$v = \sqrt{\frac{8Fd}{3m}} \qquad \Diamond$$

59. Suppose a solid disk of radius R is given an angular speed ω_0 about an axis through its center and then lowered to a horizontal surface and released, as in Problem 58 (Fig. P11.58). Furthermore, assume that the coefficient of friction between disk and surface is μ. (a) Show that the time it takes pure rolling motion to occur is $R\omega_0/3\mu g$.

(b) Show that the distance the disk travels before pure rolling occurs is $R^2\omega_0^2/18\mu g$.

Solution

(a)   If $v_0 = 0$, then at any time $\qquad v = at$

From Problem 58, once pure rolling occurs, $\omega = \frac{1}{3}\omega_0$, so that $v = \frac{1}{3}R\omega_0$.

Using these expressions for v and v_0 in the first equation, we find

$$\tfrac{1}{3}R\omega = at \qquad \text{where} \qquad a = \frac{F}{m} = \frac{-\mu mg}{m} = -\mu g$$

Therefore, $\qquad\qquad t = \dfrac{R\omega_0}{3\mu g} \qquad \Diamond$

(b)   The distance of travel is $\Delta x = \frac{1}{2}at^2$. Using the result from part (a), we find

$$\Delta x = \tfrac{1}{2}(\mu g)\left(\frac{\frac{1}{3}R\omega_0}{\mu g}\right)^2 = \frac{R^2\omega_0^2}{18\mu g} \qquad \Diamond$$

Chapter 12

Static Equilibrium and Elasticity

STATIC EQUILIBRIUM AND ELASTICITY

INTRODUCTION

Part of this chapter is concerned with the conditions under which a rigid object is in equilibrium. The term *equilibrium* implies either that the object is at rest or that its center of mass moves with constant velocity. We deal here with the former, which are referred to as objects in *static equilibrium*.

In Chapter 5 we stated that one necessary condition for equilibrium is that the net force on an object be zero. If the object is treated as a particle, this is the only condition that must be satisfied for equilibrium. That is, if the net force on the particle is zero, the particle remains at rest (if originally at rest) or moves with constant velocity (if originally in motion.)

The situation with real (extended) objects is more complex because objects cannot be treated as particles. In order for an extended object to be in static equilibrium, the net force on it must be zero *and* it must have no tendency to rotate. This second condition of equilibrium requires that *the net torque about any origin be zero*. In order to establish whether or not an object is in equilibrium, we must know its size and shape, the forces acting on different parts of it, and the points of application of the various forces.

The last section of this chapter deals with the realistic situation of objects that deform under load conditions. Such deformations are usually elastic in nature and do not affect the conditions of equilibrium. By *elastic* we mean that when the deforming forces are removed, the object returns to its original shape. Several elastic constants are defined, each corresponding to a different type of deformation.

NOTES FROM SELECTED CHAPTER SECTIONS

12.1   The Conditions of Equilibrium of a Rigid Body

A *rigid body* is defined as one that does not deform under the application of external forces. There are two necessary conditions for *equilibrium of a rigid body*: (1) the resultant external force must be zero and (2) the resultant external torque must be zero about any axis. Two forces are equal if, and only if, they have equal magnitudes and they have equal torques about any specified axis.

12.2 More on the Center of Gravity

In order to compute the torque due to the weight force, all of the weight of a body can be considered to be concentrated at a point called the *center of gravity*. At the point of the center of gravity, a force of magnitude mg and directed opposite the force of gravity will balance the body if no other external forces are acting. The center of gravity of an object coincides with its center of mass if the object is in a uniform gravitational field.

12.4 Elastic Properties of Solids

The elastic properties of solids are described in terms of *stress* and *strain*. Stress is a quantity that is proportional to the force causing a deformation of the object. Strain is a measure of the degree of the resulting deformation.

The *elastic modulus* of a material is the ratio of stress to strain for that material. There is an elastic modulus for each of three types of deformation: *Young's modulus* which measures resistance to change in length, *Shear modulus* which measures resistance to relative motion of the planes of a solid, and *Bulk modulus* which measures the resistance to a change in volume.

EQUATIONS AND CONCEPTS

In general, *an object at rest or one moving with constant velocity will only do so if the resultant force on it is zero.* This is a statement of the *first condition of equilibrium*, and corresponds to the condition of translational equilibrium.

$$\Sigma F = 0 \tag{12.1}$$

Since Equation 12.1 is a *vector sum* of all *external forces* acting on the body, this necessarily implies that the sum of the $x, y,$ and z components separately must be zero.

$$\Sigma F_x = 0$$
$$\Sigma F_y = 0$$
$$\Sigma F_z = 0$$

The *second condition of equilibrium* of a rigid body requires that the *vector sum of the torques relative to any origin must be zero.* This is the condition of rotational equilibrium.

$$\Sigma \tau = 0 \tag{12.2}$$

If all the forces acting on a rigid body lie in a common plane, say the x-y plane, then there is no z component of force. In this case, we only have to deal with three equations--two of which correspond to the first condition of equilibrium, the third coming from the second condition. In this case, the torque vector lies along the z axis. All problems in this chapter fall into this category.

$$\Sigma F_x = 0 \qquad (12.3)$$

$$\Sigma F_y = 0$$

$$\Sigma \tau_z = 0$$

In order to compute the torque due to the weight (gravitational force), all the weight can be considered to be concentrated at a single point called the center of gravity. When an object is in a uniform gravitational field, the center of gravity is located at the center of mass.

$$x_{cg} = \frac{m_1 x_1 + m_2 x_2 + m_3 x_3 + \cdots}{m_1 + m_2 + m_3 + \cdots} \qquad (12.4)$$

$$= \frac{\Sigma m_i x_i}{\Sigma m_i}$$

Young's modulus Y is a measure of the resistance of a body to elongation, and is equal to the ratio of the tensile stress (the force per unit area) to the tensile strain (the change in length over the original length).

$$Y = \frac{F/A}{\Delta L / L_0} \qquad (12.6)$$

The *Shear modulus S* is a measure of the deformation which occurs when a force is applied to one surface, while the opposite surface is held fixed. The shear modulus equals the ratio of the shearing stress to the shear strain.

$$S = \frac{F/A}{\Delta x / h} \qquad (12.7)$$

The *Bulk modulus B* is a parameter which characterizes the response of a body to uniform pressure on all sides. It is defined as the ratio of the volume stress (the pressure) to the volume strain ($\Delta V/V$).

$$B = -\frac{\Delta P}{\Delta V / V} \qquad (12.8)$$

SUGGESTIONS, SKILLS, AND STRATEGIES

Since this chapter represents the application of Newton's laws to a special situation, namely, rigid bodies in static equilibrium, it is important that you understand and follow the procedures for analyzing such problems. The following skills must be mastered in this regard:

- The need to recognize all *external* forces acting on the body, and the construction of an accurate free-body diagram.

- Resolving the external forces into their rectangular components, and applying the first condition of equilibrium $\Sigma F_x = 0$ and $\Sigma F_y = 0$.

- You must choose a convenient origin for calculating the net torque on the body. The choice of this origin is arbitrary. (The torque equation gives information which is not offered by applying $\Sigma F = 0$.)

- Solving the set of simultaneous equations obtained from the two conditions of equilibrium.

The following procedure is recommended when analyzing a body in equilibrium under the action of several external forces:

- Make a sketch of the object under consideration.

- Draw a free-body diagram and label all external forces acting on the object. Try to guess the correct direction for each force. If you select an incorrect direction that leads to a negative sign in your solution for a force, do not be alarmed; this merely means that the direction of the force is the opposite of what you assumed.

- Resolve all forces into rectangular components, choosing a convenient coordinate system. Then apply the first condition for equilibrium, which balances forces. Remember to keep track of the signs of the various force components.

- Choose a convenient axis for calculating the net torque on the object. Remember that the choice of the origin for the torque equation is *arbitrary*; therefore, choose an origin that will simplify your calculation as much as possible. Becoming adept at this is a matter of practice.

- The first and second conditions of equilibrium give a set of linear equations with several unknowns. All that is left is to solve the simultaneous equations for the unknowns in terms of the known quantities.

REVIEW CHECKLIST

▷ Describe the two necessary conditions of equilibrium for a rigid body.

▷ Locate the center of gravity of a system of particles or a rigid body and understand the subtle difference between center of gravity and center of mass.

▷ Analyze problems of rigid bodies in static equilibrium using the procedures presented in Section 12.3 of the text.

SOLUTIONS TO SELECTED END-OF-CHAPTER PROBLEMS

3. A uniform beam of weight w and length L has weights w_1 and w_2 at two positions, as in Figure P12.3. The beam is resting at two points. For what value of x will the beam be balanced at P such that the normal force at O is zero?

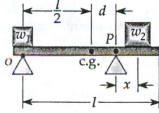

Figure P12.3

Solution Refer to the free-body diagram, and take torques about P.

$$\Sigma\tau_p = -n_0\left[\frac{l}{2}+d\right]+w_1\left[\frac{l}{2}+d\right]+wd-w_2x=0$$

We want to find x for which $n_0 = 0$. Let $n_0 = 0$ and solve for x.

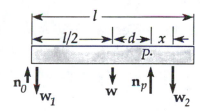

$$w_1\left[\frac{l}{2}+d\right]+wd-w_2x=0$$

$$x=\frac{w_1\left(\frac{l}{2}+d\right)+wd}{w_2} \qquad \lozenge$$

5. A ladder of weight 400 N and length 10.0 m is placed against a smooth vertical wall. A person weighing 800 N stands on the ladder 2.00 m from the bottom as measured along the ladder. The foot of the ladder is 8.00 m from the bottom of the wall. Calculate the force exerted by the wall, and the normal force exerted by the floor on the ladder.

Solution   The smooth wall exerts a normal force we call **P**, and no frictional force. The ladder makes an angle with the horizontal,

$$\theta = \cos^{-1}\left(\frac{8.00\ m}{10.0\ m}\right) = 36.9°$$

We choose to take torques about an axis at the foot of the ladder. The 800-N force has a lever arm of

$$(2.00)\cos 36.9° = 1.60\ m.$$

If the ladder is uniform, the weight of 400 N has a lever arm of $(5.00)\cos 36.9° = 4.00\ m.$

The force **P** has lever arm of $(10.0\ m)\sin 36.9° = 6.00\ m.$

The equations for equilibrium are

$$\sum F_x = 0 \qquad -f + P = 0$$

$$\sum F_y = 0 \qquad +n - 800\ N - 400\ N = 0$$

$$\sum \tau = 0 \qquad f(0) + n(0) + (800\ N)(1.6\ m) + (400\ N)(4.00\ m) - P(6.00\ m) = 0$$

The torque equation gives $P = \dfrac{2880\ N \cdot m}{6.00\ m} = 480\ N$ ◊

The vertical force equator gives $n = 1200\ N$ ◊

---

9. Consider the following mass distribution: 5.0 kg at (0, 0) m, 3.0 kg at (0, 4.0) m, and 4.0 kg at (3.0, 0) m. Where should a fourth mass of 8.0 kg be placed so that the center of gravity of the four-mass arrangement will be at (0, 0)?

Solution

$$r_{cg} = \frac{\Sigma m_i r_i}{\Sigma m_i} \qquad \text{or} \qquad r_{cg}(\Sigma m_i) = \Sigma m_i r_i$$

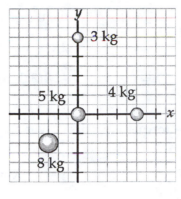

We require the center of mass to be at the origin; this simplifies the situation, leaving

$$\Sigma m_i x_i = 0 \qquad \text{and} \qquad \Sigma m_i y_i = 0$$

To find the x-coordinate:

$$[5.0 \text{ kg}][0 \text{ m}] + [3.0 \text{ kg}][0 \text{ m}] + [4.0 \text{ kg}][3.0 \text{ m}] + [8.0 \text{ kg}]x = 0$$

and $x = -1.50$ m.

Likewise, to find the y-coordinate, we solve:

$$[5.0 \text{ kg}][0 \text{ m}] + [3.0 \text{ kg}][4.0 \text{ m}] + [4.0 \text{ kg}][3.0 \text{ m}] + [8.0 \text{ kg}]y = 0$$

and $y = -1.50$ m.

Therefore, a fourth mass of 8.0 kg should be located at

$$r_4 = (-1.50i - 1.50j) \text{ m} \quad \lozenge$$

---

15. A 1500-kg automobile has a wheel base (the distance between the axles) of 3.0 m. The center of mass of the automobile is on the center line at a point 1.2 m behind the front axle. Find the force exerted by the ground on each wheel.

Solution   The car's weight is

$$w = mg = (1500 \text{ kg})(9.80 \text{ m/s}^2) = 14700 \text{ N}$$

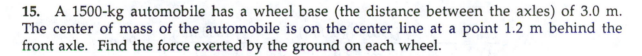

Call **F** the force of the ground on each of the front wheels and **R** the normal force on each of the rear wheels.

If we take torques around the front axle, the equations are as follows:

$$\Sigma F_x = 0 \qquad 0 = 0$$

$$\Sigma F_y = 0 \qquad 2R - 14700\ \text{N} + 2F = 0$$

$$\Sigma \tau = 0 \qquad -2R(3.0\ \text{m}) + 14700\ \text{N}(1.2\ \text{m}) + 2F(0) = 0$$

The torque equation gives :

$$R = \frac{17640\ \text{N}\cdot\text{m}}{6.0\ \text{m}} = 2940\ \text{N} = 2.94\ \text{kN} \quad \Diamond$$

Then, from the second force equation,

$$2(2.94\ \text{kN}) - 14.7\ \text{kN} + 2F = 0$$

and

$$F = 4.41\ \text{kN} \quad \Diamond$$

---

23. A 200-kg load is hung on a wire of length 4.0 m, cross-sectional area 0.20×10^{-4} m^2, and Young's modulus 8.0×10^{10} N/m^2. What is its increase in length?

Solution The load force is $(200\ \text{kg})(9.80\ \text{m/s}^2) = 1960\ \text{N}$.

Now

$$Y = \frac{F/A}{\Delta L/L_0},$$

so

$$\Delta L = \frac{FL_0}{AY} = \frac{(1960\ \text{N})(4.0\ \text{m})}{(0.20\times10^{-4}\ \text{m}^2)(8.0\times10^{10}\ \text{N/m}^2)}\left(\frac{1000\ \text{mm}}{\text{m}}\right) = 4.90\ \text{mm} \quad \Diamond$$

---

29. If the shear stress in steel exceeds about 4.0×10^8 N/m^2, the steel ruptures. Determine the shearing force necessary to (a) shear a steel bolt 1.0 cm in diameter and (b) punch a 1.0-cm-diameter hole in a 0.50-cm-thick steel plate.

Solution

(a) We do not need the equation $S = $ stress/strain. Rather, we use just the definition of stress $\sigma = F/A$, where A is the area of one of the layers sliding over each other:

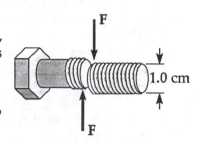

$$F = \sigma \cdot A = \pi \left(4.0 \times 10^8 \text{ N/m}^2\right)\left(0.50 \times 10^{-2} \text{ m}\right)^2 = 31.4 \text{ kN} \quad \lozenge$$

(b) Now the area of the molecular layers sliding over each other is the curved lateral surface area of the cylinder punched out, a cylinder of radius 0.50 cm and height 0.50 cm.

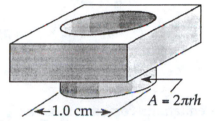

So,

$$F = \sigma \cdot A = (\sigma)(h)(2\pi r)$$

$$= 2\pi \left(4.0 \times 10^8 \text{ N/m}^2\right)\left(0.50 \times 10^{-2} \text{ m}\right)\left(0.50 \times 10^{-2} \text{ m}\right)$$

$$= 62.8 \text{ kN} \quad \lozenge$$

---

31. When water freezes, it expands about 9%. What would be the pressure increase inside your automobile engine block if the water in it froze? (The bulk modulus of ice is $2.0 \times 10^9 \text{ N/m}^2$.)

Solution V represents the original volume; $0.09V$ is the change in volume that happens if the block cracks open. Imagine squeezing the ice back down ($\Delta V = -0.09V$) to its original volume according to

$$B = -\frac{\Delta P}{\Delta V/V}$$

$$\Delta P = -\frac{B(\Delta V)}{V} = -\frac{\left(2.0 \times 10^9 \text{ N/m}^2\right)(-0.09V)}{V} = 1.80 \times 10^8 \text{ N/m}^2 \quad \lozenge$$

---

33. A bridge of length 50 m and mass 8.0×10^4 kg is supported at each end as in Figure P12.33. A truck of mass 3.0×10^4 kg is located 15 m from one end. What are the forces on the bridge at the points of support?

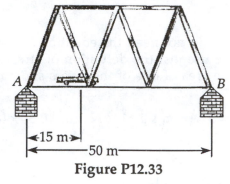

Figure P12.33

Solution   Let n_A and n_B be the normal forces at the points of support. Choosing the origin at point A, we find:

$\Sigma F_y = 0$: $n_A + n_B - (8.0 \times 10^4 \text{ kg})g - (3.0 \times 10^4 \text{ kg})g = 0$

$\Sigma \tau = 0$: $-(3.0 \times 10^4 \text{ kg})(15 \text{ m})g - (8.0 \times 10^4 \text{ kg})(25 \text{ m})g + n_B(50 \text{ m}) = 0$

The equations combine to give

$$n_A = 5.98 \times 10^5 \text{ N} \quad \Diamond$$

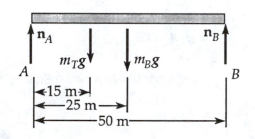

$$n_B = 4.80 \times 10^5 \text{ N} \quad \Diamond$$

39. A uniform sign of weight w and width $2l$ hangs from a light, horizontal beam, hinged at the wall and supported by a cable (Fig. P12.39). Determine (a) the tension in the cable and (b) the components of the reaction force exerted by the wall on the beam in terms of w, d, l, and θ.

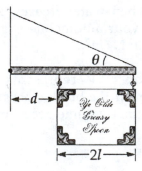

Figure P12.39

Solution   Choose the beam for analysis, and draw a free-body diagram as shown. We know the direction of the tension force at the right end is along the cable, at an angle of θ above the horizontal.

Taking torques about the left end,

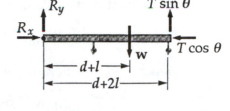

$$\Sigma F_x = 0 \qquad +R_x - T \cos \theta = 0$$

$$\Sigma F_y = 0 \qquad +R_y - w + T \sin \theta = 0$$

$$\Sigma \tau = 0 \qquad R_x(0) + R_y(0) - w(d + l) + (0)(T \cos \theta) + (d + 2l)(T \sin \theta) = 0$$

(a) The torque equation gives $\qquad T = \dfrac{w(d+l)}{(d+2l)\sin \theta} \quad \lozenge$

(b) Now from the force equations,

$$R_x = \frac{w(d+l)}{(d+2l)\tan \theta} \quad \lozenge \qquad \text{and} \qquad R_y = w - \frac{w(d+l)}{d+2l} = \frac{wl}{d+2l} \quad \lozenge$$

---

41. A 15-m uniform ladder weighing 500 N rests against a frictionless wall. The ladder makes a 60.0° angle with the horizontal. (a) Find the horizontal and vertical forces the ground exerts on the base of the ladder when an 800-N firefighter is 4.00 m from the bottom. (b) If the ladder is just on the verge of slipping when the firefighter is 9.00 m up, what is the coefficient of static friction between ladder and ground?

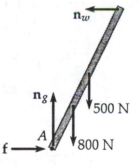

Solution   Refer to the free-body diagram.

(a) $\Sigma F_x = f - n_w = 0$

$\Sigma F_y = n_g - 800 \text{ N} - 500 \text{ N} = 0$

Taking torques about an axis at the foot of the ladder,

$$-(800 \text{ N})(4.00 \text{ m})\sin 30° - (500 \text{ N})(7.50 \text{ m})\sin 30° + n_w (15.0 \text{ m})\cos 30° = 0$$

Solving the torque equation for n_w,

$$n_w = \frac{\left[(4.00 \text{ m})(800 \text{ N}) + (7.50 \text{ m})(500 \text{ N})\right]\tan\ 30°}{15.0 \text{ m}} = 267.5 \text{ N}$$

Next substitute this value into the F_x equation to find

$$f = n_w = 268 \text{ N in the positive } x \text{ direction} \quad \lozenge$$

Solving the equation $\Sigma F_y = 0$ gives

$$n_g = 1300 \text{ N in the positive } y \text{ direction} \quad \lozenge$$

(b)   In this case, the torque equation $\Sigma \tau_A = 0$ gives

$$- (9.00 \text{ m})(800 \text{ N})\sin 30° - (7.50 \text{ m})(500 \text{ N})\sin 30° + (15.0 \text{ m})(n_w)\sin 60° = 0$$

or $n_w = 421 \text{ N}$

Since $f = n_w = 421 \text{ N}$ and $f = f_{max} = \mu n_g$,   we find

$$\mu = \frac{f_{max}}{n_g} = \frac{421 \text{ N}}{1300 \text{ N}} = 0.324 \quad \lozenge$$

---

43. A 10 000-N shark is supported by a cable attached to a 4.00-m rod that can pivot at the base. Calculate the cable tension needed to hold the system in the position shown in Figure P12.43. Find the horizontal and vertical forces exerted on the base of the rod. (Neglect the weight of the rod.)

Solution   The angle **T** makes with the rod is

$$\theta = 60° + 20° = 80°,$$

and the perpendicular component of **T** is $T\sin 80°$.

Summing torques around the base of the rod,

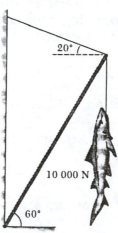

10 000 N

Figure P12.43

$$\Sigma \tau = -(4.00 \text{ m})(10000 \text{ N})\cos \ 60° + T(4.00 \text{ m})\sin 80° = 0$$

$$T = \frac{(10000 \text{ N})\cos \ 60°}{\sin 80°} = 5.08 \times 10^3 \text{ N} \quad \lozenge$$

$\Sigma F_x = 0$ yields $\qquad F_H - T \cos 20° = 0$

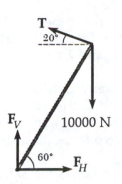

$$F_H = T \cos 20° = 4.77 \times 10^3 \text{ N} \qquad \lozenge$$

$\Sigma F_y = 0$ gives $\qquad F_V + T \sin 20° - 10000 \text{ N} = 0$

and $\quad F_V = (10000 \text{ N}) - T \sin 20° = 8.26 \times 10^3 \text{ N} \qquad \lozenge$

---

47. A force acts on a rectangular block weighing 400 N as in Figure P12.47. (a) If the block slides with constant speed when $F = 200$ N and $h = 0.400$ m, find the coefficient of sliding friction and the position of the resultant normal force. (b) If $F = 300$ N, find the value of h for which the block just begins to tip.

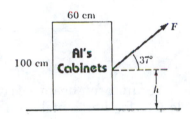

Figure P12.47

Solution

(a) Think of the normal force as acting at distance x from the lower left corner. Moving with constant speed, the block is in equilibrium:

$\Sigma F_x = 0 \qquad -f + (200 \text{ N}) \cos 37° = 0$

$\Sigma F_y = 0 \qquad -400 \text{ N} + n + (200 \text{ N}) \sin 37° = 0$

We have already $f = 160$ N and $n = 280$ N,

so $\qquad \mu_k = f/n = 0.571 \quad \lozenge$

Take torques about the lower left corner; $\Sigma \tau = 0$ gives

$$f(0) - (400 \text{ N})(30.0 \text{ cm}) + nx + (200 \text{ N}) \sin 37° \ (60.0 \text{ cm}) - (200 \text{ N}) \cos 37° \ (40.0 \text{ cm}) = 0$$

Substituting $n = 280$ N (and converting cm to m, and back) gives

$$x = \frac{120 \text{ N} \cdot \text{m} - 72.2 \text{ N} \cdot \text{m} + 63.9 \text{ N} \cdot \text{m}}{280 \text{ N}} = 39.9 \text{ cm} \quad \lozenge$$

(b)   When the block is just about to tip,   the normal force is located at the lower right corner, and $\Sigma\tau = 0$ is still true.

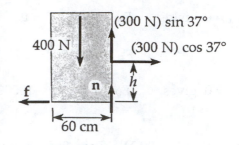

Because most of the forces are directed through the lower right corner, we choose to take torques about that point.   This leaves only two forces to deal with.

$$\Sigma\tau = 0:\quad -(300\text{ N})(\cos 37°)h + (400\text{ N})(30.0\text{ cm}) = 0$$

Solving for h,
$$h = \frac{120\text{ N}\cdot\text{m}}{240\text{ N}} = 50.1\text{ cm} \quad \lozenge$$

---

49.   A uniform beam of weight w is inclined at an angle θ to the horizontal with its upper end supported by a horizontal rope tied to a wall and its lower end resting on a rough floor (Fig. P12.49).   (a) If the coefficient of static friction between beam and floor is μ_S, determine an expression for the maximum weight W that can be suspended from the top before the beam slips. (b) Determine the magnitude of the reaction force at the floor and the magnitude of the force exerted by the beam on the rope at P in terms of w, W, and μ_S.

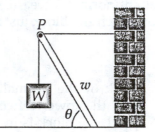

Figure P12.49

Solution   Call the length of the beam L:

(a)   We can use $\Sigma F_x = \Sigma F_y = \Sigma\tau = 0$ with the origin at the point of contact on the floor.

On the verge of slipping, the friction $f = \mu_s n$, and

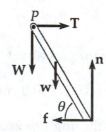

$\Sigma F_x = 0$ gives $T - \mu_s n = 0$

$\Sigma F_y = 0$ gives $n - W - w = 0$

Solving these two equations, $T = \mu_s[W + w]$

From $\Sigma\tau = 0$,

$$W(\cos\theta)L + w(\cos\theta)\frac{L}{2} - T(\sin\theta)L = 0$$

Substituting for T and solving, we get:

$$W = \frac{w}{2}\left[\frac{2\mu_s\sin\theta - \cos\theta}{\cos\theta - \mu_s\sin\theta}\right] \quad \Diamond$$

(b) At the floor, we see that the normal force is in the y direction and frictional force is in the x direction. The reaction force then is

$$R = \sqrt{n^2 + (\mu_s n)^2} = (W + w)\sqrt{1 + \mu_s^2} \quad \Diamond$$

At point P, the force of the beam on the rope is

$$F = \sqrt{T^2 + W^2} = \sqrt{W^2 + \mu_s^2(W + w)^2} \quad \Diamond$$

---

51. A stepladder of negligible weight is constructed as shown in Figure P12.51. A painter of mass 70.0 kg stands on the ladder 3.00 m from the bottom. Assuming the floor is frictionless, find (a) the tension in the horizontal bar connecting the two halves of the ladder, (b) the normal forces at A and B, and (c) the components of the reaction force at the hinge C that the left half of the ladder exerts on the right half. (*Hint:* Treat each half of the ladder separately.)

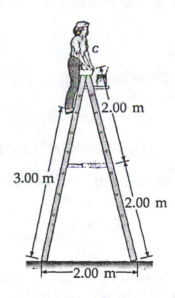

3.00 m

2.00 m

2.00 m

2.00 m

Figure P12.51

Solution   If we think of the whole ladder, we can solve part (b).

The painter is 3/4 of the way up the ladder, so the lever arm of her weight about A is

$$(3/4)(1.00\ \text{m}) = 0.75\ \text{m}.$$

Now, $\Sigma F_x = 0$ $0 = 0$

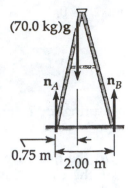

$\Sigma F_y = 0$ $+n_A - 686 \text{ N} + n_B = 0$

$\Sigma \tau_A = 0$ $n_A(0) - (686 \text{ N})(0.75 \text{ m}) + n_B (2.0 \text{ m}) = 0$

Thus

(b) $n_B = 257 \text{ N}$ and $n_A = 686 \text{ N} - 257 \text{ N} = 429 \text{ N}$ ◊

Now consider the left half of the ladder. We know the direction of the bar tension, and we make guesses for the directions of the components of the hinge force. If a guess is wrong, the answer will be negative.

The side rails make an angle with the horizontal

$$\theta = \cos^{-1}(1/4) = 75.5°$$

$\Sigma F_x = 0$ $T - C_x = 0$

$\Sigma F_y = 0$ $429 \text{ N} - 686 \text{ N} + C_y = 0$

Taking torques about the top of the ladder, we have:

$\Sigma \tau_c = 0$ $(-429 \text{ N})(1.00 \text{ m}) + T(2.00 \text{ m} \sin 75.5°) + (686 \text{ N})(0.250 \text{ m}) + C_y (0) + C_x (0) = 0$

(a)   From the torque equation, $T = \dfrac{257 \text{ N} \cdot \text{m}}{1.94 \text{ m}} = 133 \text{ N}$ ◊

(c)   From the force equations,

$C_x = T = 133 \text{ N}$ to the left $C_y = 686 \text{ N} - 429 \text{ N} = 257 \text{ N}$ up

The force that the left half exerts on the right half has opposite components:

133 N to the right, and 257 N down.   ◊

---

56. A wire of length L, Young's modulus Y, and cross-sectional area A is stretched elastically by an amount ΔL. By Hooke's law, the restoring force is $-k \, \Delta L$. (a) Show that $k = YA/L$. (b) Show that the work done in stretching the wire by an amount ΔL is

$$\text{Work} = \tfrac{1}{2}\frac{YA}{L}(\Delta L)^2$$

Solution

(a) According to Hooke's law $|F| = k \Delta L$. Young's modulus is defined as

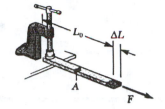

Figure 12.14

$$Y = \frac{F/A}{\Delta L/L} = k \frac{L}{A}$$

or $$k = \frac{YA}{L} \quad \lozenge$$

(b) $$W = -\int_0^{\Delta L} F \, dx = -\int_0^{\Delta L} (-kx) \, dx = \frac{YA}{L} \int_0^{\Delta L} x \, dx = \tfrac{1}{2}\frac{YA}{L}(\Delta L)^2 \quad \lozenge$$

---

59. (a) Estimate the force with which a karate master strikes a board if the hand's speed at time of impact is 10.0 m/s, decreasing to 1.0 m/s during a 0.0020 s time-of-contact with the board. The mass of coordinated hand and arm is 1.0 kg. (b) Estimate the shear stress if this force is exerted on a 1.0-cm-thick pine board that is 10 cm wide. (c) If the maximum shear stress a pine board can receive before breaking is 3.6×10^6 N/m^2, will the board break?

Solution   The impulse-momentum theorem describes the force of board on his hand:

$$Ft = mv_f - mv_0$$

$$F(0.0020 \text{ s}) = (1.0 \text{ kg})(1.0 \text{ m/s} - 10.0 \text{ m/s})$$

and $$F = -4500 \text{ N}$$

(a) Therefore, the force of his hand on board is 4500 N $\quad \lozenge$

That force produces a shear stress on the area that is exposed when the board snaps:

(b) $$\text{Stress} = \frac{F}{A} = \frac{4500 \text{ N}}{(10.0^{-1} \text{ m})(10.0^{-2} \text{ m})} = 4.50 \times 10^6 \text{ N/m}^2 \quad \lozenge$$

(c) *Yes;* this suffices to break the board. $\quad \lozenge$

Chapter 13

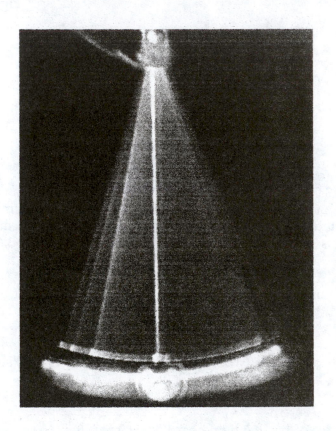

Oscillatory Motion

OSCILLATORY MOTION

INTRODUCTION

A very special kind of motion occurs when the force on a body is proportional to the displacement of the body from equilibrium. If this force always acts toward the equilibrium position of the body, there is a repetitive back-and-forth motion about this position. Such motion is an example of what is called *periodic* or *oscillatory* motion.

You are most likely familiar with several examples of periodic motion, such as the oscillations of a mass on a spring, the motion of a pendulum, and the vibrations of a stringed musical instrument.

Most of the material in this chapter deals with *simple harmonic motion.* In this type of motion, an object oscillates between two spatial positions for an indefinite period of time with no loss in mechanical energy. In real mechanical systems, retarding (frictional) forces are always present and these forces are considered in an optional section at the end of the chapter.

NOTES FROM SELECTED CHAPTER SECTIONS

13.1 Simple Harmonic Motion

Oscillatory motions are exhibited by many physical systems such as a mass attached to a spring, a pendulum, atoms in a solid, stringed musical instruments, and electrical circuits driven by a source of alternating current. *Simple harmonic motion* of a mechanical system corresponds to the oscillation of an object between two points for an indefinite period of time, with no loss in mechanical energy.

An object exhibits simple harmonic motion if the net external force acting on it is a *linear restoring force.*

The value of the phase constant ϕ depends on the initial displacement and initial velocity of the body. Two special cases are discussed in Section 13.1. Figure 13.1 on the following page represents plots of the displacement, velocity, and acceleration versus time assuming that at $t = 0$, $x_o = A$ and $v_o = 0$. In this case, one finds that $\phi = 0$. Note that the velocity is 90° out of phase with the displacement. That is, when $|x|$ is a maximum, v is zero; while $|v|$ is a maximum when x is zero. Furthermore, note that the acceleration is 180° out of phase with the displacement. That is, when x is a maximum and positive, $|a|$ is a maximum, but a is negative. In other words, a is proportional to x, but in the *opposite* direction.

It is necessary to define a few terms relative to harmonic motion:

- The amplitude, *A*, is the *maximum distance that an object moves away from its equilibrium position.* In the absence of friction, an object will continue in simple harmonic motion. During each cycle, it will reach a maximum displacement on each side of the equilibrium position equal to the amplitude.

- The period, *T*, *is the time it takes the object to execute one complete cycle of the motion.*

- The frequency, *f*, *is the number of cycles or vibrations per unit of time.*

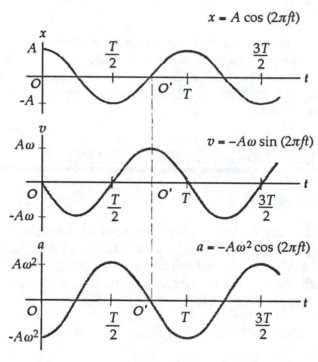

Figure 13.1

Representation of (a) the displacement, (b) the velocity, and (c) the acceleration as a function of time for an object moving with simple harmonic motion.

13.2 Mass Attached to a Spring

The most common system which undergoes simple harmonic motion is the mass-spring system shown in Figure 13.2. The mass is assumed to move on a horizontal, *frictionless* surface. The point $x = 0$ is the equilibrium position of the mass; that is, the point where the mass would reside if left undisturbed. In this position, there is no horizontal force on the mass. When the mass is displaced a distance x from its equilibrium position, the spring produces a linear restoring force given by Hooke's law, $F = -kx$, where k is the force constant of the spring, and has SI units of N/m. The minus sign means that F is to the *left* when the displacement x is positive, whereas F is to the *right* when x is negative. In other words, the direction of the force F is *always* towards the equilibrium position.

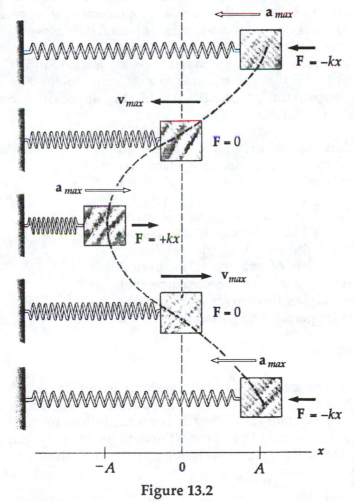

Figure 13.2

Oscillating motion of a mass on the end of a spring.

13.3   Energy of the Simple Harmonic Oscillator

You should study carefully the comparison between the motion of the mass-spring system and that of the simple pendulum. In particular, notice that when the displacement is a maximum, the energy of the system is entirely potential energy; whereas, when the displacement is zero, the energy is entirely kinetic energy. This is consistent with the fact that $v = 0$ when $|x| = A$, while $v = v_{max}$ when $x = 0$. For an arbitrary value of x, the energy is the sum of K and U.

13.4   The Pendulum

A *simple pendulum consists* of a mass m attached to a light string of length L as shown in Figure 13.3. When the angular displacement θ is small during the entire motion (less than about 15°), the pendulum exhibits simple harmonic motion. In this case, the resultant force acting on the mass m equals the component of weight *tangent* to the circle, and has a magnitude $mg \sin\theta$. Since this force is always directed towards $\theta = 0$, it corresponds to a restoring force. For small θ, we use the small angle approximation $\sin\theta \cong \theta$. In this approximation, the equation of motion reduces to Equation 13.21.

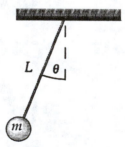

This equation is *identical* in form to Equation 13.14 for the mass-spring system. Its solution is therefore of the general form $\theta = \theta_0 \cos(\omega t + \phi)$, where ω is given by Equation 13.22. The period of motion is given by Equation 13.23. In other words, the period depends only on the length of the pendulum and the acceleration of gravity. The period *does not* depend on mass, so we conclude that *all* simple pendulums of equal length oscillate with the same frequency and period.

Figure 13.3

13.6   Damped Oscillations

Damped oscillations occur in realistic systems in which retarding forces such as friction are present. These forces will reduce the amplitudes of the oscillations with time, since mechanical energy is continually lost by the system. When the retarding force is assumed to be proportional to the velocity, but small compared to the restoring force, the system will still oscillate, but the amplitude will decrease exponentially with time.

It is possible to compensate for the energy lost in a damped oscillator by adding an additional driving force that does positive work on the system. This additional energy supplied to the system must at least equal the energy lost due to friction to maintain constant amplitude. The energy transferred to the system is a maximum when the driving force is in phase with the velocity of the system. The amplitude is a maximum when the frequency of the driving force matches the natural (resonance) frequency of the system.

EQUATIONS AND CONCEPTS

The force exerted by a spring on a mass attached to the spring and displaced a distance x from the unstretched position is given by Hooke's law. The force constant, k, is always positive and has a value which corresponds to the relative stiffness of the spring. The negative sign means that the force exerted on the mass is always directed opposite the displacement--the force is a restoring force, always directed toward the equilibrium position.

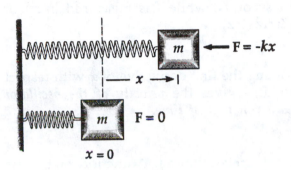

An object exhibits simple harmonic motion when the net force along the direction of motion is proportional to the displacement and in the opposite direction.

$$F = -kx \qquad (13.13)$$

Applying *Newton's second law* to motion in the x direction gives $F = ma_x = -kx$. Since $a_x = d^2x/dt^2$, this is equivalent to Equation 13.14.

$$\frac{d^2x}{dt^2} = -\omega^2 x \qquad (13.14)$$

The general solution to Equation 13.14 represents the *displacement versus time*, $x(t)$, provided that $\omega^2 = k/m$. In this expression, A represents the *amplitude* of the motion, $\omega t + \phi$ is the *phase*, ω is the *angular frequency* (rad/s), and ϕ is the *phase constant*.

$$x(t) = A\cos(\omega t + \phi) \qquad (13.1)$$

with $\omega^2 = \dfrac{k}{m}$

The *period of motion T* equals the time it takes the mass to complete *one* oscillation; that is, the time it takes the mass to return to its original position for the first time. The *frequency* of the motion, *f*, numerically equals the inverse of the period and represents the number of oscillations per unit time. *T* is measured in seconds, while *f* is measured in s⁻¹ or Hertz (Hz).

$$T = \frac{2\pi}{\omega} = 2\pi\sqrt{\frac{m}{k}} \qquad (13.15)$$

$$f = \frac{1}{T} = \frac{\omega}{2\pi} \qquad (13.3)$$

Taking the first derivative of x with respect to time gives the *velocity of the oscillator as a function of time.*

$$v = \frac{dx}{dt} = -\omega A \sin(\omega t + \phi) \qquad (13.5)$$

The acceleration as a function of time is equal to the time derivative of the velocity (or the second derivative of the displacement). Note from Equation 13.7 that the *acceleration* (and hence the force) *is always proportional to and opposite the displacement.*

$$a = \frac{dv}{dt} = -\omega^2 A \cos(\omega t + \phi) \qquad (13.6)$$

or

$$a = -\omega^2 x \qquad (13.7)$$

The kinetic energy of a simple harmonic oscillator is given by $\frac{1}{2}mv^2$, while the potential energy is equal to $\frac{1}{2}kx^2$. Using Equations 13.1 and 13.5, together with $\omega^2 = k/m$, gives the *total* energy E of the oscillator. Note that E remains constant since we have assumed there are no nonconservative forces acting on the system. The total energy of the simple harmonic oscillator is a constant of the motion and is proportional to the square of the amplitude.

$$E = \frac{1}{2}mv^2 + \frac{1}{2}kx^2$$

or

$$E = \frac{1}{2}kA^2 \qquad (13.19)$$

Energy conservation can be used to obtain an expression for velocity as a function of position.

$$v = \pm\sqrt{\frac{k}{m}\left(A^2 - x^2\right)} \qquad (13.20)$$

or

$$v = \pm\omega\sqrt{\left(A^2 - x^2\right)}$$

The speed of an object in simple harmonic motion is a maximum at $x = 0$; the speed is zero when the mass is at the points of maximum displacement $(x = \pm A)$.

The equation of motion for the simple pendulum assumes a small displacement so that $\sin\theta \cong \theta$.

$$\frac{d^2\theta}{dt^2} = -\frac{g}{L}\theta \qquad (13.21)$$

The period and frequency of a simple pendulum depend only on the length of the supporting string and the value of the acceleration due to gravity.

$$\omega = \sqrt{\frac{g}{L}} \qquad (13.22)$$

$$T = 2\pi\sqrt{\frac{L}{g}} \qquad (13.23)$$

SUGGESTIONS, SKILLS, AND STRATEGIES

Most of this chapter deals with simple harmonic motion, and the properties of the displacement expression

$$x(t) = A\cos(\omega t + \phi) \qquad (13.1)$$

In order to obtain the velocity $v(t)$ and acceleration $a(t)$ of the system, one must be familiar with the derivative operation as applied to trigonometric functions. In particular, note that

$$\frac{d}{dt}\cos(\omega t + \phi) = -\omega \sin(\omega t + \phi)$$

$$\frac{d}{dt}\sin(\omega t + \phi) = \omega \cos(\omega t + \phi)$$

Using these results, and $x(t)$ from Equation 13.1, we see that

$$v(t) = \frac{dx(t)}{dt} = -A\omega \sin(\omega t + \phi) \qquad (13.5)$$

and

$$a(t) = \frac{dv(t)}{dt} = -A\omega^2 \cos(\omega t + \phi) \qquad (13.6)$$

By direct substitution, you should be able to show that Equation 13.1 represents a general solution to the equation of motion for the mass-spring system (a second-order homogeneous differential equation) given by

$$\frac{d^2x}{dt^2} + \frac{k}{m}x = 0$$

where

$$\omega = \sqrt{\frac{k}{m}}$$

In treating the motion of the simple pendulum, we made use of the small angle approximation $\sin\theta \cong \theta$. This approximation enables us to reduce the equation of motion to that of the simple harmonic oscillator. The small angle approximation for $\sin\theta$ follows from inspecting the series expansion for $\sin\theta$, where θ is in *radians*:

$$\sin\theta = \theta - \frac{\theta^3}{3!} + \frac{\theta^5}{5!} - \ldots$$

For small values of θ, the higher order terms in θ^3, θ^5 ... are *small* compared to θ, so it follows that $\sin\theta \cong \theta$. The difference between $\sin\theta$ and θ is less than 1% for $0 < \theta < 15°$ (where $15° \cong 0.26$ rad).

REVIEW CHECKLIST

▷ Describe the general characteristics of simple harmonic motion, and the significance of the various parameters which appear in the expression for the displacement versus time, $x = A \cos(\omega t + \phi)$.

▷ Start with the expression for the displacement versus time for the simple harmonic oscillator, and obtain equations for the velocity and acceleration as functions of time.

▷ Understand the phase relations between displacement, velocity, and acceleration for simple harmonic motion, noting that acceleration is proportional to the displacement, but in the opposite direction.

▷ Describe and understand the conditions of simple harmonic motions executed by the mass-spring system (where the frequency depends on k and m) and the simple pendulum (where the frequency depends on L and g).

▷ Discuss the relationship between simple harmonic motion and the motion of a point on a circle moving with uniform angular velocity.

SOLUTIONS TO SELECTED END-OF-CHAPTER PROBLEMS

1. The displacement of a particle at $t = 0.25$ s is given by the expression $x = (4.0 \text{ m})\cos(3.0t + \pi)$, where x is in meters and t is in seconds. Determine (a) the frequency and period of the motion, (b) the amplitude of the motion, (c) the phase constant, and (d) the displacement of the particle at $t = 0.25$ s.

Solution   The particular displacement function $x = (4.0 \text{ m})\cos(3.0t + \pi)$ and the general one $x = A\cos(\omega t + \phi)$ have a specially powerful kind of equality called functional equality. They must give the same x value for all values of the variable t. This requires, then, that all parts be the same:

(a) $\omega = 3.0 = 2\pi f$ or $f = \dfrac{1.5}{\pi} = 0.477$ Hz $T = \dfrac{1}{f} = 2.09$ s   ◊

(b) $A = 4.00$ m   ◊

(c) $\phi = \pi$ rad   ◊

(d) $x(t = 0.25 \text{ s}) = (4.0 \text{ m})\cos(1.75\pi \text{ rad}) = (4.0 \text{ m})\cos(5.50 \text{ rad})$. Note that this is *not* 5.50°.

Instead, $x = (4.0 \text{ m})\cos(5.50 \text{ rad}) = (4.0 \text{ m})\cos 315° = 2.83$ m   ◊

5. The displacement of an object is $x = (8.0 \text{ cm})\cos(2.0t + \pi/3)$, where x is in centimeters and t is in seconds. Calculate (a) the speed and acceleration at $t = \pi/2$ s, (b) the maximum speed and the earliest time $(t > 0)$ at which the particle has this speed, and (c) the maximum acceleration and the earliest time $(t > 0)$ at which the particle has this acceleration.

Solution $\qquad x = (8.0 \text{ cm})\cos\left(2.0t + \dfrac{\pi}{3}\right)$

(a) $\quad v = \dfrac{dx}{dt} = -(16.0 \text{ cm})\sin\left(2.0t + \dfrac{\pi}{3}\right)$

$$\text{at} \quad t = \frac{\pi}{2} \text{ s,} \qquad v = 13.9 \text{ cm/s} \quad \lozenge$$

$$a = \frac{dv}{dt} = -\left(32.0 \text{ cm/s}^2\right)\cos\left(2.0t + \frac{\pi}{3}\right)$$

$$\text{at} \quad t = \frac{\pi}{2} \text{ s,} \qquad a = 16.0 \text{ cm/s}^2 \quad \lozenge$$

(b) $\quad v_{max} = 16.0$ cm/s $\quad \lozenge \qquad$ This occurs when $t = \frac{1}{2}\left[\sin^{-1}(1) - \dfrac{\pi}{3}\right] = 0.262$ s $\quad \lozenge$

(c) $\quad a_{max} = 32.0$ cm/s^2 $\quad \lozenge \qquad$ This occurs when $t = \frac{1}{2}\left[\cos^{-1}(-1) - \dfrac{\pi}{3}\right] = 1.05$ s $\quad \lozenge$

---

7. A particle moving along the x axis in simple harmonic motion starts from the origin at $t = 0$ and moves to the right. If the amplitude of its motion is 2.00 cm and the frequency is 1.50 Hz, (a) show that its displacement is given by $x = (2.00 \text{ cm})\sin(3.00 \pi t)$. Determine (b) the maximum speed and the earliest time $(t > 0)$ at which the particle has this speed, (c) the maximum acceleration and the earliest time $(t > 0)$ at which the particle has this acceleration, and (d) the total distance traveled between $t = 0$ and $t = 1$ s.

Solution

(a) At $t = 0$, $x = 0$ and v is positive (to the right). The sine function is zero and the cosine is positive at $\theta = 0$, so this situation corresponds to $x = A\sin\omega t$ and $v = v_o \cos\omega t$..

Since $f = 1.50$ Hz, $\omega = 2\pi f = 3\pi$

Also, $A = 2.00$ cm, so that $x = (2.00 \text{ cm}) \sin(3\pi t)$ ◊

This is equivalent to writing $x = A\cos(\omega t + \phi)$

with $A = 2.00$ cm, $\omega = 3.00\pi/\text{s}$ and $\phi = -90° = -\dfrac{\pi}{2}$. Note $T = \dfrac{1}{f} = \left(\dfrac{2}{3}\right)$ s.

(b) The velocity is $v = \dfrac{dx}{dt} = 2.00(3\pi)\cos(3\pi t)$ cm/s

The maximum speed is $v_{max} = v_0 = A\omega = 2.00(3\pi)$ cm/s $= 6\pi$ cm/s ◊

The particle has this speed at $t = 0$, when $\cos(3\pi t) = +1$,

and next at $t = \dfrac{T}{2} = \dfrac{1}{3}$ s ◊ when $\cos\left(3\pi \dfrac{1}{3}\right) = -1$

(c) Again, $a = \dfrac{dv}{dt} = (-2.00 \text{ cm})(3\pi/\text{s})^2 \sin(3\pi t)$

Its maximum value is

$$a_{max} = A\omega^2 = (2.00 \text{ cm})(3\pi/\text{s})^2 = 18\pi^2 \text{ cm/s}^2 \quad ◊$$

The acceleration has this positive value for the first time at

$$t = \dfrac{3T}{4} = 0.500 \text{ s} \quad ◊ \qquad \text{when } a = -2(3\pi)^2 \sin\left(\dfrac{3\pi}{2}\right) = +18\pi^2$$

(d) Since $T = \dfrac{2}{3}$ s and $A = 2.00$ cm, the particle will travel 8.00 cm in this time.

Hence, in $(1 \text{ s}) = \left(\dfrac{3T}{2}\right)$ the particle will travel 8.00 cm + 4.00 cm = 12.0 cm ◊

11. A 7.00-kg mass is hung from the bottom end of a vertical spring fastened to an overhead beam. The mass is set into vertical oscillations having a period of 2.60 s. Find the force constant of the spring.

Solution   A mass hanging from a vertical spring moves with SHM just like a mass moving without friction attached to a horizontal spring.

$$T = 2\pi\sqrt{\frac{m}{k}}, \text{ so} \qquad k = \frac{4\pi^2 m}{T^2} = \frac{4\pi^2(7.00 \text{ kg})}{(2.60 \text{ s})^2} = 40.9 \text{ kg/s}^2$$

Therefore, $\qquad\qquad\qquad k = 40.9 \text{ N/m}$ ◊

---

15. A 0.50 kg mass attached to a spring of force constant 8.0 N/m vibrates in simple harmonic motion with an amplitude of 10 cm. Calculate (a) the maximum value of its speed and acceleration, (b) the speed and acceleration when the mass is 6.0 cm from the equilibrium position, and (c) the time it takes the mass to move from $x = 0$ to $x = 8.0$ cm.

Solution $$\omega = \sqrt{\frac{k}{m}} = \sqrt{\frac{8.0 \text{ N/m}}{0.50 \text{ kg}}} = 4.0 \text{ s}^{-1}$$

Therefore, position is given by $x = (10 \text{ cm})\sin[(4.0 \text{ s}^{-1})t]$. From this we find that

(a) $$v = \frac{dx}{dt} = (40 \text{ cm/s})\cos(4.0t) \qquad\qquad v_{max} = 40 \text{ cm/s}$$ ◊

$$a = \frac{dv}{dt} = -(160 \text{ cm/s}^2)\sin(4.0t) \qquad\qquad a_{max} = 160 \text{ cm/s}^2$$ ◊

(b) $\quad t = \frac{1}{4}\sin^{-1}\left(\dfrac{x}{10 \text{ cm}}\right)$

When $x = 6.0$ cm, $t = 0.161$ s and we find

$$v = (40 \text{ cm/s})\cos[4.0(0.161)] = 32 \text{ cm/s}$$ ◊

$$a = -(160 \text{ cm/s}^2)\sin[(4.0 \text{ s}^{-1})(0.161 \text{ s})] = -96 \text{ cm/s}^2$$ ◊

(c) Using $t = \frac{1}{4}\sin^{-1}\left(\frac{x}{10}\right)$

When $x = 0$, $t = 0$ and when $x = 8.0$ cm, $t = 0.232$ s. Therefore $\Delta t = 0.232$ s ◊

---

19. An automobile having a mass of 1000 kg is driven into a brick wall in a safety test. The bumper behaves like a spring of constant 5.0×10^6 N/m, and compresses 3.16 cm as the car is brought to rest. What was the speed of the car before impact, assuming no energy is lost during impact with the wall?

Solution Assuming no energy lost during impact with the wall, the initial energy (kinetic) equals the final energy (elastic potential):

$$K_i = U_f \quad \text{or} \quad \tfrac{1}{2}mv^2 = \tfrac{1}{2}kx^2$$

$$v = \sqrt{\frac{k}{m}}x = \sqrt{\frac{5.0 \times 10^6 \text{ N/m}}{1000 \text{ kg}}}(3.16 \times 10^{-2} \text{ m})$$

$$v = 2.23 \text{ m/s} ◊$$

---

23. A particle executes simple harmonic motion with an amplitude of 3.00 cm. At what displacement from the midpoint of its motion does its speed equal one half of its maximum speed?

Solution From energy considerations, $v^2 + \omega^2 x^2 = \omega^2 A^2$

$$v_{max} = \omega A \quad \text{and} \quad v = \frac{v_{max}}{2} = \frac{\omega A}{2} \quad \text{so} \quad \tfrac{1}{2}\omega^2 A^2 + \omega^2 x^2 = \omega^2 A^2$$

From this we find $x^2 = \frac{3A^2}{4}$

where $A = 3.0$ cm, $x = \pm\frac{A\sqrt{3}}{2} = \pm\frac{3\sqrt{3}}{2} = \pm 2.60$ cm ◊

29. A simple pendulum has a mass of 0.250 kg and a length of 1.00 m. It is displaced through an angle of 15.0° and then released. What are (a) the maximum speed? (b) the maximum angular acceleration? (c) the maximum restoring force?

Solution We can solve this problem by either of two methods.

METHOD ONE: Since 15.0° is small enough that (in radians) (sin θ = 0.259) = (θ = 0.262) within 1%, we may treat the motion as SHM. The constant angular frequency characterizing the motion is

$$\omega = \sqrt{\frac{g}{L}} = \sqrt{\frac{9.80 \text{ m/s}^2}{1.00 \text{ m}}} = 3.13 \text{ rad/s}$$

The amplitude as a distance is $\quad A = L\theta = (1.00 \text{ m})(0.262) = 0.262 \text{ m}$

(a) The maximum linear speed is

$$v_{max} = \omega A = \left(3.13 \text{ s}^{-1}\right)(0.262 \text{ m}) = 0.820 \text{ m/s} \quad \Diamond$$

(b) Similarly, $\quad a_{max} = \omega^2 A = (3.13 \text{ 1/s})^2 (0.262 \text{ m}) = 2.57 \text{ m/s}^2$

This implies maximum angular acceleration

$$\alpha = \frac{a}{r} = \frac{2.57 \text{ m/s}^2}{1.00 \text{ m}} = 2.57 \text{ rad/s}^2 \quad \Diamond$$

(c) $\quad \Sigma F = ma = (0.250 \text{ kg})(2.57 \text{ m/s}^2) = 0.641 \text{ N} \quad \Diamond$

METHOD TWO: We may work out slightly more precise answers by using ideas we studied before. At release, the pendulum has height above its equilibrium position.

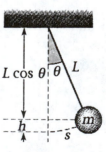

$$h = L - L \cos 15.0° = 1.00 \text{ m}(1 - \cos 15.0°) = 0.0341 \text{ m}$$

(a) Its energy is constant as it swings down:

$$(K + U)_{top} = (K + U)_{bottom}$$

$$0 + mgh = \tfrac{1}{2}mv_{max}^2 + 0$$

$$v_{max} = \sqrt{2gh} = \sqrt{2(9.80 \text{ m/s}^2)0.0341 \text{ m}} = 0.817 \text{ m/s} \quad \Diamond$$

(c)   The restoring force at release is

$$mg\sin 15.0° = (0.250\ kg)(9.80\ m/s^2)\sin 15.0°$$

$$= 0.634\ N \quad ◊$$

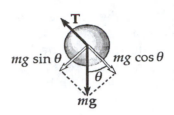

(b)   This produces linear acceleration

$$a = \frac{\Sigma F}{m} = \frac{0.634N}{0.250\ kg} = 2.54\ m/s^2$$

and angular acceleration

$$\alpha = \frac{a}{r} = \frac{2.54\ m/s^2}{1.00\ m} = 2.54\ rad/s^2 \quad ◊$$

---

33.   A physical pendulum in the form of a planar body moves in simple harmonic motion with a frequency of 0.450 Hz. If the pendulum has a mass of 2.20 kg and the pivot is located 0.350 m from the center of mass, determine the moment of inertia of the pendulum.

Solution

$$f = 0.450\ Hz, \quad d = 0.350\ m, \quad and \quad m = 2.20\ kg$$

Using equation 13.25,

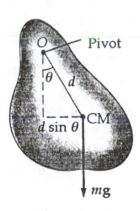

$$T = 2\pi\sqrt{\frac{I}{mgd}} \qquad T^2 = \frac{4\pi^2 I}{mgd}$$

$$I = \frac{T^2 mgd}{4\pi^2} = \left(\frac{1}{f}\right)^2 \frac{mgd}{4\pi^2} = \frac{(2.2\ kg)(9.80\ m/s^2)(0.35\ m)}{(0.450\ s^{-1})^2(4\pi^2)}$$

$$I = 0.944\ kg\cdot m^2 \quad ◊$$

---

47. When the simple pendulum illustrated in Figure P13.47 makes an angle θ with the vertical, its speed is v. (a) Calculate the total mechanical energy of the pendulum as a function of v and θ. (b) Show that when θ is small, the potential energy can be expressed as

$$\tfrac{1}{2}mgL\theta^2 = \tfrac{1}{2}m\omega^2 s^2$$

$$\left[\textit{Hint}: \text{In part (b), approximate } \cos\theta \text{ by } \cos\theta \approx 1 - \frac{\theta^2}{2}. \right]$$

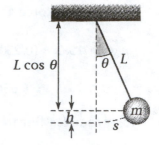

Figure P13.47
(modified)

Solution

(a) $E = K + U = \tfrac{1}{2}I\omega^2 + mgh$

where $I = mL^2$ and $\omega = \dfrac{v}{L}$

In this case (see Figure P13.47), $h = L(1 - \cos\theta)$.

So, $E = \tfrac{1}{2}mv^2 + mgL(1 - \cos\theta)$ ◊

(b) $U = mgL(1 - \cos\theta)$ and for small θ,

$$U \approx mgL\left[1 - \left(1 - \frac{\theta^2}{2} \right) \right] = \frac{mgL\theta^2}{2}$$

Since $L\theta = s$ and $\omega^2 = \dfrac{g}{L}$, we have $U = \dfrac{m\omega^2 s^2}{2}$ ◊

---

49. A particle of mass m slides inside a hemispherical bowl of radius R. Show that, for small displacements from equilibrium, the particle moves in simple harmonic motion with an angular frequency equal to that of a simple pendulum of length R. That is, $\omega = \sqrt{g/R}$.

Solution   Locate the center of curvature C of the bowl. We can measure the excursion of the object from equilibrium by the angle θ between the radial line to C and the vertical. The distance the object moves from equilibrium is, $s = R\theta$.

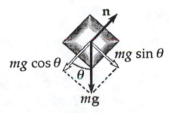

$\Sigma F_s = ma$ becomes $-mg\sin\theta = \dfrac{m\,d^2s}{dt^2}$

For small angles $\theta \cong \sin\theta$,   so by substitution,

$$-mg\theta = \frac{m\,d^2s}{dt^2}$$

$$-mg\,\frac{s}{R} = \frac{m\,d^2s}{dt^2}$$

$$\frac{d^2s}{dt^2} = -\left(\frac{g}{R}\right)s$$

The acceleration is proportional to the displacement and in the opposite direction, so we have SHM. ◊

We identify its angular frequency by comparing our equation to Equation 13.14:

$$\frac{d^2x}{dt^2} = -\omega^2 x$$

Now x and s both measure displacement, so

$$\omega^2 = \frac{g}{R} \qquad\qquad \omega = \sqrt{\frac{g}{R}} \quad ◊$$

---

51. A mass M is attached to the end of a uniform rod of mass M and length L, that is pivoted at the top (Fig. P13.51). (a) Determine the tensions in the rod at the pivot and at the point P when the system is stationary. (b) Calculate the period of oscillation for small displacements from equilibrium, and determine this period for $L = 2.00$ m. (*Hint:* Assume the mass at the end of the rod is a point mass and use Equation 13.26.)

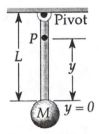

Figure P13.51

Solution (a) At the pivot, $T = Mg + Mg = 2Mg$ ◊

At P, a fraction of the rod's mass (y/L) pulls down as well as does the ball.

Therefore,
$$T = Mg\left(\frac{y}{L}\right) + Mg = Mg\left(1 + \frac{y}{L}\right) \quad ◊$$

(b) Relative to the pivot,
$$I = \tfrac{1}{3}ML^2 + ML^2$$

For a physical pendulum,
$$T = 2\pi\sqrt{\frac{I}{mgd}}$$

where, in this case, $m = 2M$ and d is the distance from the pivot to the center of mass.

$$d = \frac{\left(\dfrac{ML}{2} + ML\right)}{(M + M)} = \frac{3L}{4} \qquad \text{and we have} \qquad T = \frac{4\pi}{3}\sqrt{\frac{2L}{g}} \quad ◊$$

For $L = 2.00$ m,
$$T = \frac{4\pi}{3}\sqrt{\frac{2(2.00 \text{ m})}{9.80 \text{ m}/\text{s}^2}} = 2.68 \text{ s} \quad ◊$$

---

53. A small, thin disk of radius r and mass m is attached rigidly to the face of a second thin disk of radius R and mass M as shown in Figure P13.53. The center of the small disk is located at the edge of the large disk. The large disk is mounted at its center on a frictionless axle. The assembly is rotated through an angle θ from its equilibrium position and released. (a) Show that the speed of the center of the small disk as it passes through the equilibrium position is

$$v = \sqrt{\frac{Rg(1 - \cos\theta)}{(M/m) + (r/R)^2 + 2}}$$

(b) Show that the period of the motion is

$$T = 2\pi\sqrt{\frac{(M + 2m)R^2 + mr^2}{2mgR}}$$

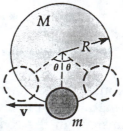

Figure P13.53

Solution

(a) $\Delta K + \Delta U = 0$, thus $K_{top} + U_{top} = K_{bot} + U_{bot}$, where $K_{top} = U_{bot} = 0$.

Therefore, $mgh = \frac{1}{2}I\omega^2$

But $h = R - R\cos\theta = R(1 - \cos\theta)$, $\omega = \dfrac{v}{R}$ and $I = \frac{1}{2}MR^2 + \frac{1}{2}mr^2 + mR^2$

Substituting, we find

$$mgR(1 - \cos\theta) = \frac{1}{2}\left(\frac{1}{2}MR^2 + \frac{1}{2}mr^2 + mR^2\right)\frac{v^2}{R^2}$$

$$mgR(1 - \cos\theta) = \left(\frac{1}{4}M + \frac{1}{4}\frac{r^2}{R^2}m + \frac{1}{2}m\right)v^2$$

and

$$v^2 = \frac{4gR(1 - \cos\theta)}{\dfrac{M}{m} + \dfrac{r^2}{R^2} + 2}$$

so

$$v = 2\sqrt{\frac{Rg(1 - \cos\theta)}{\dfrac{M}{m} + \dfrac{r^2}{R^2} + 2}} \qquad \lozenge$$

(b) $T = 2\pi\sqrt{\dfrac{I}{M_T g d_{cm}}}$ $\qquad M_T = M + m \qquad d_{cm} = \dfrac{mR + M(0)}{m + M}$

$$T = 2\pi = \sqrt{\frac{\dfrac{1}{2}MR^2 + \dfrac{1}{2}mr^2 + mR^2}{mgR}} = 2\pi\sqrt{\frac{(M + 2m)R^2 + mr^2}{2mgR}} \qquad \lozenge$$

55. A pendulum of length L and mass M has a spring of force constant k connected to it at a distance h below its point of suspension (Fig. P13.55). Find the frequency of vibration of the system for small values of the amplitude (small θ). (Assume the vertical suspension of length L is rigid, but neglect its mass.)

Solution   For the pendulum (see sketch), we have

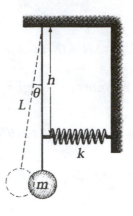

Figure P13.55

$$\Sigma\tau = I\alpha \quad \text{and} \quad \frac{d^2\theta}{dt^2} = -\alpha$$

The negative sign appears because positive θ is measured clockwise in the picture. We take torque around the point of suspension:

$$\Sigma\tau = MgL\sin\theta + kxh\cos\theta = I\alpha$$

For small amplitude vibrations, use the approximations:

$$\sin\theta \approx \theta, \quad \cos\theta \approx 1, \quad \text{and} \quad x \approx s = h\theta$$

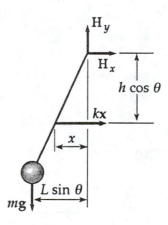

Therefore, with $I = mL^2$,

$$\frac{d^2\theta}{dt^2} = -\left[\frac{MgL + kh^2}{I}\right]\theta = -\left[\frac{MgL + kh^2}{ML^2}\right]\theta$$

This is of the form $\frac{d^2\theta}{dt^2} = -\omega^2\theta$ required for SHM,

with angular frequency,

$$\omega = \sqrt{\frac{MgL + kh^2}{ML^2}} = 2\pi f$$

The ordinary frequency is

$$f = \frac{\omega}{2\pi} = \frac{1}{2\pi}\sqrt{\frac{MgL + kh^2}{ML^2}} \quad \lozenge$$

57. A large block P executes horizontal simple harmonic motion by sliding across a frictionless surface with a frequency $f = 1.5$ Hz. Block B rests on it, as shown in Figure P13.57, and the coefficient of static friction between the two is $\mu_s = 0.60$. What maximum amplitude of oscillation can the system have if the block is not to slip?

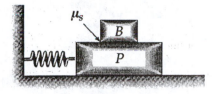

Figure P13.57

Solution   If the block B does not slip, its motion is SHM with the same amplitude and frequency as those of P, and with its acceleration caused by the static friction force exerted on it by P. Think of the block when it is just ready to slip at a turning point in its motion:

$\Sigma F = ma$ becomes

$$f_{max} = \mu_s n = \mu_s mg = ma_{max} = mA\omega^2$$

Then $A = \dfrac{\mu_s g}{\omega^2} = \dfrac{0.60(9.80 \text{ m/s}^2)}{[2\pi(1.5/s)]^2}$

$A = 6.62$ cm   ◊

---

59. A simple pendulum having a length of 2.23 m and a mass of 6.74 kg is given an initial speed of 2.06 m/s at its equilibrium position. Assume it undergoes simple harmonic motion and determine its (a) period, (b) total energy and (c) maximum angular displacement.

Solution   (a) $T = \dfrac{2\pi}{\omega} = 2\pi\sqrt{\dfrac{L}{g}} = 2\pi\sqrt{\dfrac{2.23 \text{ m}}{9.8 \text{ m/s}^2}} = 3.00$ s   ◊

(b) $E = \tfrac{1}{2}mv^2 = \tfrac{1}{2}(6.74 \text{ kg})(2.06 \text{ m/s})^2 = 14.3$ J   ◊

(c)   At maximum angular displacement, $mgh = \tfrac{1}{2}mv^2$, and $h = \dfrac{v^2}{2g} = 0.217$ m

$h = L - L\cos\theta = L(1 - \cos\theta)$,

so $\cos\theta = 1 - \dfrac{h}{L}$ $\theta = 25.5°$ ◊

Alternatively, we could write $v_{max} = \omega A$

$$A = \frac{v_{max}}{\omega} = \frac{2.06 \text{ m/s}}{2.10/\text{s}} = 0.983 \text{ m}$$

$$\theta = \frac{A}{L} = \frac{0.983 \text{ m}}{2.23 \text{ m}} = 0.441 \text{ rad} = 25.2°$$

Our two answers are not precisely equal because the pendulum does not move with precisely simple harmonic motion.

---

65. A mass m is connected to two rubber bands of length L, each under tension T, as in Figure P13.65. The mass is displaced vertically by a small distance y. Assuming the tension does not change, show that (a) the restoring force is $-(2T/L)y$ and (b) the system exhibits simple harmonic motion with an angular frequency $\omega = \sqrt{2T/mL}$.

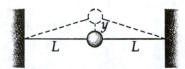

Figure P13.65

Solution

(a) $\Sigma F = -2T \sin\theta \mathbf{j}$ where $\theta = \tan^{-1}\left(\frac{y}{L}\right)$

Since for a small displacement, $\sin\theta \approx \tan\theta = \frac{y}{L}$

and the resulting force is $F = \left(-\frac{2Ty}{L}\right)\mathbf{j}$ ◊

(b) For a spring system, $F = -kx$ becomes $F = -\left(\frac{2T}{L}\right)y$

Therefore, $\omega = \sqrt{\frac{k}{m}} = \sqrt{\frac{2T}{mL}}$ ◊

Chapter 14

The Law of Gravity

Chapter 14

THE LAW OF GRAVITY

INTRODUCTION

In this chapter we study the law of gravity. Emphasis is placed on describing the motion of the planets, because astronomical data provide an important test of the validity of the law of gravity. We show that the laws of planetary motion developed by Johannes Kepler follow from the law of gravity and the concept of the conservation of angular momentum. A general expression for the gravitational potential energy is derived, and the energetics of planetary and satellite motion are treated. The law of gravity is also used to determine the force between a particle and an extended body.

NOTES FROM SELECTED CHAPTER SECTIONS

14.1 Newton's Law of Gravity

Newton's law of gravitation states that every particle in the Universe attracts every other particle with a force that is directly proportional to the product of their masses and inversely proportional to the square of their distance of separation. The gravitational force is an *action-at-a-distance force* which always exists between two particles regardless of the medium which separates them.

14.4 Kepler's Laws

Using astronomical data provided by Brahe, Kepler deduced the following empirical laws as they apply to our Solar System:

- All planets move in elliptical orbits with the Sun at one of the focal points.

- The radius vector drawn from the Sun to any planet sweeps out equal areas in equal times.

- The square of the orbital period of any planet is proportional to the cube of the semimajor axis for the elliptical orbit.

Kepler's second law is a consequence of the central nature of the gravitational force, which leads to conservation of angular momentum. Kepler's third law follows from the inverse square nature of the gravitational force.

14.7 Gravitational Potential Energy

The *gravitational potential energy* for any pair of particles varies as $1/r$. The potential energy is negative since the force is one of attraction and the potential energy is taken to be zero when the distance of separation between the two particles is infinity. The absolute value of the potential energy is the *binding energy* of the system.

14.8 Energy Considerations in Planetary and Satellite Motion

Both the total energy and the total angular momentum of a planet-sun system are *constants of the motion.*

The *escape velocity* of an object is independent of the mass of the object and is independent of the direction of the velocity.

14.10 Gravitational Force Between a Particle and a Spherical Mass

You should be aware of some interesting special cases of gravitational force between a particle and a spherically symmetric mass distribution.

For the case of a *spherical shell*:

- If a particle of mass m is located *outside* a spherical shell of mass M, the spherical shell attracts the particle as though the mass of the shell were concentrated at its center.

- If the particle is located *inside* the spherical shell, the force on it is zero.

Note that the shell of mass does *not* act as a gravitational shield. The particle may experience forces due to other masses outside the shell.

For the case of a *solid sphere*:

- If a particle of mass m is located *outside* a homogeneous solid sphere of mass M, the sphere attracts the particle as though the mass of the sphere were concentrated at its center. This follows from case (1) above, since a solid sphere can be considered a collection of concentric spherical shells.

- If a particle of mass m is located (at a distance r from the center) *inside* a homogeneous solid sphere of mass M and radius R, the force on m is due *only* to the mass M' contained within the sphere of radius $r < R$.

$$F = -\frac{GmM}{R^3}r\hat{\mathbf{r}} \quad \text{for } r < R$$

That is, the force goes to zero at the center of the sphere.

- If a particle is located *inside* a solid sphere having a density ρ that is spherically symmetric but *not* uniform, then M' is given by an integral of the form $M' = \int \rho \, dV$, where the integration is taken over the volume contained *within* the dotted surface. This integral can be evaluated if the radial variation of ρ is given. The integral is easily evaluated if the mass distribution has spherical symmetry; that is, if ρ is a function of r only. In this case, we take the volume element dV as the volume of a spherical shell of radius r and thickness dr, so that $dV = 4\pi r^2 \, dr$.

EQUATIONS AND CONCEPTS

The *universal law of gravity* states that any pair of particles *attract* each other with a force that is proportional to the product of their masses and inversely proportional to the square of their separation.

$$F_g = G\frac{m_1 m_2}{r^2} \tag{14.1}$$

The constant G is called the *universal constant of gravity*.

$$G = 6.672 \times 10^{-11} \frac{\text{N} \cdot \text{m}^2}{\text{kg}^2} \tag{14.2}$$

The acceleration due to gravity, g', decreases with increasing altitude, h, measured from the Earth's surface.

$$g' = \frac{GM_E}{(R_E + h)^2} \tag{14.5}$$

A *gravitational field* $\mathbf{g}$ exists at some point in space if a particle of mass m experiences a gravitational force $\mathbf{F} = m\mathbf{g}$ at that point. That is, the gravitational field represents the ratio of the gravitational force experienced by the mass divided by that mass.

$$\mathbf{g} \equiv \frac{\mathbf{F}_g}{m} \tag{14.8}$$

The gravitational field at a distance r from the center of the Earth points radially inward toward the center of the Earth. Over a small region near the Earth's surface, $\mathbf{g}$ is an approximately uniform downward field.

$$\mathbf{g} = -\frac{GM_E}{r^2}\hat{\mathbf{r}} \qquad (14.9)$$

Since the gravitational force is conservative, we can define a gravitational energy function corresponding to that force. As a mass m moves from one position to another in the presence of the Earth's gravity, its potential energy changes by an amount given by Equation 14.11, where r_i and r_f are the initial and final distances of the mass from the center of the Earth.

$$U_f - U_i = -GM_E m \left(\frac{1}{r_f} - \frac{1}{r_i} \right) \qquad (14.11)$$

The *gravitational potential energy* associated with *any pair* of particles of masses m_1 and m_2 separated by a distance r is given by Equation 14.13. The negative sign in this expression corresponds to the attractive nature of the gravitational force. An external agent must do positive work to increase the separation of the particles.

$$U = -\frac{Gm_1 m_2}{r} \qquad (14.13)$$

In this expression it is assumed that $U_i = 0$ at $r_i = \infty$, and the equation is valid for an Earth-particle system when $r > R_E$.

As a body of mass m moves in a circular orbit around a very massive body of mass M (where $M \gg m$), the *total energy* of the system is the sum of the kinetic energy of m (taking the massive body to be at rest) and the potential energy of the system.

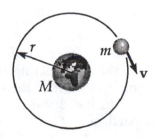

When the two contributions are evaluated, one finds that the *total energy* E is negative, and given by Equation 14.18. This arises from the fact that the (positive) kinetic energy is equal to one half of the magnitude of the (negative) potential energy.

$$E = -\frac{GMm}{2r} \qquad (14.18)$$

The *escape velocity* is defined as the *minimum* velocity a body must have, when projected from the Earth whose mass is M_E and radius is R_E, in order to escape the Earth's gravitational field (that is, to just reach $r = \infty$ with zero speed). Note that v_{esc} does not depend on the mass of the projected body.

$$v_{esc} = \sqrt{\frac{2GM_E}{R_E}} \qquad (14.20)$$

The *potential energy* associated with a particle of mass m and an *extended* body of mass M can be evaluated using Equation 14.21, where the extended body is divided into segments of mass dM, and r is the distance from dM to the particle.

$$U = -Gm \int \frac{dM}{r} \qquad (14.21)$$

If a particle of mass m is *outside* a *uniform* solid sphere or spherical shell of radius R, the sphere attracts the particle as though the mass of the sphere were concentrated at its center.

$$\mathbf{F}_g = -\frac{GMm}{r^2}\hat{\mathbf{r}} \qquad (r \geq R) \qquad (14.23a)$$

If a particle is located *inside a uniform spherical shell*, the force acting on the particle is *zero*.

$$\mathbf{F}_g = 0 \qquad (r < R) \qquad (14.23b)$$

If a particle is *inside* a *homogeneous solid sphere* of radius R, the force acting on the particle acts toward the center of the sphere and is proportional to the distance r from the center to the particle.

$$\mathbf{F}_g = -\frac{GmM}{R^3}r\hat{\mathbf{r}} \qquad (r < R) \qquad (14.25)$$

SUGGESTIONS, SKILLS, AND STRATEGIES

In this chapter we made use of the definite integral in evaluating the potential energy function associated with the conservative gravitational force. You should be familiar with the following type of definite integral:

$$\int_{x_1}^{x_2} x^n dx = \frac{x^{n+1}}{n+1}\Bigg]_{x_1}^{x_2} = \frac{x_2^{n+1} - x_1^{n+1}}{n+1}\,(n \neq -1)$$

For example, in deriving Equation 14.11, we used the above expression as follows:

$$\int_{r_i}^{r_f} \frac{dr}{r^2} = \int_{r_i}^{r_f} r^{-2}\,dr = \frac{r_f^{-1} - r_i^{-1}}{-2+1} = -\left(\frac{1}{r_f} - \frac{1}{r_i}\right)$$

REVIEW CHECKLIST

▷ State Kepler's three laws of planetary motion and recognize that the laws are empirical in nature; that is, they are based on astronomical data. Describe the nature of Newton's universal law of gravity, and the method of deriving Kepler's third law ($T^2 \alpha r^3$) from this law for circular orbits. Recognize that Kepler's second law is a consequence of conservation of angular momentum and the central nature of the gravitational force.

▷ Understand the concepts of the gravitational field and the gravitational potential energy, and know how to derive the expression for the potential energy for a pair of particles separated by a distance r.

▷ Describe the total energy of a planet or Earth satellite moving in a circular orbit about a large body located at the center of motion. Note that the total energy is negative, as it must be for any closed orbit.

▷ Understand the meaning of escape velocity, and know how to obtain the expression for v_{esc} using the principle of conservation of energy.

▷ Learn the method for calculating the gravitational force between a particle and an extended object. In particular, you should be familiar with the force between a particle and a spherical body when the particle is located outside and inside the spherical body.

SOLUTIONS TO SELECTED END-OF-CHAPTER PROBLEMS

6. When a falling meteor is at a distance $d = 3R_E$ above the Earth's surface, what is its free-fall acceleration?

Solution   The acceleration of gravity on Earth, $g = Gm/r^2$, follows an inverse-square law. At the surface (at distance one Earth-radius R_E from the center), it is 9.80 m/s².

At an altitude $3R_E$ above the surface (at distance $4R_E$ from the center), the acceleration of gravity will be $4^2 = 16$ times smaller:

$$g = \frac{GM_E}{(4R_E)^2} = \frac{GM_E}{16R_E^2} = \frac{9.80 \text{ m/s}^2}{16} = 0.612 \text{ m/s}^2 \text{ down} \quad \Diamond$$

11. Plaskett's binary system consists of two stars that revolve in a circular orbit about a center of gravity midway between them. This means that the masses of the two stars are equal (Figure P14.11). If the orbital speed of each star is 220 km/s and the orbital period of each is 14.4 days, find the mass M of each star. (For comparison, the mass of our Sun is 2×10^{30} kg.)

Solution

The circumference of the stars' common orbit is

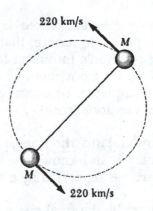

Figure P14.11

$$2\pi r = vT = (220 \times 10^3 \text{ m/s})(14.4 \text{ d})\left(\frac{86\,400 \text{ s}}{\text{d}}\right)$$

$$2\pi r = 2.74 \times 10^{11} \text{ m}$$

so each is at a distance from the center of

$$r = 2.74 \times 10^{11} \text{ m}/2\pi = 4.36 \times 10^{10} \text{ m}$$

while the distance between the stars is $2r = 8.72 \times 10^{10}$ m.

The gravitational force of each star on the other is the source of the central force about the center:

$$\Sigma F = ma \qquad \text{becomes} \qquad \frac{GMM}{(2r)^2} = \frac{Mv^2}{r}$$

$$M = \frac{v^2 4r}{G} = \frac{(220 \times 10^3 \text{ m/s})^2 (4)(4.36 \times 10^{10} \text{ m})}{6.67 \times 10^{-11} \text{ N} \cdot \text{m}^2 / \text{kg}^2} = 1.26 \times 10^{32} \text{ kg} \quad \lozenge$$

(or about 63 solar masses for each star).

14. The *Explorer VIII* satellite, placed into orbit November 3, 1960, to investigate the ionosphere, had the following orbit parameters: perigee 459 km and apogee 2289 km (both distances above the Earth's surface); period 112.7 min. Find the ratio v_p / v_a.

Solution

The satellite moves with constant angular momentum. Equating the angular momentum at apogee and perigee gives

$$mr_a v_a \sin 90.0° = mr_p v_p \sin 90.0°$$

$$\frac{v_p}{v_a} = \frac{r_a}{r_p} = \frac{6370 \text{ km} + 2289 \text{ km}}{6370 \text{ km} + 459 \text{ km}} = 1.27 \quad \lozenge$$

15. Io, a small moon of Jupiter, has an orbital period of 1.77 days and an orbital radius of 4.22×10^5 km. From these data, determine the mass of Jupiter.

Solution   The gravitational force of Jupiter on Io is the central force on Io:

$$\Sigma F_{Io} = M_{Io} a$$

$$\frac{GM_J M_{Io}}{r^2} = \frac{M_{Io} v^2}{r} = \frac{M_{Io}}{r}\left(\frac{2\pi r}{T}\right)^2 = \frac{4\pi^2 r M_{Io}}{T^2}$$

Thus,

$$M_J = \frac{4\pi^2 r^3}{GT^2} = \frac{4\pi^2 (4.22 \times 10^8 \text{ m})^3 \text{ kg}^2}{(6.67 \times 10^{-11} \text{ N} \cdot \text{m}^2)(1.77 \text{ d})^2}\left(\frac{1 \text{ d}}{86\,400 \text{ s}}\right)^2\left(\frac{\text{N} \cdot \text{s}}{\text{kg} \cdot \text{m}}\right)$$

and $$M_J = 1.90 \times 10^{27} \text{ kg} \quad \Diamond$$

---

19. A synchronous satellite, which always remains above the same point on a planet's equator, is put in orbit around Jupiter to study the famous red spot. Jupiter rotates once every 9.9 h. Use the data of Table 14.2 to find the altitude of the satellite.

Solution   The gravitational force is the central force:

$$\frac{GM_s M_J}{r^2} = \frac{M_s v^2}{r} = \left(\frac{M_s}{r}\right)\left(\frac{2\pi r}{T}\right)^2$$

$$GM_J T^2 = 4\pi^2 r^3$$

$$r = \left(\frac{GM_J T^2}{4\pi^2}\right)^{1/3} = \left(\frac{(6.67 \times 10^{-11} \text{ N} \cdot \text{m}^2)(1.90 \times 10^{27} \text{ kg})(9.90 \times 3600 \text{ s})^2}{(\text{kg}^2)(4\pi^2)}\right)^{1/3}$$

$$r = 15.98 \times 10^7 \text{ m}$$

$$\text{Altitude} = (15.98 \times 10^7 \text{ m}) - (6.99 \times 10^7 \text{ m}) = 8.99 \times 10^7 \text{ m} \quad \Diamond$$

---

21. Compute the magnitude and direction of the gravitational field at a point P on the perpendicular bisector of two equal masses separated by a distance $2a$ as shown in Figure P14.21.

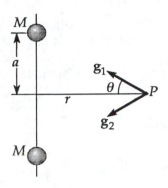

Solution   We must add the vector fields created by each mass:

$$\mathbf{g} = \mathbf{g}_1 + \mathbf{g}_2 \quad \text{where}$$

$$\mathbf{g}_1 = \frac{GM}{r^2 + a^2} \quad \text{to the left and upward at } \theta$$

Figure P14.21
(modified)

and $\mathbf{g}_2 = \dfrac{GM}{r^2 + a^2} \quad$ to the left and downward at θ

Therefore,

$$\mathbf{g} = \frac{GM}{r^2 + a^2}\cos\theta(-\mathbf{i}) + \frac{GM}{r^2 + a^2}\sin\theta(\mathbf{j})$$

$$+ \frac{GM}{r^2 + a^2}\cos\theta(-\mathbf{i}) + \frac{GM}{r^2 + a^2}\sin\theta(-\mathbf{j})$$

$$\mathbf{g} = \frac{2GM}{r^2 + a^2}\frac{r}{\sqrt{r^2 + a^2}}(-\mathbf{i}) + 0\mathbf{j} = \frac{-2GMr}{(r^2 + a^2)^{3/2}}\mathbf{i} \quad \lozenge$$

27. After it exhausts its nuclear fuel, the ultimate fate of our Sun is possibly to collapse to a *white dwarf*, which is a star that has approximately the mass of the Sun, but the radius of the Earth. Calculate (a) the average density of the white dwarf, (b) the free-fall acceleration at its surface, and (c) the gravitational potential energy of a 1.00-kg object at its surface.

Solution

(a) $\rho = \dfrac{M_s}{V} = \dfrac{M_s}{\left(\frac{4}{3}\right)\pi R_E^3} = \dfrac{1.99\times10^{30}\ \text{kg}}{\left(\frac{4}{3}\right)\pi(6.37\times10^6\ \text{m})^3}$

$\rho = 1.84\times10^9\ \text{kg/m}^3$ ◊

(b) For an object of mass m on its surface,

$$mg = \dfrac{GM_s m}{R_E^2}$$

So $\qquad g = \dfrac{GM_s}{R_E^2} = \dfrac{(6.67\times10^{-11}\ \text{N}\cdot\text{m}^2)(1.99\times10^{30}\ \text{kg})}{(\text{kg}^2)(6.37\times10^6\ \text{m})^2}$

$g = 3.27\times10^6\ \text{m/s}^2$ ◊

(c) $U_g = \dfrac{-GM_s m}{R_E}$

$U_g = \dfrac{(-6.67\times10^{-11}\ \text{N}\cdot\text{m}^2)(1.99\times10^{30}\ \text{kg})(1\ \text{kg})}{(\text{kg}^2)(6.37\times10^6\ \text{m})}$

$U_g = -2.08\times10^{13}\ \text{J}$ ◊

29. A spaceship is fired from the Earth's surface with an initial speed of 2.00×10^4 m/s. What will its speed be when it is very far from the Earth? (Neglect friction.)

Solution Energy is conserved between surface and the distant point:

$$(K + U_g)_i + W_{nc} = (K + U_g)_f$$

$$\tfrac{1}{2}mv_i^2 - \frac{GM_E m}{R_E} + 0 = \tfrac{1}{2}mv_f^2 - \frac{GM_E m}{\infty}$$

$$v_f^2 = v_i^2 - \frac{2GM_E}{R_E}$$

$$v_f^2 = (2.00 \times 10^4 \text{ m/s})^2 - \frac{2(6.67 \times 10^{-11} \text{ N} \cdot \text{m}^2)(5.98 \times 10^{24} \text{ kg})}{\text{kg}^2 \, (6.37 \times 10^6 \text{ m})}$$

$$v_f^2 = 4.00 \times 10^8 \text{ m}^2/\text{s}^2 - 1.25 \times 10^8 \text{ m}^2/\text{s}^2 = 2.75 \times 10^8 \text{ m}^2/\text{s}^2$$

$$v_f = 1.66 \times 10^4 \text{ m/s} \quad \Diamond$$

---

33. A satellite moves in a circular orbit just above the surface of a planet. Show that the orbital speed v and escape speed of the satellite are related by the expression $v_{esc} = \sqrt{2}v$.

Solution Call M the mass of the planet and R its radius. For the orbiting "treetop satellite," $\Sigma F = ma$ becomes

$$\frac{GMm}{R^2} = \frac{mv^2}{R}$$

$$v = \sqrt{\frac{GM}{R}}$$

Applying conservation of energy for an object launched with escape velocity gives

$$\tfrac{1}{2}mv_{esc}^2 - \frac{GMm}{R} = 0$$

$$v_{esc} = \sqrt{\frac{2GM}{R}}$$

Thus,
$$v_{esc} = \sqrt{2}\,v \quad \lozenge$$

38. A uniform rod of mass M is in the shape of a semicircle of radius R (Fig. P14.38). Calculate the force on a point mass m placed at the center of the semicircle.

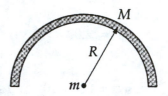

Solution Consider one segment of arc of the curved rod, subtending angle $d\theta$ at the center. Its length is $ds = R\,d\theta$. Since the whole rod has length πR and mass M, this incremental element has mass

Figure P14.38

$$dM = \left(\frac{M}{\pi R}\right)R\,d\theta = \frac{M\,d\theta}{\pi}$$

This mass element exerts a force $d\mathbf{F}$ on the point mass at the center.

From $F = \dfrac{GM_1M_2}{r^2} = G\dfrac{Mm}{r^2}$,

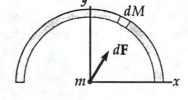

$$d\mathbf{F} = \frac{Gm\,dM}{R^2} = \frac{Gm}{R^2}\left(\frac{M}{\pi}\,d\theta\right)$$

directed at an angle θ above the x-axis

or
$$d\mathbf{F} = \frac{GmM}{\pi R^2}\,d\theta\,(\cos\theta\,\mathbf{i} + \sin\theta\,\mathbf{j})$$

To find the net force on the point mass, we must add (integrate) the contributions for all elements from $\theta = 0$ to $\theta = 180.0°$:

$$\mathbf{F} = \int\limits_{\text{all } m} d\mathbf{F} = \int_{\theta=0}^{180°} \left(\frac{GMm\, d\theta}{\pi R^2}\right)(\cos\theta\, \mathbf{i} + \sin\theta\, \mathbf{j})$$

$$= \left(\frac{GMm\, \mathbf{i}}{\pi R^2}\right)\int_0^{180°}\cos\theta\, d\theta + \left(\frac{GMm\, \mathbf{j}}{\pi R^2}\right)\int_0^{180°}\sin\theta\, d\theta$$

$$= \left(\frac{GMm\, \mathbf{i}}{\pi R^2}\right)\sin\theta\Big|_0^{180°} + \left(\frac{GMm\, \mathbf{j}}{\pi R^2}\right)[-\cos\theta]_0^{180°}$$

$$= \left(\frac{GMm\, \mathbf{i}}{\pi R^2}\right)(0-0) + \left(\frac{GMm\, \mathbf{j}}{\pi R^2}\right)[-(-1)+1]$$

$$= 0\mathbf{i} + \left(2G\frac{Mm}{\pi R^2}\right)\mathbf{j} \quad \lozenge$$

The direction of the force on the point mass at the center is vertically upward in the plane of the figure.

---

41. A 500-kg uniform solid sphere has a radius of 0.400 m. Find the magnitude of the gravitational force exerted by the sphere on a 50.0-g particle located (a) 1.50 m from the center of the sphere, (b) at the surface of the sphere, and (c) 0.200 m from the center of the sphere.

Solution

(a) $\quad F = \dfrac{GmM}{r^2} = \dfrac{(6.67\times10^{-11}\ \text{N}\cdot\text{m}^2/\text{kg}^2)(0.0500\ \text{kg})(500\ \text{kg})}{(1.50\ \text{m})^2}$

Therefore $\qquad F = 7.41\times10^{-10}\ \text{N} \quad \lozenge$

(b) $F = \dfrac{\left(6.67 \times 10^{-11}\ \text{N} \cdot \text{m}^2 / \text{kg}^2\right)(0.0500\ \text{kg})(500\ \text{kg})}{(0.400\ \text{m})^2}$

$F = 1.04 \times 10^{-8}\ \text{N} \quad \Diamond$

(c) In this case the mass m is a distance r from a sphere of mass,

$M = (500\ \text{kg})(0.200\ \text{m} / 0.400\ \text{m})^3 = 62.5\ \text{kg}$ and

$F = \dfrac{\left(6.67 \times 10^{-11}\ \text{N} \cdot \text{m}^2 / \text{kg}^2\right)(0.0500\ \text{kg})(62.5\ \text{kg})}{(0.200\ \text{m})^2} = 5.21 \times 10^{-9}\ \text{N} \quad \Diamond$

---

45. A cylindrical habitat in space 6.0 km in diameter and 30 km long has been proposed (by G. K. O'Neill, 1974). Such a habitat would have cities, land, and lakes on the inside surface and air and clouds in the center. This would all be held in place by rotation of the cylinder about its long axis. How fast would the cylinder have to rotate to imitate the Earth's gravitational field at the walls of the cylinder?

Solution For a 6.0-km diameter cylinder, $r = 3000$ m. To simulate $1g = 9.8$ m/s^2, we have

$$g = \frac{v^2}{r} = \omega^2 r \qquad \text{or} \qquad \omega = \sqrt{\frac{g}{r}} = 0.0572\ \text{rad} / \text{s}$$

But $\quad 1\ \text{rad}/\text{s} = \dfrac{30}{\pi}\ \text{rev/min}$

Therefore, the required rotation rate would be

$$(0.0572\ \text{rad} / \text{s})\left(\frac{30}{\pi} \frac{\text{rev} / \text{min}}{\text{rad} / \text{s}}\right) = 0.546\ \text{rev} / \text{min}$$

or $\quad\quad\quad\quad\quad\quad\quad\quad\quad\quad$ 1 rev/110 s $\quad \Diamond$

---

55. Two hypothetical planets of masses m_1 and m_2 and radii r_1 and r_2, respectively, are at rest when they are an infinite distance apart. Because of their gravitational attraction, they head toward each other on a collision course. (a) When their center-to-center separation is d, find the speed of each planet and their relative velocity. (b) Find the kinetic energy of each planet just before they collide if $m_1 = 2.0 \times 10^{24}$ kg, $m_2 = 8.0 \times 10^{24}$ kg, $r_1 = 3.0 \times 10^6$ m, and $r_2 = 5.0 \times 10^6$ m. (*Hint:* Both energy and momentum are conserved.)

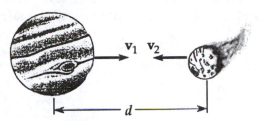

Solution

(a)   At infinite separation, $U = 0$; and at rest, $K = 0$. Since energy is conserved, we have

$$0 = \tfrac{1}{2}m_1v_1^2 + \tfrac{1}{2}m_2v_2^2 - \frac{Gm_1m_2}{d} \qquad (1)$$

The initial momentum is zero and momentum is conserved.

Therefore, $\qquad\qquad 0 = m_1v_1 - m_2v_2 \qquad\qquad (2)$

Combine (1) and (2) to find

$$v_1 = m_2\sqrt{\frac{2G}{d(m_1 + m_2)}} \quad \text{and} \quad v_2 = m_1\sqrt{\frac{2G}{d(m_1 + m_2)}}$$

Relative velocity $\qquad v_r = v_1 - (-v_2) = \sqrt{\dfrac{2G(m_1 + m_2)}{d}} \qquad \Diamond$

(b)   Substitute the given numerical values into the equation found for v_1 and v_2 in part (a) to find $v_1 = 1.03 \times 10^4$ m/s and $v_2 = 2.58 \times 10^3$ m/s.

Therefore, $\qquad\qquad K_1 = \tfrac{1}{2}m_1v_1^2 = 1.07 \times 10^{32}$ J $\qquad \Diamond$

and $\qquad\qquad K_2 = \tfrac{1}{2}m_2v_2^2 = 2.67 \times 10^{31}$ J $\qquad \Diamond$

59. X-ray pulses from Cygnus X-1, a celestial x-ray source, have been recorded during high-altitude rocket flights. The signals can be interpreted as originating when a blob of ionized matter orbits a black hole with a period of 5 ms. If the blob were in a circular orbit about a black hole whose mass is $20M_{Sun}$, what is the orbit radius?

Solution   The gravity outside a black hole is ordinary gravity. For the orbiting blob, $\Sigma F = ma$ becomes

$$\frac{GM_h M_b}{r^2} = \frac{M_b v^2}{r} = \frac{M_b}{r}\left(\frac{2\pi r}{T}\right)^2$$

So we have Kepler' third law: $T^2 GM_h = 4\pi^2 r^3$

$$r = \left(\frac{T^2 GM_h}{4\pi^2}\right)^{1/3}$$

$$r = \left(\frac{(5.0\times 10^{-3}\text{ s})^2(6.67\times 10^{-11}\text{ N}\cdot\text{m}^2)(20)(1.99\times 10^{30}\text{ kg})}{(4\pi^2)(\text{kg}^2)}\right)^{1/3}$$

$r = 119\text{ km}$ ◊

---

63. A sphere of mass M and radius R has a nonuniform density that varies with r, the distance from its center, according to the expression $\rho = Ar$, for $0 \le r \le R$.   (a) What is the constant A in terms of M and R? (b) Determine the force on a particle of mass m placed outside the sphere. (c) Determine the force on the particle if it is inside the sphere. (*Hint:* See Section 14.10 and note that the distribution is spherically symmetric.)

Solution

(a)   If we consider a hollow shell in the sphere with radius r and thickness dr, then $dM = \rho dV = \rho(4\pi r^2 dr)$. The total mass is then

$$M = \int_0^R \rho dV = \int_0^R (Ar)\left(4\pi r^2 dr\right) = \pi A R^4$$

and $$A = \frac{M}{\pi R^4}$$ ◊

(b)   The total mass of the sphere acts as if it were at the center of the sphere and $F = GmM/r^2$ directed toward the center of the sphere.

(c)   Inside the sphere at a distance r from the center, $dF = (Gm/r^2)dM$ where dM is just the mass of a shell enclosed within the radius r.

$$F = \frac{Gm}{r^2}\int_0^r dM = \frac{Gm}{r^2}\int_0^r Ar\left(4\pi r^2\right)dr = \frac{Gm}{r^2}\frac{M4\pi}{\pi R^4}\frac{r^4}{4} = \frac{GmMr^2}{R^4} \qquad \lozenge$$

64.   Two stars of masses M and m, separated by a distance d, revolve in circular orbits about their center of mass (Fig. P14.64). Show that each star has a period given by

$$T^2 = \frac{4\pi^2}{G(M+m)}d^3$$

(*Hint:* Apply Newton's second law to each star, and note that the center-of-mass condition requires that $Mr_2 = mr_1$, where $r_1 + r_2 = d$.)

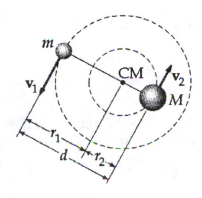

Figure P14.64

Solution   For the star of mass M and orbital radius r_2, $\Sigma F = ma$ gives

$$\frac{GMm}{d^2} = \frac{Mv_2^2}{r_2} = \frac{M}{r_2}\left(\frac{2\pi r_2}{T}\right)^2$$

For the star of mass m, $\Sigma F = ma$ gives

$$\frac{GMm}{d^2} = \frac{mv_1^2}{r_1} = \frac{m}{r_1}\left(\frac{2\pi r_1}{T}\right)^2$$

Clearing of fractions, we then obtain simultaneous equations:

$$GmT^2 = 4\pi^2 d^2 r_2$$

$$GMT^2 = 4\pi^2 d^2 r_1$$

Adding, we find

$$G(M+m)T^2 = 4\pi^2 d^2(r_1 + r_2) = 4\pi^2 d^3$$

$$T^2 = \frac{4\pi^2 d^3}{G(M+m)}$$

In a visual binary star system T, d, r_1, and r_2 can be measured, so the mass of each component can be computed.

---

67. A satellite is in a circular orbit about a planet of radius R. If the altitude of the satellite is h and its period is T, (a) show that the density of the planet is

$$\rho = \frac{3\pi}{GT^2}\left(1+\frac{h}{R}\right)^3$$

(b) Calculate the average density of the planet if the period is 200 min and the satellite's orbit is close to the planet's surface.

Solution

(a) The density is just $\rho = \dfrac{M}{V} = \dfrac{3M}{4\pi R^3}$.

From Kepler's 3rd law, $\qquad\qquad\qquad T^2 = \left(\dfrac{4\pi^2}{GM}\right)r^3 \qquad$ where $r = R + h$

Combining these two equations, we find $\qquad \rho = \dfrac{3\pi}{GT^2}\left(1+\dfrac{h}{R}\right)^3 \quad \Diamond$

(b) If the satellite is close to the surface, then $h/R \ll 1$ and

$$\rho \cong \frac{3\pi}{GT^2} = \frac{3\pi}{\left(6.67\times10^{-11}\ \text{N}\cdot\text{m}^2/\text{kg}^2\right)\left(1.20\times10^4\ \text{s}\right)^2} = 981\ \text{kg}/\text{m}^3 \quad \Diamond$$

---

69. A particle of mass m is located inside a uniform solid sphere of radius R and mass M. If the particle is at a distance r from the center of the sphere, (a) show that the gravitational potential energy of the system is $U = (GmM/2R^3)r^2 - 3GmM/2R$. (b) How much work is done by the gravitational force in bringing the particle from the surface of the sphere to its center?

Solution
$$U = \int F\,dr$$

(a) Initially take the particle from ∞ and move it to the sphere's surface. Then

$$U = \int_{\infty}^{R}\left(\frac{GmM}{r^2}\right)dr = -\frac{GmM}{R}$$

Now move it to a position r from the center of the sphere. The force in this case is a function of the mass enclosed by r at any point. Since

$$\rho = \frac{M}{\frac{4}{3}\pi R^3}$$

we have

$$U = \int_{R}^{r}\frac{Gm(4\pi r^3)\rho}{3r^2}\,dr = \frac{GmM}{R^3}\left(\frac{r^2 - R^2}{2}\right)$$

and the total gravitational potential is

$$U = \frac{GmM}{2R^3}(r^2 - 3R^2) = \left(\frac{GmM}{2R^3}\right)r^2 - \frac{3GmM}{2R}$$

(b) $U(R) = -\frac{GMm}{R}$ and $U(0) = -\frac{3GmM}{2R}$, so

$$W_g = -[U(0) - U(R)] = \frac{GMm}{2R} \quad \Diamond$$

Chapter 15

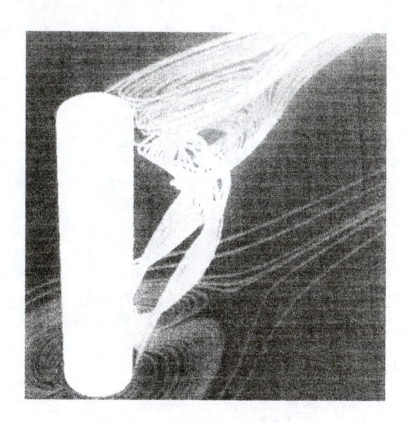

Fluid Mechanics

FLUID MECHANICS

INTRODUCTION

In our treatment of the mechanics of fluids (liquids and gases), we shall see that no new physical principles are needed to explain such effects as the buoyant force on a submerged object and the dynamic lift on an airplane wing. First, we consider a fluid at rest and derive an expression for the pressure exerted by the fluid as a function of its density and depth. We then treat fluids in motion, an area of study called *fluid dynamics*. A fluid in motion can be described by a model in which certain simplifying assumptions are made. We use this model to analyze some situations of practical importance. An underlying principle known as the *Bernoulli principle* enables us to determine relationships among the pressure, density, and velocity at every point in a fluid. As we shall see, the Bernoulli principle is a result of conservation of energy applied to an ideal fluid. We conclude the chapter with a brief discussion of internal friction in a fluid and turbulent motion.

NOTES FROM SELECTED CHAPTER SECTIONS

15.1 Pressure

The *density*, ρ, of a substance of uniform composition is defined as its *mass per unit volume* and has units of kilograms per cubic meter (kg/m^3) in the SI system.

The *specific gravity* of a substance is a dimensionless quantity which is the ratio of the density of the substance to the density of water.

15.2 Variation of Pressure with Depth

In a *fluid at rest*, all points at the same depth are at the same pressure. *Pascal's law* states that a change in pressure applied to an *enclosed* fluid is transmitted undiminished to every point in the fluid and the walls of the containing vessel.

15.3 Pressure Measurements

The *absolute pressure* of a fluid is the sum of the *gauge pressure* and atmospheric pressure. The SI unit of pressure is the Pascal (Pa). Note that $1\ Pa \equiv 1\ N/m^2$.

15.4 Buoyant Forces and Archimedes' Principle

Any object partially or completely submerged in a fluid experiences a buoyant force equal in magnitude to the weight of the fluid displaced by the object and acting vertically upward through the point which was the center of gravity of the displaced fluid.

15.5 Fluid Dynamics

When fluid is in motion, its flow can be characterized as being one of two main types. The flow is said to be steady if each particle of the fluid follows a smooth path, and the paths of each particle do not cross each other. Above a certain critical speed, fluid flow becomes nonsteady or turbulent, characterized by small whirlpool-like regions.

The term *viscosity* is commonly used in fluid flow to characterize the degree of internal friction in the fluid. This internal friction is associated with the resistance to two adjacent layers of the fluid to move relative to each other. Because of viscosity, part of the kinetic energy of a fluid is converted to thermal energy.

In our model of an ideal fluid, we make the following four assumptions:

- *Nonviscous fluid.* In a nonviscous fluid, internal friction is neglected. An object moving through a nonviscous fluid would experience no retarding viscous force.

- *Steady flow.* In a steady flow, we assume that the velocity of the fluid at each point remains constant in time.

- *Incompressible fluid.* The density of the fluid is assumed to remain constant in time.

- *Irrotational flow.* Fluid flow is irrotational if there is no angular momentum of the fluid about any point. If a small wheel placed anywhere in the fluid does not rotate about its center of mass, the flow would be considered irrotational. (If the wheel were to rotate, as it would if turbulence were present, the flow would be rotational.)

15.6 Streamlines and the Equation of Continuity

The path taken by a fluid particle under *steady flow* is called a *streamline*. The velocity of the fluid particle is always tangent to the streamline at that point and no two streamlines can cross each other. A set of streamlines forms a *tube of flow*. In steady flow, particles of the fluid cannot flow into or out of a tube of flow.

EQUATIONS AND CONCEPTS

The *density* of a homogeneous substance is defined as its ratio of mass per unit volume. The value of density is characteristic of a particular type of material and independent of the total quantity of material in the sample.

$$\rho \equiv \frac{m}{V}$$

The SI units of density are kg per cubic meter.

$$1\,g/cm^3 = 1000\ kg/m^3$$

The (average) pressure of a fluid is defined as the normal force per unit area acting on a surface immersed in the fluid.

$$P \equiv \frac{F}{A}$$ (15.1)

Atmospheric pressure is often expressed in other units: atmospheres, mm of mercury (Torr), or pounds per square inch.

$$1\,atm = 1.013 \times 10^5\ Pa$$

$$1\,Torr = 133.3\ Pa$$

$$1\,lb/in^2 = 6.9 \times 10^3\ Pa$$

The SI units of pressure are newtons per square meter, or Pascal (Pa).

$$1\,Pa \equiv 1\ N/m^2$$ (15.3)

The absolute pressure, P, at a depth, h, below the surface of a liquid which is open to the atmosphere is greater than atmospheric pressure, P_0, by an amount which depends on the depth below the surface.

$$P = P_0 + \rho g h$$ (15.4)

The quantity $P_g = \rho g h$ is called the gauge pressure and P is the absolute pressure. Therefore,

$$P_{absolute} = P_{atmosphere} + P_g$$

Pascal's law states that pressure applied to an enclosed fluid (liquid or gas) is transmitted undiminished to every point within the fluid and over the walls of the vessel which contain the fluid.

$$P = P_a + \rho g h$$

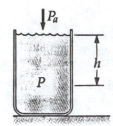

Archimedes' principle states that when an object is partially or fully submersed in a fluid, the fluid exerts an upward buoyant force on the object. The magnitude of the buoyant force depends on the density of the fluid and the volume of displaced fluid, V. In particular, note that B equals the weight of the displaced fluid.

Archimedes' principle.

$$B = \rho_f V g \qquad (15.5)$$

Fluid dynamics, the treatment of fluids in motion, is greatly simplified under the assumption that the fluid is ideal with the following characteristics:

- nonviscous - internal friction between adjacent fluid layers is negligible

- incompressible - the density of the fluid is constant throughout

- steady - the velocity, density, and pressure at each point in the fluid are constant in time

- irrotational (without turbulence) - there are no eddy currents within the fluid (each element of the fluid has zero angular velocity about its center)

The equation of continuity is really just a statement of conservation of mass: *the mass that flows into a tube equals the mass that flows out.*

$$\rho A_1 v_1 = \rho A_2 v_2$$

For an *incompressible fluid* (ρ = constant), the equation of continuity can be written as Equation 15.7. The product Av is called the flow rate. Therefore, the flow rate at any point along a pipe carrying an incompressible fluid is constant.

$$A_1 v_1 = A_2 v_2 \qquad (15.7)$$

Bernoulli's equation is the most fundamental law in fluid mechanics. The equation is a statement of the law of conservation of mechanical energy as applied to a fluid. Bernoulli's equation states that the sum of pressure, kinetic energy per unit volume, and potential energy per unit volume remains constant along a streamline of an ideal fluid.

$$P + \tfrac{1}{2}\rho v^2 + \rho g y = \text{constant} \qquad (15.9)$$

The maximum power per unit area of a wind machine will double if the wind velocity increases by 26%.

$$\frac{\text{Maximum power}}{A} = \frac{8}{27}\rho v^3 \qquad (15.12)$$

The *coefficient of viscosity* for a fluid is defined as the ratio of the shearing stress to the rate of change of the shear strain.

$$\eta = \frac{Fl}{Av} \qquad (15.13)$$

Equation 15.13 is valid when the speed gradient of the fluid is uniform, that is, when the speed varies linearly with the position in the fluid.

REVIEW CHECKLIST

▷ Understand the concept of pressure at a point in a fluid, and the variation of pressure with depth. Understand the relationships among absolute, gauge, and atmospheric pressure values; and know the several different units commonly used to express pressure.

▷ Understand the origin of buoyant forces; and state and explain Archimedes' principle.

▷ State the simplifying assumptions of an ideal fluid moving with streamline flow.

▷ State and understand the physical significance of the *equation of continuity* (constant flow rate) and *Bernoulli's equation* for fluid flow (relating *flow velocity*, *pressure*, and *pipe elevation*).

SOLUTIONS TO SELECTED END-OF-CHAPTER PROBLEMS

1. Calculate the mass of a solid iron sphere that has a diameter of 3.0 cm.

Solution The definition of density $\rho = \dfrac{m}{V}$ is often written as $m = \rho V$.

Here $V = \left(\dfrac{4}{3}\right)\pi r^3$.

So, $m = \rho\left(\dfrac{4}{3}\right)\pi\left(\dfrac{d}{2}\right)^3 = \left(7.86\times 10^3 \text{ kg}/\text{m}^3\right)\left(\dfrac{4}{3}\right)\pi(1.50 \text{ cm})^3\left(\dfrac{1 \text{ m}^3}{1\,000\,000 \text{ cm}^3}\right) = 111 \text{ g}$ ◊

3. Estimate the density of the *nucleus* of an atom. What does this result suggest concerning the structure of matter? (Use the fact that the mass of a proton is 1.67×10^{-27} kg and its radius is approximately 10^{-15} m.)

Solution N = sum of number of protons and neutrons in the nucleus,

$$m = Nm_p, \quad \text{and} \quad V = N\left(\frac{4\pi r^3}{3}\right)$$

Therefore, $\rho = \dfrac{m}{V} = \dfrac{Nm_p}{N\left(\dfrac{4\pi r^3}{3}\right)} = \dfrac{1.67 \times 10^{-27} \text{ kg}}{\dfrac{4}{3}\pi\left(10^{-15} \text{ m}\right)^3} = 3.99 \times 10^{17} \text{ kg/m}^3$ ◊

---

5. A 50.0-kg woman balances on one heel of a pair of high-heel shoes. If the heel is circular with radius 0.500 cm, what pressure does she exert on the floor?

Solution The area of the circular base of the heel is

$$\pi r^2 = \pi(0.500 \text{ cm})^2\left(\frac{1 \text{ m}^2}{10\,000 \text{ cm}^2}\right) = 7.85 \times 10^{-5} \text{ m}^2$$

The force she exerts is her weight, $mg = (50.0 \text{ kg})(9.8 \text{ m/s}^2) = 490 \text{ N}.$

Then $P = \dfrac{F}{A} = \dfrac{490 \text{ N}}{7.85 \times 10^{-5} \text{ m}^2} = 6.24 \times 10^6 \text{ Pa}$ ◊

---

11. The spring of the pressure gauge shown in Figure 15.2 has a force constant of 1000 N/m, and the piston has a diameter of 2.0 cm. Find the depth in water for which the spring compresses by 0.50 cm.

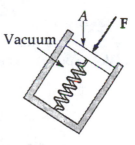

Solution

$$F_{\text{spring}} = F_{\text{fluid}} \quad \text{or} \quad kx = \rho g h A$$

Figure 15.2

and $h = \dfrac{kx}{\rho g A} = \dfrac{\left(1000 \text{ N/m}^2\right)(0.0050 \text{ m})}{\left(10^3 \text{ kg/m}^3\right)\left(9.80 \text{ m/s}^2\right)(0.010 \text{ m})^2 \pi} = 1.62 \text{ m}$ ◊

---

13. What must be the contact area between a suction cup (completely exhausted) and a ceiling in order to support the weight of an 80.0-kg student?

Solution   "Suction" is not a new kind of force. Familiar forces hold the cup in equilibrium. The ceiling pushes down on the cup with some normal force. The student pulls down with a force of magnitude

$$w = mg = 80.0 \text{ kg } 9.80 \text{ m/s}^2 = 784 \text{ N}$$

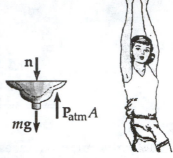

The vacuum between cup and ceiling exerts no force on either. The air below the cup pushes up on it with a force $(P_{atm})(A)$. If the cup barely supports the student, the normal force of the ceiling is nearly zero, and

$$\Sigma F_y = 0 + P_{atm}A - mg = 0$$

$$A = \frac{mg}{P_{atm}} = \frac{784 \text{ N}}{1.013 \times 10^5 \text{ N/m}^2} = 7.74 \times 10^{-3} \text{ m}^2 \quad \Diamond$$

---

17. In some places, the Greenland ice sheet is 1.0 km thick. Estimate the pressure on the ground underneath the ice. ($\rho_{ice} = 920 \text{ kg/m}^3$.)

Solution   Pressure equals atmospheric pressure plus the weight per unit area of the ice sheet:

$$P = P_{atm} + P_{ice} = P_{atm} + \frac{mg}{A}$$

where m is the mass of a block of ice of cross-sectional area A and depth d. Also, $m = \rho(\text{Vol}) = \rho(Ad)$. Therefore,

$$P = P_{atm} + \frac{\rho Adg}{A} = P_{atm} + \rho dg$$

or $$P = 1.013 \times 10^5 \text{ N/m}^2 + (920 \text{ kg/m}^3)(1.0 \times 10^3 \text{ m})(9.80 \text{ m/s}^2)$$

$$P = 9.12 \times 10^6 \text{ N/m}^2$$

or $$P = 9.12 \times 10^3 \text{ kPa} \quad \Diamond$$

---

21. Blaise Pascal duplicated Torricelli's barometer using (as a Frenchman would) a red Bordeaux wine as the working liquid (Fig. P15.21). The density of the wine he used was 0.984×10^3 kg/m³. What was the height h of the wine column for normal atmospheric pressure? Would you expect the vacuum above the column to be as good as for mercury?

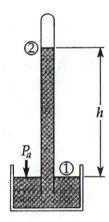

Figure P15.21
(modified)

Solution　　In Bernoulli's equation,

$$P_1 + \tfrac{1}{2}\rho v_1{}^2 + \rho g y_1 = P_2 + \tfrac{1}{2}\rho v_2{}^2 + \rho g y_2$$

Take point 1 at the wine surface in the pan, where $P_1 = P_{atm}$, and point 2 at the wine surface up in the tube. Here we approximate $P_2 = 0$, although some alcohol and water will evaporate. The vacuum is not as good as with mercury. Unless you are careful, a lot of dissolved oxygen or carbon dioxide may come bubbling out.

Now $P_1 = P_2 + \rho g(y_2 - y_1)$.

You could alternatively think of Equation 15.4 as the starting-point.

$$1 \text{ atm} = 0 + (984 \text{ kg/m}^3)(9.80 \text{ m/s}^2)(y_2 - y_1)$$

$$y_2 - y_1 = \frac{1.013 \times 10^5 \text{ N/m}^2}{9640 \text{ N/m}^3} = 10.5 \text{ m} \quad \lozenge$$

A water barometer in a stairway of a three-story building is a nice display. Red wine makes the fluid level easier to see. On television, Mister Wizard uses grape juice.

---

27. A cube of wood 20 cm on a side and having a density of 0.65×10^3 kg/m³ floats on water. (a) What is the distance from the top face of the cube to the water level? (b) How much lead weight has to be placed on top of the cube so that its top is just level with the water? (Assume its top face remains parallel to the water's surface.)

Solution　　Set h equal to the distance from the top of the cube to the water level.

(a)　According to Archimedes' principle,

$$B = \rho_w V g = \left(1.0 \text{ g/cm}^3\right)\left[(20 \text{ cm})^2(20 \text{ cm} - h \text{ cm})\right]g$$

But $\quad B = \text{Weight of block} = Mg = \rho_{\text{wood}}V_{\text{wood}}g = \left(0.65 \text{ g/cm}^3\right)(20 \text{ cm})^3 g$

Setting these two equations equal,

$$(0.65 \text{ g/cm}^3)(20 \text{ cm})^3 g = (1 \text{ g/cm}^3)(20 \text{ cm})^2(20 \text{ cm} - h \text{ cm}) g$$

$$20 - h = 20(0.65)$$

$$h = 20(1.0 \text{ cm} - 0.65 \text{ cm}) = 7.00 \text{ cm} \quad \Diamond$$

(b) $\quad B = w + Mg \quad$ where $\quad M = \text{mass of lead}$

$$1(20)^3 g = (0.65)(20)^3 g + Mg$$

$$M = (20 \text{ cm})^3 \left(1.00 \text{ g/cm}^3 - 0.65 \text{ g/cm}^3\right) = (20 \text{ cm})^3 \left(0.35 \text{ g/cm}^3\right) = 2800 \text{ g} = 2.80 \text{ kg} \quad \Diamond$$

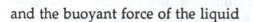

29. A plastic sphere floats in water with 50% of its volume submerged. This same sphere floats in oil with 40% of its volume submerged. Determine the densities of the oil and the sphere.

Solution The forces on the ball are its weight

$$w = mg = \rho_{\text{plastic}}V_{\text{ball}}g$$

and the buoyant force of the liquid

$$B = \rho_{\text{fluid}}V_{\text{immersed}}g.$$

When floating in water, $\Sigma F_y = 0$;

$$-\rho_{\text{plastic}}V_{\text{ball}}g + \rho_{\text{water}}0.50V_{\text{ball}}g = 0$$

$$\rho_{\text{plastic}} = 0.50\rho_{\text{water}} = 500 \text{ kg/m}^3 \quad \Diamond$$

When floating in oil, $\Sigma F_y = 0$;

$$-\rho_{\text{plastic}} V_{\text{ball}} g + \rho_{\text{oil}} 0.40 V_{\text{ball}} g = 0$$

$$\rho_{\text{plastic}} = 0.40 \rho_{\text{oil}}$$

$$\rho_{\text{oil}} = \frac{500 \text{ kg/m}^3}{0.40} = 1250 \text{ kg/m}^3 \quad \Diamond$$

This oil would sink in water.

35. How many cubic meters of helium are required to lift a balloon with a 400-kg payload to a height of 8000 m? ($\rho_{\text{He}} = 0.18 \text{ kg/m}^3$.) Assume the balloon maintains a constant volume and that the density of air decreases with altitude z according to the expression $\rho_{\text{air}} = \rho_0 e^{-z/8000}$, where z is in meters, and ρ_0 (= 1.25 kg/m^3) is the density of air at sea level.

Solution At $z = 8000$ m, the density of air is

$$\rho_{\text{air}} = \rho_0 e^{-z/8000} = \left(1.25 \text{ kg/m}^3\right)e^{-1} = \left(1.25 \text{ kg/m}^3\right)(0.368)$$

$$\rho_{\text{air}} = 0.460 \text{ kg/m}^3$$

Think of the balloon reaching equilibrium at this height. The weight of its payload is

$$(400 \text{ kg})(9.80 \text{ m/s}^2) = 3920 \text{ N}$$

and the weight of the helium in it is

$$mg = \rho_{\text{He}} V g$$

$\Sigma F_y = 0$ becomes

$$+\rho_{\text{air}} V g - 3920 \text{ N} - \rho_{\text{He}} V g = 0$$

Solving,

$$(\rho_{\text{air}} - \rho_{\text{He}}) V g = 3920 \text{ N}$$

and

$$V = \frac{400 \text{ kg}}{\rho_{\text{air}} - \rho_{\text{He}}} = \frac{400 \text{ kg}}{(0.460 - 0.18) \text{ kg/m}^3} = 1.43 \times 10^3 \text{ m}^3 \quad \Diamond$$

37. A large storage tank filled with water develops a small hole in its side at a point 16 m below the water level. If the rate of flow from the leak is 2.5×10^{-3} m^3/min, determine (a) the speed at which the water leaves the hole and (b) the diameter of the hole.

Solution   Assuming the top is open to the atmosphere, then $P_1 = P_a$.

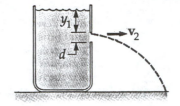

(a) $P_1 + \frac{1}{2}\rho v_1^2 + \rho g y_1 = P_2 + \frac{1}{2}\rho v_2^2 + \rho g y_2$

$A_1 \gg A_2$, so $v_1 \ll v_2$

Assuming $v_1 \approx 0$ and $P_1 = P_2 = P_0$,

$$v_2 = \sqrt{2gy_1} = \sqrt{2(9.80 \text{ m/s}^2)(16 \text{ m})} = 17.7 \text{ m/s} \quad \Diamond$$

(b) Flow rate = 2.5×10^{-3} m^3/min = 4.167×10^{-5} m^3/s

Flow rate = $A_2 v_2 = \left(\dfrac{\pi d^2}{4}\right)(17.7 \text{ m/s}) = 4.167 \times 10^{-5}$ m^3/s

Thus, $d = 1.73 \times 10^{-3}$ m = 1.73 mm $\quad \Diamond$

---

41. Water flows through a fire hose of diameter 6.35 cm at a rate of 0.0120 m^3/s. The fire hose ends in a nozzle of inner diameter 2.20 cm. What is the speed with which the water exits the nozzle?

Solution   Take point 1 inside the hose and point 2 at the surface of the water stream leaving the nozzle. The volume flow rate is constant:

$$0.0120 \text{ m}^3/\text{s} = A_1 v_1 = A_2 v_2$$

$$v_1 = \frac{0.0120 \text{ m}^3/\text{s}}{\pi(6.35 \times 10^{-2} \text{ m}/2)^2} = 3.79 \text{ m/s} \quad \text{and} \quad v_2 = \frac{0.0120 \text{ m}^3/\text{s}}{\pi(1.10 \times 10^{-2} \text{ m})^2} = 31.6 \text{ m/s} \quad \Diamond$$

Related Calculation   Find the pressure at both points.

At point 2, the water pressure is 101.3 kPa. The air exerts this much force per unit area on the water. By Newton's third law, the water exerts the same force on the air. If we have a nonviscous incompressible fluid in steady nonturbulent flow,

$$P_1 + \tfrac{1}{2}\rho v_1^2 + \rho g y_1 = P_2 + \tfrac{1}{2}\rho v_2^2 + \rho g y_2$$

$$P_1 + \tfrac{1}{2}\left(1000 \text{ kg}/\text{m}^3\right)(3.79 \text{ m}/\text{s})^2 + 0 = 1.013 \times 10^5 \text{ N}/\text{m}^2 + \tfrac{1}{2}\left(1000 \text{ kg}/\text{m}^3\right)(31.6 \text{ m}/\text{s})^2 + 0$$

$$P_1 = 101.3 \text{ kPa} + 498.3 \text{ kPa} - 7.18 \text{ kPa} = 592 \text{ kPa} \quad \lozenge$$

---

49. A large storage tank is filled to a height h_0. If the tank is punctured at a height h from the bottom of the tank (Fig. P15.49), how far from the tank will the stream land?

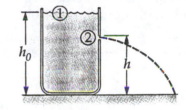

Figure P15.49
(modified)

Solution   Take point 1 at the top liquid surface. Since the tank is large, the fluid level falls only slowly and $v_1 \cong 0$. Take point 2 at the surface of the water stream leaving the hole. At both points the pressure is one atmosphere because the water can push no more or less strongly than the air pushes on it, as described by Newton's third law.

$$P_1 + \tfrac{1}{2}\rho v_1^2 + \rho g y_1 = P_2 + \tfrac{1}{2}\rho v_2^2 + \rho g y_2$$

$$P_a + 0 + \rho g h_0 = P_a + \tfrac{1}{2}\rho v_2^2 + \rho g h$$

$$v_2 = \sqrt{2g(h_0 - h)}$$

Now each drop of water moves as a projectile. Its velocity v_2, which we now call its original velocity, has zero vertical component. Its time of fall is given by $y = v_{y0}t + \tfrac{1}{2}a_y t^2$:

$$-h = 0 - \tfrac{1}{2}g t^2$$

$$t = \sqrt{\frac{2h}{g}}$$

and its horizontal displacement is

$$x = v_{x0}t + \tfrac{1}{2}a_x t^2 = \sqrt{2g(h_0 - h)}\sqrt{2h/g} + 0$$

$$x = \sqrt{4h(h_0 - h)} \quad \lozenge$$

---

53. Consider a windmill with blades of cross-sectional area A, as in Figure P15.53, and assume the mill is facing directly into the wind. (a) If the wind speed is v, show that the kinetic energy of the air that passes through the blades in a time Δt is $K = \tfrac{1}{2}\rho A v^3 \Delta t$. (b) What is the maximum available power according to this model? Compare your result with Equation 15.10.

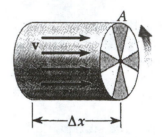

Figure P15.53

Solution

(a) In time Δt, all of the air in a cylinder of base area A and length $\Delta x = v\,\Delta t$ will pass through the windmill. Its volume is $A\,\Delta x = Av\,\Delta t$. It mass is $m = \rho V = \rho A v\,\Delta t$, and its kinetic energy is

$$\tfrac{1}{2}mv^2 = \tfrac{1}{2}\rho A v\,\Delta t\,v^2 = \tfrac{1}{2}\rho A v^3\,\Delta t \quad \lozenge$$

(b) If all of this energy could be extracted from the air, the power is

$$P = \frac{E}{\Delta t} = \tfrac{1}{2}\rho A v^3, \qquad \text{in agreement with Equation 15.10.} \quad \lozenge$$

---

55. A Ping-Pong ball has a diameter of 3.8 cm and average density of 0.084 g/cm^3. What force would be required to hold it completely submerged under water?

Solution At equilibrium, $\Sigma F = 0$ or

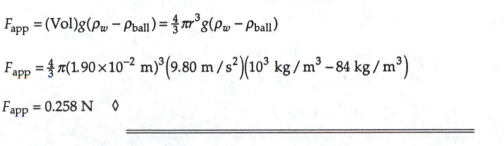

$$F_{app} + mg - B = 0 \quad \text{where } B \text{ is the buoyant force.}$$

The applied force, $F_{app} = B - mg$ when $B \equiv (\text{Vol})\rho_w g$

and $m = (\text{Vol})\rho_{ball}$. So,

$$F_{app} = (\text{Vol})g(\rho_w - \rho_{ball}) = \tfrac{4}{3}\pi r^3 g(\rho_w - \rho_{ball})$$

$$F_{app} = \tfrac{4}{3}\pi(1.90\times10^{-2}\text{ m})^3(9.80\text{ m/s}^2)(10^3\text{ kg/m}^3 - 84\text{ kg/m}^3)$$

$$F_{app} = 0.258\text{ N} \quad \lozenge$$

---

61. The true weight of a body is its weight when measured in a vacuum where there are no buoyant forces. A body of volume V is weighed in air on a balance using weights of density ρ. If the density of air is ρ_a and the balance reads w', show that the true weight w is

$$w = w' + \left(V - \frac{w'}{\rho g}\right)\rho_a g$$

Solution The "balanced" condition is one in which the apparent weight of the body equals the apparent weight of the weights. This condition can be written as:

$$w - w_b = w' - w_b',$$

where w_b and w_b' are the buoyant forces on the body and weights respectively. The buoyant force experienced by an object of volume V in air is

$$B = (\text{Volume of object})\rho_a g$$

so we have $\qquad w_b = V\rho_a g \qquad$ and $\qquad w'_b = \left(\frac{w'}{\rho g}\right)\rho_a g$

Therefore, $\qquad\qquad w = w' + \left(V - \frac{w'}{\rho g}\right)\rho_a g \qquad \lozenge$

---

63. A pipe carrying water has a diameter of 2.5 cm. Estimate the maximum flow speed if the flow is to be laminar. Assume the temperature is 20°C.

Solution

For laminar flow we require a Reynolds number less than 2000:

$$\frac{\rho v d}{\eta} < 2000$$

$$v < \frac{2000\,\eta}{\rho d} = \frac{2000\left(1.00\times10^{-3}\ \text{N}\cdot\text{s}/\text{m}^2\right)}{1000\ (\text{kg}/\text{m}^3)(2.50\times10^{-2}\ \text{m})}\left(\frac{\text{kg}\cdot\text{m}}{\text{N}\cdot\text{s}^2}\right)$$

$$v < 8.00\ \text{cm/s} \quad \Diamond$$

---

67. With reference to Figure 15.5, show that the total torque exerted by the water behind the dam about an axis through O is $\frac{1}{6}\rho g w H^3$. Show that the effective line of action of the total force exerted by the water is at a distance $\frac{1}{3}H$ above O.

Solution

The torque is $\tau = \int d\tau = \int r\,dF$.

From Figure 15.5, we have

$$\tau = \int_0^H y\left[\rho g(H-y)w\right]dy = \frac{1}{6}\rho g w H^3$$

The total force is given as $\frac{1}{2}\rho g w H^2$.

If this were applied at a height y_{eff} such that the torque remains unchanged,

$$\frac{1}{6}\rho g w H^3 = y_{\text{eff}}\left[\frac{1}{2}\rho g w H^2\right] \quad \text{and} \quad y_{\text{eff}} = \frac{1}{3}H \quad \Diamond$$

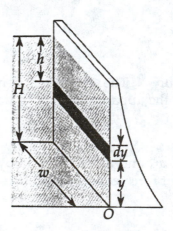

Figure 15.5

69. In 1983, the United States began coining the cent piece out of copper-clad zinc rather than pure copper. If the mass of the old copper cent is 3.083 g while that of the new cent is 2.517 g, calculate the percent of zinc (by volume) in the new cent. The density of copper is 8.960 g/cm^3 and that of zinc is 7.133 g/cm^3. The new and old coins have the same volume.

Solution Let f represent the fraction of the volume V occupied by zinc in the new coin. We have $m = \rho V$ for both coins:

$$3.083 \text{ g} = (8.960 \text{ g/cm}^3)V$$

$$2.517 \text{ g} = (7.133 \text{ g/cm}^3)(fV) + (8.960 \text{ g/cm}^3)(1-f)V$$

By substitution,

$$2.517 \text{ g} = (7.133 \text{ g/cm}^3)fV + 3.083 \text{ g} - (8.960 \text{ g/cm}^3)fV$$

$$fV = \frac{3.083 \text{ g} - 2.517 \text{ g}}{8.960 \text{ g/cm}^3 - 7.133 \text{ g/cm}^3}$$

$$f = \frac{0.566 \text{ g}}{1.827 \text{ g/cm}^3}\left(\frac{8.960 \text{ g/cm}^3}{3.083 \text{ g}}\right) = 0.9004 = 90.04\% \quad \Diamond$$

Chapter 16

Wave Motion

WAVE MOTION

INTRODUCTION

In this chapter we are going to study the properties of mechanical waves. In the case of mechanical waves, what we interpret as a wave corresponds to the disturbance of a body or medium through which the wave travels. Therefore, we can consider a wave to be the *motion of a disturbance.*

The mathematics used to describe wave phenomena is common to all waves. In general, we shall find that mechanical wave motion is described by specifying the positions of all points of the disturbed medium as a function of time.

The mechanical waves discussed in this chapter require (1) some source of disturbance, (2) a medium that can be disturbed, and (3) some physical connection through which adjacent portions of the medium can influence each other. We shall find that all waves carry energy. The amount of energy transmitted through a medium and the mechanism responsible for that transport of energy differ from case to case.

NOTES FROM SELECTED CHAPTER SECTIONS

16.1 Introduction

The production of *mechanical waves* requires: (1) an *elastic medium* which can be disturbed, (2) an *energy source* to provide a disturbance or deformation in the medium, and (3) a physical mechanism by way of which adjacent portions on the medium can *influence* each other. The three parameters important in characterizing waves are (1) wavelength, (2) frequency, and (3) wave velocity.

16.2 Types of Waves

A transverse wave is one in which the particles of the disturbed medium oscillate back and forth along a direction perpendicular to the direction of the wave velocity.

A longitudinal wave is one in which the particles of the medium undergo a displacement (oscillate back and forth) along a direction parallel to the direction of the wave velocity (direction along which the pulse travels).

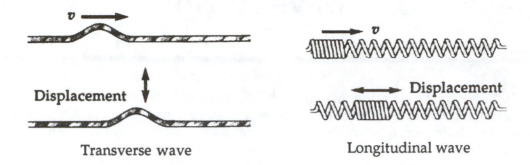

Transverse wave   Longitudinal wave

16.3 One-Dimensional Traveling Waves

A one-dimensional traveling wave can be described mathematically by its *wave function* $y(x, t) = f(x \mp vt)$. If the wave is assumed to be traveling along the x direction, then $y(x, t)$ represents the y coordinate of any point on the string at any time, t.

In the expression for the wave function, the *negative sign* describes a wave pulse traveling *toward the right* and the *positive sign* describes the wave function for a pulse traveling *toward the left*.

In the case of a wave on a string at a *fixed value of x*, the wave function represents the y coordinate of a particular *point as a function of time*. On the other hand, if *t is fixed*, the wave function defines a curve showing the *shape of the wave pulse* at a given time.

16.4 Superposition and Interference of Waves

If two or more waves are moving through a medium, the *resultant wave function* is the *algebraic sum* of the wave functions of the individual waves. Two traveling waves can pass through each other without being destroyed or altered.

16.5 The Speed of Waves on Strings

For linear waves, the *velocity* of *mechanical waves* depends only on the physical properties of the medium through which the disturbance travels. In the case of waves on a *string*, the velocity depends on the tension in the string and the mass per unit length (linear mass density).

16.6 Reflection and Transmission of Waves

The following general rules apply to reflected waves: *When a wave pulse travels from medium A to medium B and $v_A > v_B$ (that is, when B is more massive than A), the reflected part of the pulse is inverted upon reflection. When a wave pulse travels from medium A to medium B and $v_A < v_B$ (A is more massive than B), the reflected pulse is not inverted.*

16.7 Sinusoidal Waves

A sinusoidal wave is one whose shape is sinusoidal at every instant of time. Sinusoidal waves which differ in phase are shown in the figure below.

The wave amplitude, A, is the maximum possible value of the displacement of a particle of the medium away from its equilibrium position.

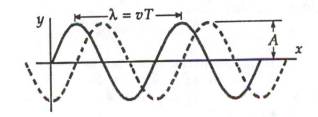

The wavelength, λ, is the minimum distance between two points which are the same distance from their equilibrium positions and are moving in the same direction (such a pair of points are said to be in phase).

The period of the wave is the time required for the disturbance (or pulse) to travel along the direction of propagation a distance equal to the wavelength. The period is also the time required for any point in the medium to complete one complete cycle in its harmonic motion about its equilibrium point.

16.8 Energy Transmitted by Sinusoidal Waves on Strings

As waves propagate through a medium, they transport energy; this occurs without any net transfer of matter. The power transmitted by any sinusoidal wave is proportional to the square of the angular frequency and to the square of the amplitude.

EQUATIONS AND CONCEPTS

The *wave speed* or *phase velocity* is in general the rate at which the profile of the disturbance moves along the direction of travel (e.g. the x axis).

$$v = \frac{dx}{dt}$$ (16.3)

In the special case of a transverse pulse moving along a stretched string, the wave speed depends on the tension in the string F and the linear density μ of the string (mass per unit length).

$$v = \sqrt{\frac{F}{\mu}}$$ (16.4)

The displacement repeats itself when x is increased by an integral multiple of λ and the wave moves to the right a distance of vt in a time t.

$$y = A\sin\left[\frac{2\pi}{\lambda}(x - vt)\right]$$ (16.6)

It is convenient to define three additional characteristic wave quantities: *the wave number k*, the *angular frequency ω*, and the *harmonic frequency f*.

$$k \equiv \frac{2\pi}{\lambda}$$ (16.9)

$$\omega \equiv \frac{2\pi}{T}$$ (16.10)

$$f = \frac{1}{T}$$ (16.12)

The expression for the *wave function* can be written in a more compact form in terms of the parameters defined above. If the transverse displacement is not zero at $x = 0$ and $t = 0$, it is necessary to include a phase constant, ϕ.

$$y = A\sin(kx - \omega t)$$ (16.11)

$$y = A\sin(kx - \omega t - \phi)$$ (16.15)

The *wave speed v* or phase velocity can also be expressed in alternative forms.

$$v = \frac{\omega}{k}$$ (16.13)

$$v = f\lambda$$

The *transverse velocity* v_y of a point on a harmonic wave is out of phase with the *transverse acceleration* a_y of that point by $\pi/2$ radians.

$$v_y = -\omega A \cos(kx - \omega t)$$ (16.16)

$$a_y = -\omega^2 A \sin(kx - \omega t)$$ (16.17)

The *power* transmitted by any harmonic wave is proportional to the square of the frequency and the square of the amplitude, where μ is the mass per unit length of the string.

$$\text{Power} = \frac{1}{2}\mu\omega^2 A^2 v$$ (16.21)

Any wave function having the form $y = f(x \pm vt)$ satisfies the *linear wave equation*.

$$\frac{\partial^2 y}{\partial x^2} = \frac{1}{v^2}\frac{\partial^2 y}{\partial t^2}$$ (16.26)

REVIEW CHECKLIST

▷ Recognize whether or not a given function is a possible description of a traveling wave.

▷ Express a given harmonic wave function in several alternative forms involving different combinations of the wave parameters: wavelength, period, phase velocity, wave number, angular frequency, and harmonic frequency.

▷ Given a specific wave function for a harmonic wave, obtain values for the characteristic wave parameters: A, ω, k, λ, f, and ϕ.

▷ Calculate the rate at which energy is transported by harmonic waves in a string.

Chapter 16

SOLUTIONS TO SELECTED END-OF-CHAPTER PROBLEMS

1. At $t = 0$, a transverse wave pulse in a wire is described by the function

$$y = \frac{6}{x^2 + 3}$$

where x and y are in meters. Write the function $y(x, t)$ that describes this wave if it is traveling in the positive x direction with a speed of 4.5 m/s.

Solution $y(x, t)$ must allow x to vary with t, but must become $y = \frac{6}{x^2 + 3}$ when $t = 0$.

To guarantee the same form, we substitute the term $x = (x' + ut)$, and solve for y:

$$y = \frac{6}{(x' + ut)^2 + 3}$$

Note that as t increases, x' must decrease by $u\Delta t$; aside from that, the equation remains the same.

At $t = 0$, $x = 0$.

So $$x' + u[0] = x = 0 \tag{1}$$

In order to cause the wave to appear to move to the right, we need to force our reference point (x) to the left. Therefore, 1 second later, at $t = 1$, the wave has moved 4.5 m in the $+x$ direction, but x moves 4.5 m in the $-x$ direction.

$$x = x' + ut[1] = -4.5 \text{ m} \tag{2}$$

Subtracting equations (1) and (2), $u = -4.5$ m/s, and our new equation is:

$$y(x, t) = \frac{6}{(x - 4.5t)^2 + 3} \quad \lozenge$$

In general, we can cause any waveform to move along the x axis at a velocity v_x by substituting $(x - v_x t)$ for x. The same principle applies to motion in other directions.

7. Two sinusoidal waves in a string are defined by the functions

$$y_1 = (2.0 \text{ cm}) \sin (20x - 30t) \quad \text{and} \quad y_2 = (2.0 \text{ cm}) \sin (25x - 40t)$$

where y and x are in centimeters and t is in seconds. (a) What is the phase difference between these two waves at the point $x = 5.0$ cm at $t = 2.0$ s? (b) What is the positive x value closest to the origin for which the two phases differ by $\pm\pi$ at $t = 2.0$ s? (This is where the two waves add to zero.)

Solution

(a) At $x = 5.0$ cm and $t = 2.0$ s,

$$\phi_1 = (20 \text{ rad/cm})(5.0 \text{ cm}) - (30 \text{ rad/s})(2.0 \text{ s}) = 40 \text{ rad}$$

$$\phi_2 = (25 \text{ rad/cm})(5.0 \text{ cm}) - (40 \text{ rad/s})(2.0 \text{ s}) = 45 \text{ rad}$$

$$\Delta\phi = \phi_2 - \phi_1 = 5.0 \text{ rad} = 286° = -74° \quad \Diamond$$

(b) $\Delta\phi = 20x - 30t - (25x - 40t)$

At $t = 2.0$ s and when $\Delta\phi = \pm\pi \pm 2n\pi$, the expression for $\Delta\phi$ becomes

$$-5x + 20 = \pm\pi \pm 2n\pi \quad \text{or} \quad 5x = 20 \pm \pi \pm 2n\pi$$

The smallest positive value of x is

$$x = 0.858 \text{ cm} \quad \Diamond$$

---

9. Two pulses traveling on the same string are described by

$$y_1 = \frac{5}{(3x - 4t)^2 + 2} \quad \text{and} \quad y_2 = \frac{-5}{(3x + 4t - 6)^2 + 2}$$

(a) In which direction does each pulse travel? (b) At what time do the two cancel? (c) At what point do the two waves always cancel?

Solution

(a) Set a constant phase: $\phi_{const} = 3x - 4t$ or $x = \dfrac{\phi_{const} + 4t}{3}$

As t increases, x increases, so the first wave moves to the right.   ◊

In a similar manner, in the second case $x = \dfrac{\phi_{const} - 4t + 6}{3}$

As t increases, x must decrease, so the second wave moves to the left.   ◊

(b) We require that $y_1 + y_2 = 0$: $\dfrac{5}{(3x - 4t)^2 + 2} + \dfrac{-5}{(3x + 4t - 6)^2 + 2} = 0$

This can be written as $(3x - 4t)^2 = (3x + 4t - 6)^2$

Solving for the positive root, $t = \dfrac{3}{4}$ s   ◊

(c) The negative root yields: $(3x - 4t) = -(3x + 4t - 6)$

We find that the t terms cancel, leaving $x = 1$ m   ◊

At this point, the waves *always* cancel.

---

13. Transverse waves travel with a speed of 20.0 m/s in a string under a tension of 6.00 N. What tension is required for a wave speed of 30.0 m/s in the same string?

Solution   If the linear density remains constant, $v_1 = \sqrt{\dfrac{F_1}{\mu}}$ and $v_2 = \sqrt{\dfrac{F_2}{\mu}}$

Dividing, $\dfrac{v_2}{v_1} = \sqrt{\dfrac{F_2}{F_1}}$

and $F_2 = \left(\dfrac{v_2}{v_1}\right)^2 F_1 = \left(\dfrac{30.0 \text{ m/s}}{20.0 \text{ m/s}}\right)^2 (6.00 \text{ N}) = 13.5 \text{ N}$ ◊

---

17. A 30.0-m steel wire and a 20.0-m copper wire, both with 1.00-mm diameters, are connected end to end and stretched to a tension of 150 N. How long does it take a transverse wave to travel the entire length of the two wires?

Solution   The total time of travel is the sum of the two times.

In each wire, $t = \dfrac{L}{v} = L\sqrt{\dfrac{\mu}{T}}$ where $\mu = \rho A = \dfrac{\pi \rho d^2}{4}$,   so $t = L\sqrt{\dfrac{\pi \rho d^2}{4T}}$

For copper, $t_1 = (20.0 \text{ m})\sqrt{\dfrac{\pi(8920 \text{ kg}/\text{m}^3)(0.00100 \text{ m})^2}{4(150 \text{ kg} \cdot \text{m}/\text{s}^2)}} = 0.1366 \text{ s}$

For steel, $t_2 = (30.0 \text{ m})\sqrt{\dfrac{\pi(7860 \text{ kg}/\text{m}^3)(0.001 \text{ m})^2}{4(150 \text{ kg} \cdot \text{m}/\text{s}^2)}} = 0.1925 \text{ s}$

The total time is $(0.137 \text{ s}) + (0.192 \text{ s}) = 0.329 \text{ s}$ ◊

---

21. A sinusoidal wave is traveling along a rope. The oscillator that generates the wave completes 40.0 vibrations in 30.0 s. Also, a given maximum travels 425 cm along the rope in 10.0 s. What is the wavelength?

Solution $f = \dfrac{40.0 \text{ waves}}{30.0 \text{ s}} = 1.33 \text{ s}^{-1}$ and $v = \dfrac{425 \text{ cm}}{10.0 \text{ s}} = 42.5 \text{ cm}/\text{s}$

Since $v = \lambda f$, $\lambda = \dfrac{v}{f} = \dfrac{42.5 \text{ cm}/\text{s}}{1.33 \text{ s}^{-1}} = 0.319 \text{ m}$ ◊

---

27. (a) Write the expression for y as a function of x and t for a sinusoidal wave traveling along a rope in the *negative* x direction with the following characteristics: $A = 8.00$ cm, $\lambda = 80.0$ cm, $f = 3.00$ Hz, and $y(0, t) = 0$ at $t = 0$. (b) Write the expression for y as a function of x for the wave in (a) assuming that $y(x, 0) = 0$ at the point $x = 10.0$ cm.

Solution

(a) $A = y_{max} = 8.00$ cm $= 0.0800$ m

$$k = \frac{2\pi}{\lambda} = \frac{2\pi}{0.800 \text{ m}} = 7.85 \text{ m}^{-1}$$

$$\omega = 2\pi f = 2\pi(3.00 \text{ s}^{-1}) = 6.00\pi \text{ rad/s}$$

Since $\phi = 0$, $y = A \sin (kx + \omega t)$, and

$$y = (0.0800 \text{ m}) \sin (7.85x + 6.00\pi t) \quad \Diamond$$

(b) In general, $y = (0.0800 \text{ m}) \sin (7.85x + 6.00\pi t + \phi)$.

If $y(x, 0) = 0$ at $x = 0.100$ m, then

$$0 = (0.0800 \text{ m}) \sin (0.785 + \phi), \quad \text{and} \quad \phi = -0.785 \text{ rad}$$

Therefore, $y = (0.0800 \text{ m}) \sin (7.85x + 6.00\pi t - 0.785) \quad \Diamond$

---

31. A wave is described by $y = (2.0 \text{ cm}) \sin (kx - \omega t)$, where $k = 2.11$ rad/m, $\omega = 3.62$ rad/s, x is in meters and t is in seconds. Determine the amplitude, wavelength, frequency, and speed of the wave.

Solution Given the general sinusoidal wave equation,

$$y = A \sin (kx - \omega t) \quad \text{where} \quad \phi = 0$$

By comparison, the amplitude is $A = 2.00$ cm $\Diamond$

The angular wave number $k = 2.11$ rad/m, so $\lambda = \frac{2\pi}{k} = 2.98$ m $\Diamond$

The angular frequency $\omega = 3.62$ rad/s, so $f = \frac{\omega}{2\pi} = 0.576$ Hz $\Diamond$

The speed is $v = f\lambda$, so $v = (0.576 \text{ s}^{-1})(2.98 \text{ m}) = 1.72 \text{ m/s} \quad \Diamond$

38. Sinusoidal waves 5.00 cm in amplitude are to be transmitted along a string that has a linear density of 4.00×10^{-2} kg/m. If the maximum power delivered by the source is 300 W and the string is under a tension of 100 N, what is the highest vibrational frequency at which the source can operate?

Solution   The wave speed $v = \sqrt{\dfrac{F}{\mu}} = \sqrt{\dfrac{100 \text{ N}}{4.00 \times 10^{-2} \text{ kg/m}}} = 50.0 \text{ m/s}$

From $P = \frac{1}{2}\mu\omega^2 A^2 v,$

$$\omega^2 = \frac{2P}{\mu A^2 v} = \frac{2(300 \text{ N·m/s})}{\left(4.00 \times 10^{-2} \text{ kg/m}\right)\left(5.00 \times 10^{-2} \text{ m}\right)^2 (50.0 \text{ m/s})}$$

$\omega = 346.4 \text{ rad/s}$ and $f = \dfrac{\omega}{2\pi} = 55.1 \text{ Hz}$ ◊

---

39. A sinusoidal wave on a string is described by the equation

$$y = (0.15 \text{ m}) \sin (0.80x - 50t)$$

where x and y are in meters and t is in seconds. If the mass per unit length of this string is 12 g/m, determine (a) the speed of the wave, (b) the wavelength, (c) the frequency, and (d) the power transmitted to the wave.

Solution   Compare the given wave function, $y = (0.15 \text{ m}) \sin (0.80x - 50t)$

with the general wave equation, $y = A \sin (kx - \omega t)$

(a) $v = f\lambda = \dfrac{\omega}{k} = \dfrac{50 \text{ rad/s}}{0.80 \text{ rad/m}} = 62.5 \text{ m/s}$ ◊

(b) $\lambda = \dfrac{2\pi}{k} = \dfrac{2\pi \text{ rad}}{0.80 \text{ rad/m}} = 7.85 \text{ m}$ ◊

(c) $f = \dfrac{\omega}{2\pi} = \dfrac{50 \text{ rad/s}}{2\pi \text{ rad}} = 7.96 \text{ Hz}$ ◊

(d) $P = \frac{1}{2}\mu\omega^2 A^2 v = \frac{1}{2}(0.012 \text{ kg/m})\left(50 \text{ s}^{-1}\right)^2 (0.15 \text{ m})^2 (62.5 \text{ m/s}) = 21.1 \text{ W}$ ◊

---

41. Show that the wave function $y = \ln[b(x-vt)]$ is a solution to Equation 16.25, where b is a constant.

Solution Remembering that
$$\frac{d}{dx}[\ln f(x)] = \frac{d[f(x)]/dx}{f(x)}, \qquad \frac{\partial y}{\partial x} = \frac{b}{b(x-vt)} = \frac{1}{x-vt}$$

Since
$$\frac{d}{dx}\left[\frac{1}{f(x)}\right] = -\frac{d[f(x)]/dx}{[f(x)]^2}, \qquad \frac{\partial^2 y}{\partial x^2} = -\frac{1}{(x-vt)^2}$$

In a similar manner, we find
$$\frac{\partial y}{\partial t} = \frac{-v}{(x-vt)} \quad \text{and} \quad \frac{\partial^2 y}{\partial t^2} = -\frac{v^2}{(x-vt)^2}$$

From the second-order partial derivatives, we see that
$$\frac{\partial^2 y}{\partial x^2} = \frac{1}{v^2}\frac{\partial^2 y}{\partial t^2} \quad \lozenge$$

Therefore, the given function is a solution to Equation 16.25 when $v^2 = F/\mu$.

The important thing to remember with partial derivatives is that you treat all variables as constants, except the single variable of interest.

---

45. The wave function for a linearly polarized wave on a taut string is (in SI units)

$$y(x, t) = (0.35 \text{ m}) \sin (10\pi t - 3\pi x + \pi/4)$$

(a) What are the speed and direction of travel of the wave? (b) What is the vertical displacement of the string at $t = 0$, $x = 0.10$ m? (c) What are the wavelength and frequency of the wave? (d) What is the maximum magnitude of the transverse speed of the string?

Solution We compare the given equation with $y = A \sin (kx - \omega t + \phi)$; and find

$$k = 3\pi \text{ rad/m} \quad \text{and} \quad \omega = 10\pi \text{ rad/s}$$

(a) The speed is
$$v = f\lambda = \frac{\omega}{k} = \frac{10\pi \text{ rad/s}}{3\pi \text{ rad/m}} = 3.33 \text{ m/s} \quad \lozenge$$

(b) Substituting $t = 0$ and $x = 0.10$ m,

$$y = (0.35 \text{ m}) \sin (-0.30\pi + 0.25\pi) = -0.0548 \text{ m} = -5.48 \text{ cm} \quad \lozenge$$

Note that when you take the sine of a quantity with no units, it is not in degrees, but in radians.

(c) $\quad \lambda = \dfrac{2\pi \,\text{rad}}{k} = \dfrac{2\pi \,\text{rad}}{3\pi \,\text{rad}/\text{m}} = 0.667 \text{ m}$ $\quad \Diamond \quad$ and

$\quad f = \dfrac{\omega}{2\pi \,\text{rad}} = \dfrac{10\pi \,\text{rad}/\text{s}}{2\pi \,\text{rad}} = 5.00 \text{ Hz} \quad \Diamond$

(d) $\quad v_y = \dfrac{\partial y}{\partial t} = (0.35 \text{ m})(10\pi \,\text{rad}/\text{s}) \cos \left(10\pi t - 3\pi x + \dfrac{\pi}{4} \right)$

The maximum occurs when the cosine term is 1:

$$v_{y,\text{max}} = (10\pi \,\text{rad}/\text{s})(0.35 \text{ m}/\text{s}) = 11.0 \text{ m}/\text{s} \quad \Diamond$$

Note how different the maximum particle speed is from the wave speed found in part (a).

---

52. A rope of total mass m and length L is suspended vertically. Show that a transverse wave pulse will travel the length of the rope in a time $t = 2\sqrt{L/g}$. (*Hint:* First find an expression for the wave speed at any point a distance x from the lower end by considering the tension in the rope as resulting from the weight of the segment below that point.)

Solution The tension in the rope at any point is the weight of the rope below that point.

Therefore, $\quad v = \sqrt{\dfrac{F}{\mu}} \quad$ where $F = \mu x g$, the weight of the portion of the rope below point x,

and the phase velocity, $\quad v = \sqrt{gx}$, at each point x from 0 (at the bottom) to L.

But at each point x, the wave phase progresses at a rate of $\quad v = \dfrac{dx}{dt}$.

So, $\qquad\qquad\qquad\qquad \dfrac{dx}{dt} = \sqrt{gx} \qquad \text{or} \qquad dt = dx / \sqrt{gx}$

Integrating both sides, $\qquad t = \dfrac{1}{\sqrt{g}} \int_0^L \dfrac{dx}{\sqrt{x}} = \dfrac{\left[2\sqrt{x} \right]_0^L}{\sqrt{g}} = 2\sqrt{\dfrac{L}{g}} \quad \Diamond$

---

55. An aluminum wire is clamped at each end under zero tension at room temperature (22°C). The tension in the wire is increased by reducing the temperature, which results in a decrease in the wire's equilibrium length. What strain ($\Delta L/L$) results in a transverse wave speed of 100 m/s? Take the cross-sectional area of the wire to be 5.0×10^{-6} m², the density to be 2.7×10^3 kg/m³, and Young's modulus to be 7.0×10^{10} N/m².

Solution

The expression for the elastic modulus, $\quad Y = \dfrac{F/A}{\Delta L/L}$.

becomes an equation for strain, $\qquad \dfrac{\Delta L}{L} = \dfrac{F/A}{Y}$ (1)

Combine this with the equation for the wave speed, $\quad v = \sqrt{\dfrac{F}{\mu}}$

Substitute, $\qquad \mu = \dfrac{m}{L} = \dfrac{\rho(AL)}{L} = \rho A$

and solve, $\qquad v^2 = \dfrac{F}{\mu} = \dfrac{1}{\rho}\left(\dfrac{F}{A}\right) \quad$ or $\quad \left(\dfrac{F}{A}\right) = \rho v^2$ (2)

Substiuting (2) into (1), $\quad \dfrac{\Delta L}{L} = \dfrac{\rho v^2}{Y} = \dfrac{(2.7 \times 10^3 \text{ kg/m}^3)(100 \text{ m/s})^2}{7.0 \times 10^{10} \text{ N/m}^2}$

$$\left(\dfrac{\Delta L}{L}\right) = 3.86 \times 10^{-4} \quad \lozenge$$

57. It is stated in Problem 52 that a wave pulse travels from the bottom to the top of a rope of length L in a time $t = 2\sqrt{L/g}$. Use this result to answer the following questions. (It is *not* necessary to set up any new integrations.)

(a) How long does it take for a wave pulse to travel halfway up the rope? (Give your answer as a fraction of the quantity $2\sqrt{L/g}$.)

(b) A pulse starts traveling up the rope. How far has it traveled after a time $\sqrt{L/g}$?

Solution

(a) The speed in the lower half of a rope of length L is the same function of distance (from the bottom end) as the speed along the entire length of a rope of length $\left(\dfrac{L}{2}\right)$.

Thus, the time required $= 2\sqrt{\dfrac{L'}{g}}$ with $L' = \dfrac{L}{2}$

and the time required $= 2\sqrt{\dfrac{L}{2g}} = 0.707\left(2\sqrt{\dfrac{L}{g}}\right)$ ◊

It takes the pulse more than 70% of the total time to cover 50% of the distance.

(b) By the same reasoning applied in part (a), the distance climbed in τ is given by

$$d = \frac{g\tau^2}{4}$$

For $\tau = \dfrac{t}{2} = \sqrt{\dfrac{L}{g}}$, we find the *distance climbed* $= \dfrac{L}{4}$. ◊

In half the total trip time, the pulse has climbed $\dfrac{1}{4}$ of the total length.

Chapter 17

Sound Waves

Chapter 17

SOUND WAVES

INTRODUCTION

Sound waves are the most important example of longitudinal waves. They can travel through any material medium with a speed that depends on the properties of the medium. As the waves travel, the particles in the medium vibrate to produce density and pressure changes along the direction of motion of the wave. These changes result in a series of high- and low-pressure regions called *condensations* and *rarefactions,* respectively. If the source of the sound waves vibrates sinusoidally, the pressure variations are also sinusoidal. We shall find that the mathematical description of harmonic sound waves is identical to that of harmonic string waves discussed in the previous chapter.

NOTES FROM SELECTED CHAPTER SECTIONS

17.2 Periodic Sound Waves

The motion of the medium particles is *back and forth along the direction in which the wave travels.* This is in contrast to a transverse wave, in which the vibrations of the medium are *at right angles to the direction of travel of the wave.* The *pressure amplitude* is proportional to the *displacement amplitude;* however, the pressure wave is 90° out of phase with the displacement.

17.3 Intensity of Periodic Sound Waves

The *intensity* of a wave is the rate at which sound energy flows through a unit area perpendicular to the direction of travel of the wave.

17.4 Spherical and Plane Waves

The intensity of a *spherical wave* produced by a point source is proportional to the average power emitted and inversely proportional to the square of the distance from the source.

17.5 The Doppler Effect

In general, a Doppler effect is experienced whenever there is relative motion between source and observer. When the source and observer are moving toward each other, the frequency heard by the observer is higher than the frequency of the source. When the source and observer are moving away from each other, the observer hears a frequency lower than the source frequency.

EXAMPLES OF DOPPLER EFFECT WITH OBSERVER/SOURCE IN MOTION

Observer	Source	Equation	Remark
O→	S	$f' = f\left(\dfrac{v + v_0}{v}\right)$	Observer moving toward stationary source
←O	S	$f' = f\left(\dfrac{v - v_0}{v}\right)$	Observer moving away from stationary source
O	←S	$f' = f\left(\dfrac{v}{v - v_s}\right)$	Source moving toward stationary observer
O	S→	$f' = f\left(\dfrac{v}{v + v_s}\right)$	Source moving away from stationary observer
O→	S→	$f' = f\left(\dfrac{v + v_0}{v + v_s}\right)$	Observer following moving source
←O	←S	$f' = f\left(\dfrac{v - v_0}{v - v_s}\right)$	Source following moving observer
←O	S→	$f' = f\left(\dfrac{v - v_0}{v + v_s}\right)$	Observer and source moving away from each other along opposite directions
O→	←S	$f' = f\left(\dfrac{v + v_0}{v - v_s}\right)$	Observer and source moving toward each other
O	S	$f' = f$	Observer and source both stationary

EQUATIONS AND CONCEPTS

A sound wave propagates as a *compressional wave* with a *speed v* which depends on the bulk modulus B and equilibrium density ρ of the medium in which the wave is traveling. The *Bulk Modulus* is equal to the negative ratio of the pressure variation to the fractional change in volume of the medium.

$$v = \sqrt{\frac{B}{\rho}} \qquad (17.1)$$

$$B = -\frac{\Delta P}{\Delta V / V}$$

The *speed of sound in a gas* depends on the characteristic parameter γ (the ratio of the specific heat at constant pressure to the specific heat at constant volume), and can be expressed either in terms of the pressure and density or the absolute temperature T and molar mass M.

$$v = \sqrt{\frac{\gamma P}{\rho}}$$

$$v = \sqrt{\frac{\gamma RT}{M}}$$

The *speed of sound in a solid rod* depends on the value of Young's modulus Y and the density of the material. Young's modulus is equal to the ratio of the tensile stress to the tensile strain (see Chapter 12).

$$v = \sqrt{\frac{Y}{\rho}}$$

A harmonic sound wave is produced in a gas when the source is a body vibrating in simple harmonic motion (such as a vibrating guitar string). Under these conditions, both the *displacement* of the medium s and the *pressure variations* of the gas ΔP vary harmonically in time. Note that the displacement and the pressure variation are out of phase by $\pi/2$.

$$s(x, t) = s_{max} \cos(kx - \omega t) \qquad (17.2)$$

$$\Delta P = \Delta P_{max} \sin(kx - \omega t) \qquad (17.3)$$

A sound wave may be considered as either a displacement wave or a pressure wave. The *pressure amplitude* is *proportional* to the *displacement amplitude*.

$$\Delta P_{max} = \rho v \omega s_{max} \tag{17.4}$$

The *intensity* of a harmonic sound wave is proportional to the square of the source frequency and to the square of the displacement (or pressure) amplitude.

$$I = \tfrac{1}{2} \rho (\omega s_{max})^2 v \tag{17.5}$$

$$I = \frac{\Delta P^2_{max}}{2\rho v} \tag{17.6}$$

The human ear is sensitive to a wide *range of intensities*. For this reason, a *logarithmic intensity level scale* is defined using a reference intensity of $I_0 = 10^{-12}$ W/m^2. On this scale, the unit of intensity level is the decibel (dB).

$$\beta \equiv 10 \log\left(\frac{I}{I_0}\right) \tag{17.7}$$

The *amplitude* of a periodic spherical wave varies inversely with the distance r from the source.

$$\psi(r,\, t) = \frac{s_0}{r} \sin(kr - \omega t) \tag{17.9}$$

(Spherical wave)

At large distances from the source ($r \gg \lambda$), a spherical wave can be approximated as a *plane wave*. In this case, the *amplitude* of the wave is *constant*.

$$\psi(x,\, t) = A \sin(kx - \omega t) \tag{17.10}$$

(Plane wave)

The *intensity I* of a spherical wave decreases as the square of the distance from the source. This is so because the intensity is proportional to the *square of the amplitude*, which in turn varies inversely with distance from the source.

$$I = \frac{P_{av}}{4\pi r^2} \tag{17.8}$$

The change in frequency heard by an observer whenever there is *relative motion between the source and the observer* is called the *Doppler effect*. The upper signs refer to motion of one toward the other and the lower signs apply to motion of one away from the other. In Equation 17.17, v_0 and v_s are measured *relative to the medium in which the sound travels.*

$$f' = f\left(\frac{v \pm v_0}{v \mp v_s}\right) \qquad (17.17)$$

Shock waves are produced when a sound source moves through a medium with a speed v_s which is greater than the wave speed in that medium. The shock wave front has a conical shape with a half angle which depends on the *Mach number* of the source, defined as the ratio v_s/v.

$$\sin\theta = \frac{v}{v_s}$$

SUGGESTIONS, SKILLS, AND STRATEGIES

When making calculations using Equation 17.7 which defines the intensity of a sound wave on the decibel scale, the properties of logarithms must be kept clearly in mind.

In order to determine the decibel level corresponding to two sources sounded simultaneously, you must first find the intensity, I, of each source in W/m^2; add these values, and then convert the resulting intensity to the decibel scale. As an illustration of this technique, note that if two sounds of intensity 40 dB and 45 dB are sounded together, the intensity level of the combined sources *is 46.2 dB (NOT 85 dB)*.

Your most likely error in using Equation 17.17 to calculate the Doppler frequency shift due to relative motion between a sound source and an observer is using the incorrect algebraic sign for the velocity of either the observer or the source. These sign conventions are illustrated in the chart in the section on Equations and Concepts in which the directions of motion of the observer O and source S are indicated by arrows; and the correct choice of signs used in Equation 17.17,

$$f' = f\left(\frac{v \pm v_0}{v \mp v_s}\right)$$

where f is the true frequency of the source and f' is the apparent frequency as measured by the observer.

REVIEW CHECKLIST

▷ Calculate the speed of sound in various media in terms of the appropriate elastic properties of the medium (including bulk modulus, Young's modulus, and the pressure-volume relationships of an ideal gas), and the corresponding inertial properties (usually mass density).

▷ Describe the harmonic displacement and pressure variation as functions of time and position for a harmonic sound wave. Relate the displacement amplitude to the pressure amplitude for a harmonic sound wave and calculate the wave intensity from each of these parameters.

▷ Understand the basis of the logarithmic intensity level scale (decibel scale). Determine the intensity ratio for two sound sources whose decibel levels are known. Calculate the decibel level for some combination of sources whose individual sound levels are known.

▷ Describe the wave function for spherical and planar harmonic waves. Understand the amplitude dependence on distance from the source for spherical and plane waves.

▷ Describe the various situations under which a Doppler shifted frequency is produced. Note that a Doppler shift is observed as long as there is a *relative* motion between the observer and the source.

SOLUTIONS TO SELECTED END-OF-CHAPTER PROBLEMS

1. Suppose that you hear a thunder clap 16.2 s after seeing the associated lightning stroke. The speed of sound waves in air is 343 m/s and the speed of light in air is 3.0×10^8 m/s. How far are you from the lightning stroke?

Solution

$$\Delta t = t_{\text{light}} - t_{\text{sound}} = \frac{d}{v_s} - \frac{d}{c_{\text{air}}}$$

$$d = \Delta t \left(\frac{c_{\text{air}} v_s}{c_{\text{air}} - v_s} \right)$$

$$d = (16.2 \text{ s}) \left(\frac{(3.0 \times 10^8 \text{ m/s})(343 \text{ m/s})}{3.0 \times 10^8 \text{ m/s} - 343 \text{ m/s}} \right) \cong (16.2 \text{ s})(343 \text{ m/s}) = 5.56 \text{ km} \quad \Diamond$$

13. An experimenter wishes to generate in air a sound wave that has a displacement amplitude equal to 5.5×10^{-6} m. The pressure amplitude is to be limited to 8.4×10^{-1} Pa. What is the minimum wavelength the sound wave can have?

Solution   We are given $s_{max} = 5.5 \times 10^{-6}$ m and $\Delta P_{max} = 0.84$ Pa.   The pressure amplitude is

$$\Delta P_{max} = \rho v \omega\, s_{max} = \rho v \left(\frac{2\pi v}{\lambda} \right) s_{max}$$

or

$$\lambda_{min} = \frac{2\pi \rho v^2 s_{max}}{\Delta P_{max}} = \frac{2\pi (1.20 \text{ kg/m}^3)(343 \text{ m/s})^2 (5.5 \times 10^{-6} \text{ m})}{0.84 \text{ Pa}} = 5.81 \text{ m} \quad \Diamond$$

17. Write an expression that describes the pressure variation as a function of position and time for a sinusoidal sound wave in air if $\lambda = 0.10$ m and $\Delta P_{max} = 0.20$ Pa.

Solution   We write the pressure variation as $\Delta P = \Delta P_{max} \sin(kx - \omega t)$

Noting that $k = \dfrac{2\pi}{\lambda}$, $k = \dfrac{2\pi \text{ rad}}{0.10 \text{ m}} = 62.8 \text{ rad/m}$

Likewise, $\omega = \dfrac{2\pi v}{\lambda}$,   so $\omega = \dfrac{(2\pi \text{ rad})(343 \text{ m/s})}{0.10 \text{ m}} = 2.16 \times 10^4 \text{ rad/s}$

We now can create our equation:

$$\Delta P = (0.20 \text{ Pa}) \sin(62.8x - 21600t) \quad \Diamond$$

19. Calculate the sound level in dB of a sound wave that has an intensity of $4.0\ \mu W/m^2$.

Solution   We use the equation $\beta = 10\ \log\left(\dfrac{I}{I_0}\right)$, where $I_0 = 10^{-12}$ W/m².

$$\beta = 10\ \log\left(\frac{4.0\times10^{-6}\ W/m^2}{10^{-12}\ W/m^2}\right) = 66.0\ dB \qquad \lozenge$$

24. An explosive charge is detonated at a height of several kilometers in the atmosphere. At a distance of 400 m from the explosion, the acoustic pressure reaches a maximum of 10 Pa. Assuming that the atmosphere is homogeneous over the distances considered, what will be the sound level (in dB) at 4 km from the explosion? (Sound waves in air are absorbed at a rate of approximately 7 dB/km.)

Solution   At a distance of 400 m from the explosion, $\Delta P_{max} = 10$ Pa.

At this point, $I = \dfrac{\Delta P_{max}^2}{2\rho v}$

so $I = \dfrac{(10\ N/m^2)^2}{2(1.20\ kg/m^3)(343\ m/s)} = 0.121\ W/m^2$

From equation 17.8, we can calculate the intensity and decibel level (due to distance alone) 4 km away:

$$I' = \frac{I(400\ m)^2}{(4000\ m)^2} = 1.21\times10^{-3}\ W/m^2 \quad \text{and} \quad \beta = 10\ \log\left(\frac{I}{I_0}\right) = 10\ \log\left(\frac{1.21\times10^{-3}}{10^{-12}}\right) = 90.8\ dB.$$

At a distance of 4 km from the explosion, absorption will have decreased the sound level by an additional

$$\Delta\beta = (7\ dB/km)(3.6\ km) = 25.2\ dB$$

So at 4 km, the sound level will be

$$\beta_2 = \beta - \Delta\beta = 90.8\ dB - 25.2\ dB = 65.6\ dB \qquad \lozenge$$

26. Two sources have sound levels of 75 dB and 80 dB. If they are sounding simultaneously, (a) what is the combined sound level? (b) What is their combined intensity in W/m^2?

Solution From $\beta = 10 \log\left(\dfrac{I}{I_0}\right)$, we have $I = I_0 10^{(\beta/10)}$

So $I_1 = \left(10^{-12} \dfrac{W}{m^2}\right) 10^{(75/10)} = 3.16 \times 10^{-5} \ W/m^2$

$I_2 = \left(10^{-12} \dfrac{W}{m^2}\right) 10^{(80/10)} = 1.00 \times 10^{-4} \ W/m^2$

(b) The combined intensity is $I = I_1 + I_2 = 1.32 \times 10^{-4} \ W/m^2$ ◊

(a) The combined sound level is $\beta = 10 \log\left(\dfrac{1.32 \times 10^{-4}}{10^{-12}}\right) = 81.2 \ dB$ ◊

---

29. The sound level at a distance of 3.0 m from a source is 120 dB. At what distance will the sound level be (a) 100 dB and (b) 10 dB?

Solution $\beta = 10 \log\left(\dfrac{I}{10^{-12} \ W/m^2}\right)$, so $I = \left[10^{\beta/10}\right] 10^{-12} \ W/m^2$

$I_{120} = 1 \ W/m^2$ $\qquad I_{100} = 10^{-2} \ W/m^2$ $\qquad I_{10} = 10^{-11} \ W/m^2$

(a) The power passing through any sphere around the source is $P = 4\pi r^2 I$,

so $r_1^2 I_1 = r_2^2 I_2$

and $r_2 = r_1 \sqrt{\dfrac{I_1}{I_2}} = (3.0 \ m) \sqrt{\dfrac{1}{10^{-2} \ W/m^2}} = 30 \ m$ ◊

(b) $r_2 = r_1 \sqrt{\dfrac{I_1}{I_2}} = (3.0 \ m) \sqrt{\dfrac{1}{10^{-11}}} = 9.49 \times 10^5 \ m$ ◊

---

37. Standing at a crosswalk, you hear a frequency of 560 Hz from the siren on an approaching police car. After the police car passes, the observed frequency of the siren is 480 Hz. Determine the car's speed from these observations.

Solution Approaching car: $\qquad f' = \dfrac{f}{\left(1 - \dfrac{v_s}{v}\right)}$ (Eq. 17.14)

Departing car: $\qquad f'' = \dfrac{f}{\left(1 + \dfrac{v_s}{v}\right)}$ (Eq. 17.15)

Since $f' = 560$ Hz and $f'' = 480$ Hz,

$$560\left(1 - \frac{v_s}{v}\right) = 480\left(1 + \frac{v_s}{v}\right)$$

$$1040\,\frac{v_s}{v} = 80$$

$$v = \frac{80(343)}{1040}\ \text{m/s} = 26.4\ \text{m/s} \quad \Diamond$$

---

40. A tuning fork vibrating at 512 Hz falls from rest and accelerates at 9.80 m/s². How far below the point of release is the tuning fork when waves of frequency 485 Hz reach the release point? Take the speed of sound in air to be 340 m/s.

Solution In order to solve this problem, we must first determine how fast the tuning fork is falling when its frequency is 485 Hz.

The tuning fork (source) is moving *away* from a stationary listener.

Therefore, we use the equation $f' = f\left(\dfrac{v}{v + v_s}\right)$:

$$485\ \text{Hz} = (512\ \text{Hz})\frac{340\ \text{m/s}}{340\ \text{m/s} + v_{\text{fall}}}$$

Solving,
$$v_{\text{fall}} = (340 \text{ m/s})\left(\frac{512 \text{ Hz}}{485 \text{ Hz}} - 1\right) = 18.93 \text{ m/s}$$

Now, using the kinematic equation $v_2{}^2 = v_1{}^2 + 2as$, we calculate that

$$s = \frac{v_2{}^2}{2a} = \frac{(18.93 \text{ m/s})^2}{2(9.80 \text{ m/s}^2)} = 18.28 \text{ m}$$

Since $v = at$,
$$t = \frac{v_{\text{fall}}}{a} = \frac{18.93 \text{ m/s}}{9.8 \text{ m/s}^2} = 1.932 \text{ s}$$

At this moment, the fork would appear to ring at 485 Hz *next to the fork*. However, it takes some additional time for the waves to reach the point of release.

$$\Delta t = \frac{s}{v} = \frac{18.28 \text{ m}}{340 \text{ m/s}} = 0.0540 \text{ s}$$

Over the total time $t + \Delta t$, the fork falls a distance

$$s_{\text{total}} = \tfrac{1}{2}a(t + \Delta t)^2 = \tfrac{1}{2}(9.80 \text{ m/s}^2)(1.986 \text{ s})^2 = 19.3 \text{ m} \quad \lozenge$$

43. A supersonic jet traveling at Mach 3 at an altitude of 20 000 m is directly overhead at time $t = 0$ as in Figure P17.43. (a) How long will it be before one encounters the shock wave? (b) Where will the plane be when it is finally heard? (Assume the speed of sound in air is uniform at 335 m/s.)

Solution Because the shock wave proceeds a set angle θ from the plane, we solve part (b) first.

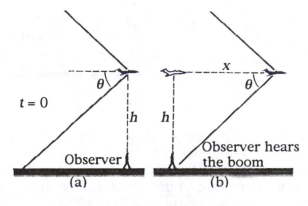

Figure P17.43

(b)   We use $\sin\theta = \dfrac{v}{v_3} = \dfrac{\text{Mach } 1}{\text{Mach } 3} = \dfrac{1}{3}$,   and solve for $\theta = 19.47°$

$$x = \frac{h}{\tan\theta} = \frac{20\,000\text{ m}}{\tan\,19.47°} = 56\,570\text{ m} = 56.6\text{ km} \quad \lozenge$$

(a)   It takes the plane $t = \dfrac{x}{v_s} = \dfrac{56\,570\text{ m}}{3(335\text{ m/s})} = 56.3\text{ s}$ to travel this distance $\lozenge$

---

45. An earthquake on the ocean floor in the Gulf of Alaska induces a *tsunami* (sometimes called a "tidal wave") that reaches Hilo, Hawaii, 4450 km distant, in a time of 9 h 30 min. Tsunamis have enormous wavelengths (100 - 200 km), and for such waves the propagation speed is $v \approx \sqrt{g\overline{d}}$, where $\overline{d}$ is the average depth of the water. From the information given, find the average wave speed and the average ocean depth between Alaska and Hawaii. (This method was used in 1856 to estimate the average depth of the Pacific Ocean long before soundings were made to give a direct determination.)

Solution

$$v = \frac{4450 \times 10^3\text{ m}}{(9.5\text{ h})(3600\text{ s/h})} = 130\text{ m/s} \quad \lozenge$$

$$\overline{d} = \frac{v^2}{g} = \frac{(130\text{ m/s})^2}{9.80\text{ m/s}^2} = 1700\text{ m} = 1.73\text{ km} \quad \lozenge$$

---

49. Two ships are moving along a line due east. The trailing vessel has a speed relative to a land-based observation point of 64 km/h, and the leading ship has a speed of 45 km/h relative to that point. The two ships are in a region of the ocean where the current is moving uniformly due west at 10 km/h. The trailing ship transmits a sonar signal at a frequency of 1200 Hz. What frequency is monitored by the leading ship? (Use 1520 m/s as the speed of sound in ocean water.)

Solution When the observer is moving in front of and in the same direction as the source,

$$f_o = f_s \left(\frac{v - v_o}{v - v_s} \right)$$

where v_o and v_s are measured relative to the *medium* in which the sound is propagated. In this case the ocean current is opposite the direction of travel of the ships and

$$v_o = 45 \text{ km/h} - (-10 \text{ km/h}) = 55 \text{ km/h} = 15.3 \text{ m/s}$$

$$v_s = 64 \text{ km/h} - (-10 \text{ km/h}) = 74 \text{ km/h} = 20.6 \text{ m/s}$$

Therefore,

$$f_o = (1200 \text{ Hz}) \left(\frac{1520 \text{ m/s} - 15.3 \text{ m/s}}{1520 \text{ m/s} - 20.6 \text{ m/s}} \right) = 1204 \text{ Hz} \quad \lozenge$$

---

51. A meteoroid the size of a truck enters the Earth's atmosphere at a speed of 20 km/s and is not significantly slowed before entering the ocean. (a) What is the Mach angle of the shock wave from the meteoroid in the atmosphere? (Use 331 m/s as the sound speed.) (b) Assuming that the meteoroid survives the impact with the ocean surface, what is the (initial) Mach angle of the shock wave that the meteoroid produces in the water? (Use the wave speed for sea water given in Table 17.1.)

Solution

(a) The Mach angle is

$$\theta_{\text{atm}} = \sin^{-1} \left(\frac{v}{v_s} \right) = \sin^{-1} \left(\frac{331 \text{ m/s}}{2.0 \times 10^4 \text{ m/s}} \right) = 0.948° \quad \lozenge$$

(b) At impact with the ocean,

$$\theta_{\text{ocean}} = \sin^{-1} \left(\frac{v}{v_s} \right) = \sin^{-1} \left(\frac{1533 \text{ m/s}}{2 \times 10^4 \text{ m/s}} \right) = 4.40° \quad \lozenge$$

---

53. By proper excitation, it is possible to produce both longitudinal and transverse waves in a long metal rod. A particular metal rod is 150 cm long and has a radius of 0.20 cm and a mass of 50.9 g. Young's modulus for the material is 6.8×10^{10} N/m^2. What must the tension in the rod be if the ratio of the speed of longitudinal waves to the speed of transverse waves is 8?

Solution

For the longitudinal wave, $v_L = \sqrt{\dfrac{Y}{\rho}}$ where $\rho = \dfrac{\text{mass}}{\text{volume}} = \dfrac{m}{\pi r^2 L}$

For the transverse wave, $v_T = \sqrt{\dfrac{F}{\mu}}$ where $\mu = \dfrac{m}{L}$

$$v_L = 8v_T, \quad \text{so} \quad F = \frac{\mu Y}{64\rho}$$

This gives $F = \dfrac{\pi r^2 Y}{64} = \dfrac{\pi(0.0020 \text{ m})^2(6.8\times10^{10} \text{ N/m}^2)}{64} = 1.34\times10^4$ N ◊

---

57. In order to be able to determine her speed, a skydiver carries a tone generator. A friend on the ground at the landing site has equipment for receiving and analyzing sound waves. While the skydiver is falling at terminal speed, her tone generator emits a steady tone of 1800 Hz. (Assume that the air is calm and that the sound speed is 343 m/s, independent of altitude.) (a) If her friend on the ground (directly beneath the skydiver) receives waves of frequency 2150 Hz, what is the skydiver's speed of descent? (b) If the skydiver were also carrying sound-receiving equipment sensitive enough to detect waves reflected from the ground, what frequency would she receive?

Solution Call f_e = 1800 Hz the emitted frequency; v_e, the speed of the skydiver; and f_g = 2150 Hz, the frequency of the wave crests reaching the ground.

(a) The skydiver source is moving toward the stationary ground, so we use the equation

$$f_g = f_e\left(\frac{v}{v - v_e}\right)$$

and
$$v_e = v\left(1 - \frac{f_e}{f_g}\right) = (343 \text{ m/s})\left(1 - \frac{1800 \text{ Hz}}{2150 \text{ Hz}}\right) = 55.8 \text{ m/s} \quad \lozenge$$

(b)   The ground now becomes a stationary source, reflecting crests with the 2150-Hz frequency at which they reach the ground, and sending them to a moving observer:

$$f_{e2} = f_g\left(\frac{v + v_e}{v}\right) = 2150\left(\frac{343 \text{ m/s} + 55.8 \text{ m/s}}{343 \text{ m/s}}\right) = 2500 \text{ Hz} \quad \lozenge$$

---

59.   Three metal rods are located relative to each other as shown in Figure P17.59, where $L_1 + L_2 = L_3$.   Values of density and Young's modulus for the three materials are $\rho_1 = 2.7 \times 10^3 \text{ kg/m}^3$, $Y_1 = 7.0 \times 10^{10} \text{ N/m}^2$, $\rho_2 = 11.3 \times 10^3 \text{ kg/m}^3$, $Y_2 = 1.6 \times 10^{10} \text{ N/m}^2$, $\rho_3 = 8.8 \times 10^3 \text{ kg/m}^3$, and $Y_3 = 11 \times 10^{10} \text{ N/m}^2$.

(a) If $L_3 = 1.5$ m, what must the ratio L_1/L_2 be if a sound wave is to travel the length of rods 1 and 2 in the same time for the wave to travel the length of rod 3?   (b) If the frequency of the source is 4.00 kHz, determine the phase difference between the wave traveling along rods 1 and 2 and the one traveling along rod 3.

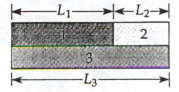

Figure P17.59

Solution

(a)   The time required for a sound pulse to travel a distance L at a speed v is given by

$$t = \frac{L}{v} = \frac{L}{\sqrt{Y/\rho}}$$

Using this expression, we find

$$t_1 = L_1\sqrt{\frac{\rho_1}{Y_1}} = L_1\sqrt{\frac{2.7 \times 10^3 \text{ kg/m}^3}{7.0 \times 10^{10} \text{ N/m}^2}} = L_1\left(1.96 \times 10^{-4} \text{ s/m}\right)$$

$$t_2 = (1.5 - L_1)\sqrt{\frac{11.3 \times 10^3 \text{ kg/m}^3}{1.6 \times 10^{10} \text{ N/m}^2}} = 1.26 \times 10^{-3} \text{ s} - \left(8.4 \times 10^{-4} \text{ s/m}\right)L_1$$

$$t_3 = 1.5 \text{ m}\sqrt{\frac{8.8 \times 10^3 \text{ kg/m}^3}{11 \times 10^{10} \text{ N/m}^2}} = 4.24 \times 10^{-4} \text{ s}$$

We require $t_1 + t_2 = t_3$,   or

$$1.96 \times 10^{-4} L_1 + 1.26 \times 10^{-3} - 8.4 \times 10^{-4} L_1 = 4.24 \times 10^{-4}$$

This gives $L_1 = 1.30$ m   and $L_2 = 1.50 - 1.30 = 0.201$ m

The ratio of lengths is $\dfrac{L_1}{L_2} = 6.45$ ◊

(b)   The ratio of lengths $\dfrac{L_1}{L_2}$ is adjusted in part (a) so that $t_1 + t_2 = t_3$.

Therefore, sound travels the two paths in equal time intervals and the phase difference, $\Delta\phi = 0$. ◊

Chapter 18

Superposition
and Standing Waves

Chapter 18

SUPERPOSITION AND STANDING WAVES

INTRODUCTION

An important aspect of waves is the combined effect of two or more of them traveling in the same medium.

In a linear medium, that is, one in which the restoring force of the medium is proportional to the displacement of the medium, the principle of superposition can be applied to obtain the resultant disturbance.

This chapter is concerned with the superposition principle as it applies to harmonic waves. If the harmonic waves that combine in a given medium have the same frequency and wavelength, and are traveling in opposite directions, one finds that a stationary pattern, called a *standing wave*, can be produced at certain frequencies.

NOTES FROM SELECTED CHAPTER SECTIONS

18.1 Superposition and Interference of Sinusoidal Waves

The *superposition principle* is that when two or more waves move in the same linear medium, the *net displacement* of the medium at any point (the resultant wave) equals the algebraic sum of the displacements of all the waves. If the individual waves are harmonic and of equal frequency, the resultant wave function is also harmonic and has the *same frequency* and *same wavelength* as the individual waves.

Figure 18.1 (on the following page) shows the resultant of two traveling harmonic waves for

(a) $\phi = 0, 2\pi, 4\pi, \ldots$ corresponding to constructive interference,

(b) $\phi = \pi, 3\pi, 5\pi, \ldots$ corresponding to destructive interference, and

(c) $0 < \phi < \pi$ for which the resultant amplitude has a value between 0 and $2A$.

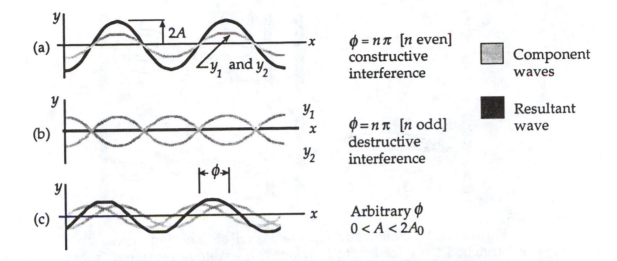

Figure 18.1 Superposition of two waves with the same amplitude and frequency, with a phase difference ϕ of (a) $n\pi$ (n even), (b) $n\pi$ (n odd), and (c) arbitrary ϕ. The gray curves represent y_1 and y_2; the solid curves represent $y = y_1 + y_2$.

18.3 Standing Waves in a String Fixed at Both Ends

Standing waves can be set up in a string by a continuous superposition of waves incident on and reflected from the ends of the string. The string has a number of natural patterns of vibration, called *normal modes*. Each normal mode has a *characteristic frequency*. The lowest of these frequencies is called the *fundamental frequency*, which together with the higher frequencies form a *harmonic series*.

Figure 18.2 (on the following page) is a schematic representation of the first three normal modes of vibration of string fixed at both ends.

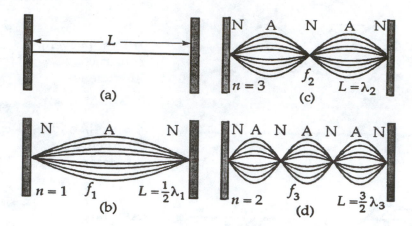

Figure 18.2 Schematic representation of standing waves on a stretched string of length L, where the envelope represents many successive vibrations. The points of zero displacement are called *nodes*; the points of maximum displacement are called *antinodes*.

18.5 Standing Waves in Air Columns

Standing waves are produced in strings by interfering *transverse* waves. Sound sources can be used to produce *longitudinal* standing waves in air columns. The phase relationship between incident and reflected waves depends on whether or not the reflecting end of the air column is open or closed. This gives rise to two sets of possible standing wave conditions.

The first three natural modes of vibration for (a) an open pipe and (b) a closed pipe are shown in Figure 18.3 on the following page.

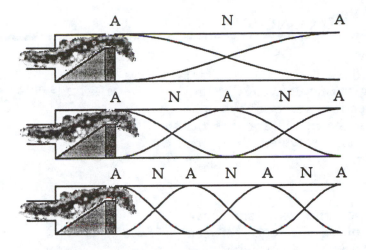

Figure 18.3 (a) Natural modes of vibration in a hollow pipe open at each end. All harmonics are present.

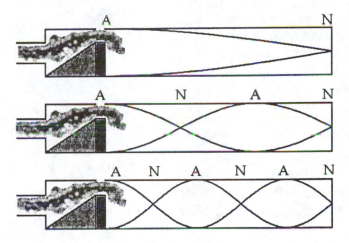

Figure 18.3 (b) Natural modes of vibration for air in a hollow pipe closed at one end. Only odd harmonics are present.

EQUATIONS AND CONCEPTS

The wave function which is the resultant of two traveling harmonic waves having the same direction, frequency, and amplitude is also harmonic and has the same frequency and wavelength as the individual waves.

$$y = 2A_0 \cos\left(\frac{\phi}{2}\right) \sin\left(kx - \omega t - \frac{\phi}{2}\right) \qquad (18.1)$$

The amplitude of the resultant wave depends on the phase difference between the two individual waves according to:

$$y_{max} = 2A_0 \cos\left(\frac{\phi}{2}\right)$$

(amplitude)

A *phase difference* can arise between two waves generated by the same source and arriving at a common point after having traveled along paths of *unequal path length*. In Equation 18.2, ϕ is the phase difference and Δr is the *path difference* between the two waves.

$$\Delta r = \frac{\lambda}{2\pi} \phi \qquad (18.2)$$

A *standing wave* can be produced in a string due to the interference of two sinusoidal waves with equal amplitude and frequency traveling in opposite directions.

$$y = (2A_0 \sin kx) \cos \omega t \qquad (18.3)$$

The amplitude of a standing wave is a function of the position x along the string. *Antinodes* are points of maximum displacement, while *nodes* are points of zero displacement.

distance N to N $= \lambda / 2$

A to A $= \lambda / 2$

N to A $= \lambda / 4$

A series of natural patterns of vibration called *normal modes* can be excited in a string fixed at both ends. Each mode corresponds to a characteristic frequency and wavelength.

$$\lambda_n = \frac{2L}{n} \qquad (18.6)$$

$$f_n = \frac{n}{2L}v \qquad (18.7)$$

$$f_n = \frac{n}{2L}\sqrt{\frac{F}{\mu}} \qquad (18.8)$$

$$n = 1, 2, 3, \ldots$$

In an "open" pipe (open at both ends), all integral multiples of the fundamental frequency can be excited. In a "closed" pipe (closed at one end), only the odd multiples of the fundamental frequency--the odd harmonics--are possible.

$$f_n = n\frac{v}{2L} \qquad n = 1, 2, 3, \ldots \qquad (18.11)$$

$$f_n = n\frac{v}{4L} \qquad n = 1, 3, 5, \ldots \qquad (18.12)$$

Beats are formed by the combination of two waves of equal amplitude but slightly different frequencies traveling in the *same* direction.

$$(18.13)$$

$$y = 2A_0 \cos\left[2\pi t\frac{f_1 - f_2}{2}\right]\cos\left[2\pi t\frac{f_1 + f_2}{2}\right]$$

The amplitude of the wave described by Equation 18.13 is time dependent. Each occurrence of maximum amplitude results in a "beat"; the *beat frequency* f_b equals the difference in the frequencies of the individual waves.

$$A = 2A_0 \cos 2\pi\left(\frac{f_1 - f_2}{2}\right)t \qquad (18.14)$$

$$f_b = |f_1 - f_2| \qquad (18.15)$$

Any *complex periodic wave form* can be represented by the combination of sinusoidal waves which form a harmonic series (combination of fundamental and various harmonics). Such a sum of sine and cosine terms is called a *Fourier series*.

$$y(t) = \sum_n \left(A_n \sin 2\pi f_n t + B_n \cos 2\pi f_n t\right) \qquad (18.16)$$

REVIEW CHECKLIST

▷ Write out the wave function which represents the superposition of the two sinusoidal waves of equal amplitude and frequency traveling in opposite directions in the same medium.

▷ Identify the angular frequency, maximum amplitude, and determine the values of x which correspond to nodal and antinodal points of a standing wave, given an expression for the wave function.

▷ Calculate the normal mode frequencies for a string under tension, and for open and closed air columns.

▷ Describe the time dependent amplitude and determine the effective frequency of vibration when two waves of slightly different frequency interfere. Also, calculate the expected beat frequency for this situation.

SOLUTIONS TO SELECTED END-OF-CHAPTER PROBLEMS

1. Two harmonic waves are described by

$$y_1 = (5.0 \text{ m})\sin[\pi(4.0x - 1200t)]$$

$$y_2 = (5.0 \text{ m})\sin[\pi(4.0x - 1200t - 0.25)]$$

where x, y_1, and y_2 are in meters and t is in seconds. (a) What is the amplitude of the resultant wave? (b) What is the frequency of the resultant wave?

Solution We can represent the waves symbolically as

$$y_1 = A_0 \sin(kx - \omega t) \quad \text{and} \quad y_2 = A_0 \sin(kx - \omega t - \phi)$$

with $A_0 = 5.0$ m, $\omega = 1200\pi \text{ s}^{-1}$, and $\phi = 0.25\pi$

The resultant wave function has the form

$$y = y_1 + y_2 = 2A_0 \cos\left(\frac{\phi}{2}\right)\sin\left(kx - \omega t - \frac{\phi}{2}\right)$$

(a)   with amplitude $A = 2A_0 \cos\left(\dfrac{\phi}{2}\right) = 2(5.0)\cos\left(\dfrac{\pi}{8}\right) = 9.24 \text{ m}$ ◊

(b)   and frequency $f = \dfrac{\omega}{2\pi} = \dfrac{1200\pi}{2\pi} = 600 \text{ Hz}$ ◊

---

7. Two speakers are driven by a common oscillator at 800 Hz and face each other at a distance of 1.25 m. Locate the points along a line joining the two speakers where relative minima would be expected. (Use $v = 343$ m/s.)

Solution   The wavelength is

$\lambda = \dfrac{v}{f} = \dfrac{343 \text{ m/s}}{800 \text{ Hz}} = 0.429 \text{ m}$

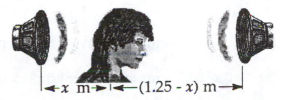

$\leftarrow x$ m $\rightarrow \leftarrow (1.25 - x)$ m $\rightarrow$

For minima, the difference in path length must be an odd number of half wavelengths. So, if x is the distance from the listener to the speaker on the left, then the path difference will be

$$(1.25 \text{ m} - x) - x = (2n+1)\dfrac{\lambda}{2}$$

where n is any integer. Use $\lambda = 0.429$ m and substitute allowed values of n to calculate corresponding values of x. Solving, gives the following results:

$n = 0$, $x = 0.518$ m	$n = -1$, $x = 0.732$ m
$n = 1$, $x = 0.303$ m	$n = -2$, $x = 0.947$ m
$n = 2$, $x = 0.0891$ m	$n = -3$, $x = 1.16$ m

The nodes are at distances of 0.0891 m, 0.304 m, 0.518 m, 0.732 m, 0.947 m, and 1.16 m from either speaker. ◊

---

9. Two harmonic waves are described by

$$y_1 = (3.0 \text{ cm})\sin\pi(x + 0.60t)$$

$$y_2 = (3.0 \text{ cm})\sin\pi(x - 0.60t)$$

where x is in centimeters and t is in seconds. Determine the *maximum* displacement of the motion at (a) $x = 0.25$ cm, (b) $x = 0.50$ cm, and (c) $x = 1.5$ cm. (d) Find the three smallest values of x corresponding to antinodes.

Solution   Adding y_1 and y_2, and using the trigonometry identity

$$\sin(\alpha + \beta) = \sin\alpha\cos\beta + \cos\alpha\sin\beta$$

We get $$y = y_1 + y_2 = (6.0 \text{ cm}) \sin(\pi x) \cos(0.60\pi t)$$

Since $\cos(0) = 1$, we can find the maximum value of y by setting $t = 0$:

$$y_{max}(x) = y_1 + y_2 = 6.0 \sin(\pi x) = (6 \text{ cm}) \sin(\pi x)$$

(a)   At $x = 0.25$ cm, $y = (6.0 \text{ cm}) \sin(\pi \times 0.25) = 4.24$ cm   ◊

(b)   At $x = 0.50$ cm, $y = (6.0 \text{ cm}) \sin(\pi \times 0.50) = 6.00$ cm   ◊

(c)   At $x = 1.5$ cm, $y = (6.0 \text{ cm}) \sin(\pi \times 1.5) = -6.00$ cm   ◊

(d)   The antinodes occur when $x = \dfrac{n\lambda}{4}$ $(n = 1, 3, 5, \ldots)$

But $k = \dfrac{2\pi}{\lambda} = \pi$,   so $\lambda = 2.0$ cm,   and

$$x_1 = \frac{\lambda}{4} = \frac{2.0 \text{ cm}}{4} = 0.50 \text{ cm}   ◊$$

$$x_2 = \frac{3\lambda}{4} = \frac{3(2.0 \text{ cm})}{4} = 1.5 \text{ cm}   ◊$$

$$x_3 = \frac{5\lambda}{4} = \frac{5(2.0 \text{ cm})}{4} = 2.5 \text{ cm}   ◊$$

13. A standing wave is formed by the interference of two traveling waves, each of which has an amplitude $A = \pi$ cm, angular wave number $k = (\pi/2)$ cm^{-1}, and angular frequency $\omega = 10\pi$ rad/s. (a) Calculate the distance between the first two antinodes. (b) What is the amplitude of the standing wave at $x = 0.25$ cm?

Solution

(a) Using the given parameters, we find the wave function to be

$$y = (2\pi \text{ cm})\sin\left(\frac{\pi x}{2}\right)\cos(10\pi t)$$

We need to find values of x for which $\sin\left(\frac{\pi x}{2}\right) = 1$

This condition requires that $\frac{\pi x}{2} = \pi\left(n + \frac{1}{2}\right)$; $n = 0, 1, 2, \ldots$

For $n = 0$, $x = 1$ cm   For $n = 1$, $x = 3$ cm

Therefore, the distance between antinodes, $\Delta x = 2.00$ cm ◊

(b) $A = (2\pi \text{ cm})\sin\left(\frac{\pi x}{2}\right)$;   when $x = 0.25$ cm, $A = 2.40$ cm   ◊

___

16. Two waves in a long string are given by

$$y_1 = (0.015 \text{ m})\cos\left(\frac{x}{2} - 40t\right) \quad \text{and} \quad y_2 = (0.015 \text{ m})\cos\left(\frac{x}{2} + 40t\right)$$

where y and x are in meters and t is in seconds. (a) Determine the positions of the nodes of the resulting standing wave. (b) What is the maximum displacement at the position $x = 0.40$ m?

Solution   The resultant standing wave is the sum of the two wave functions of equal amplitude, frequency, and wavelength.

$$y = A_o \cos(kx - \omega t) + A_o \cos(kx + \omega t)$$

$$y = 2A_o \cos kx \cos \omega t$$

$$y = (2)(0.015 \text{ m})\cos\left(\frac{x}{2}\right)\cos(40t)$$

(a)   Nodes occur where $y = 0$, or where $\cos\left(\dfrac{x}{2}\right) = 0$

This occurs when $\quad \dfrac{x}{2} = (2n+1)\dfrac{\pi}{2}$

Therefore, nodes will be present for $\quad x = (2n+1)\pi = \pi,\ 3\pi,\ 5\pi,\ ... \quad ◊$

(b)   At $x = 0.40$ m,

$$y_{max} = (0.030\text{ m})\cos\left(\frac{0.40}{2}\right) = 0.0294\text{ m} \quad ◊$$

---

21. Find the fundamental frequency and the next three frequencies that could cause a standing wave pattern on a string that is 30 m long, has a mass per unit length 9.0×10^{-3} kg / m, and is stretched to a tension of 20 N.

Solution   The wave speed is

$$v = \sqrt{\frac{F}{\mu}} = \sqrt{\frac{20\text{ N}}{9.0 \times 10^{-3}\text{ kg/m}}} = 47.1\text{ m/s}$$

For a vibrating string of length L, the wavelength of the fundamental is $\lambda = 2L = 60$ m; and the frequency is

$$f_1 = \frac{v}{\lambda} = \frac{v}{2L} = \frac{47.1\text{ m/s}}{60\text{ m}} = 0.786\text{ Hz} \quad ◊$$

The next three harmonics are

$$f_2 = 2f_1 = 1.57\text{ Hz} \quad ◊$$

$$f_3 = 3f_1 = 2.36\text{ Hz} \quad ◊$$

$$f_4 = 4f_1 = 3.14\text{ Hz} \quad ◊$$

---

25. A cello A-string vibrates in its fundamental mode with a frequency of 220 vibrations/s. The vibrating segment is 70 cm long and has a mass of 1.2 g. (a) Find the tension in the string. (b) Determine the frequency of the harmonic that causes the string to vibrate in three segments.

Solution   The length of the string L = 0.70 m, so the fundamental harmonic wavelength is

$$\lambda = 2L = 1.40 \text{ m}$$

and the velocity $v = f\lambda = \left(220 \text{ s}^{-1}\right)(1.40 \text{ m}) = 308 \text{ m/s}$

From the tension equation $v = \sqrt{\dfrac{F}{\mu}} = \sqrt{\dfrac{F}{m/L}}$ we get

(a) $F = \dfrac{v^2 m}{L} = \dfrac{(308 \text{ m/s})^2 \left(1.2 \times 10^{-3} \text{ kg}\right)}{0.70 \text{ m}} = 163 \text{ N}$ ◊

(b)   For the third harmonic state, the tension, linear density, and speed are the same. However, the string vibrates in three segments and the wavelength is one third as long as in the fundamental.

$$\lambda_3 = \lambda/3$$

From the equation $\lambda f = v$, we find that the frequency is three times as high:

$$f_3 = \frac{v}{\lambda_3} = 3\frac{v}{\lambda} = 3f = 660 \text{ Hz} \quad \Diamond$$

_____

27. A 60-cm guitar string under a tension of 50 N has a mass per unit length of 0.10 g/cm. What is the highest resonant frequency that can be heard by a person capable of hearing frequencies up to 20 000 Hz?

Solution

$L = 60 \text{ cm} = 0.60 \text{ m}, \quad F = 50 \text{ N}, \quad \mu = 0.10 \text{ g/cm} = 0.010 \text{ kg/m}$

$$f_n = n\left(\frac{v}{\lambda}\right) = n\left(\frac{v}{2L}\right) \qquad \text{where} \qquad v = \sqrt{\frac{F}{\mu}} = 70.7 \text{ m/s}$$

$$f_n = n\left(\frac{70.7 \text{ m/s}}{1.2 \text{ m}}\right) = 58.9n \text{ Hz}$$

Since $n = \dfrac{20\,000 \text{ Hz}}{58.9 \text{ Hz}} = 339.4$, the largest integer n is 339,

which corresponds to $f = 19.976$ kHz ◊

35. Calculate the length for a pipe that has a fundamental frequency of 240 Hz if the pipe is (a) closed at one end and (b) open at both ends.

Solution

(a) For the fundamental mode in a closed pipe, $\lambda = 4L$.

But $v = f\lambda$, therefore $L = \dfrac{v}{4f}$

So, $L = \dfrac{343 \text{ m/s}}{4(240/\text{s})} = 0.357 \text{ m}$ ◊

(b) For an open pipe, $\lambda = 2L$.

So, $L = \dfrac{v}{2f} = \dfrac{343 \text{ m/s}}{2(240/\text{s})} = 0.715 \text{ m}$ ◊

43. A glass tube is open at one end and closed at the other (by a movable piston). The tube is filled with 30.0°C air, and a 384-Hz tuning fork is held at the open end. Resonance is heard when the piston is 22.8 cm from the open end and again when it is 68.3 cm from the open end. (a) What speed of sound is implied by these data? (b) Where would the piston be for the next resonance?

Solution   For resonance in a closed tube, Equation 18.12 gives

$$f = n\frac{v}{4L} \quad (n = 1, 3, 5, \ldots)$$

 $f = 384$ Hz

When $n = 1$, $v = 4Lf$

(a) $v = 4(0.228 \text{ m})(384 \text{ s}^{-1}) = 350 \text{ m/s}$ ◊

(b)   For the next resonance, $n = 5$, and

$$L = \frac{5v}{4f} = \frac{5(350 \text{ m/s})}{4(384 \text{ s}^{-1})} = 1.14 \text{ m} \quad ◊$$

---

45. A shower stall measures 86 cm × 86 cm × 210 cm. When you sing in the shower, which frequencies will sound the richest (resonate), assuming the shower acts as a pipe closed at both ends (nodes at both sides)? Assume also that the human voice ranges from 130 Hz to 2000 Hz (not necessarily one person's voice, however). Let the speed of sound in the hot shower stall be 355 m/s.

Solution   For a closed box, the resonant frequencies will have nodes at both sides, so the permitted wavelengths will be defined by

$$L = \frac{n\lambda}{2}, \quad (n = 1, 2, 3, \ldots)$$

$$L = \frac{n\lambda}{2} = \frac{nv}{2f} \quad \text{or} \quad f_n = \frac{nv}{2L}$$

Therefore, with $L = 0.860$ m, the side-to-side resonant frequencies are

$$f_n = n\frac{355 \text{ m/s}}{2(0.860 \text{ m})} = n(206 \text{ Hz}), \quad \text{for each } n \text{ from 1 to 9} \quad ◊$$

With $L' = 2.10$ m, the top-to-bottom resonance frequencies are

$$f_n = n\frac{355 \text{ m/s}}{2(2.10 \text{ m})} = n(84.5 \text{ Hz}), \quad \text{for each } n \text{ from 1 to 23} \quad ◊$$

---

49. An aluminum rod 1.6 m long is held at its center. It is stroked with a rosin-coated cloth to set up longitudinal vibrations in the fundamental mode. (a) What is the frequency of the waves established in the rod? (b) What harmonics are set up in the rod held in this manner? (c) What would be the fundamental frequency if the rod were copper?

Solution Either using $v = \sqrt{\dfrac{Y}{\rho}}$ or Table 17.1, we can find the speed of sound in aluminum and copper to be:

$$v_{Al} = 5100 \text{ m/s} \qquad \text{and} \qquad v_{Cu} = 3560 \text{ m/s}$$

(a) Our first harmonic frequency, then, is

$$f_1 = \frac{v}{\lambda_1} = \frac{v}{2L} = \frac{5100 \text{ m/s}}{3.2 \text{ m}} = 1.59 \text{ kHz} \qquad \Diamond$$

(b) Since the rod is free at each end, the ends will be antinodes. Since the central clamp establishes a node at the center, the fundamental mode of vibration will be ANA. The next vibration state will be ANANANA, as shown, with a wavelength and frequency of:

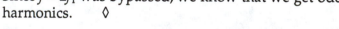

$$\lambda = \frac{2L}{3} \qquad \text{and} \qquad f = \frac{v}{\lambda} = \frac{3v}{2\lambda} = 3f_1.$$

Since $f = 2f_1$ was bypassed, we know that we get odd harmonics. $\Diamond$

(c) For a copper rod, the speed of sound changes:

$$f_1 = \frac{v}{2L} = \frac{3560 \text{ m/s}}{3.2 \text{ m}} = 1.11 \text{ kHz} \qquad \Diamond$$

50. In certain ranges of a piano keyboard, more than one string is tuned to the same note to provide extra loudness. For example, the note at 110 Hz has two strings at this pitch. If one string slips from its normal tension of 600 N to 540 N, what beat frequency will be heard when the two strings are struck simultaneously?

Solution   Combining the velocity equation $v = f\lambda$ and the tension equation $v = \sqrt{\dfrac{F}{\mu}}$

we find that

$$f = \sqrt{\frac{F}{\mu\lambda^2}}$$

and since μ and λ are constant,

$$\frac{f_1}{f_2} = \sqrt{\frac{F_1}{F_2}}$$

We solve

$$f_2 = (110 \text{ Hz})\sqrt{\frac{540 \text{ N}}{600 \text{ N}}} = 104.4 \text{ Hz}$$

and calculate the beat frequency

$$f_b = |f_1 - f_2| = 110 \text{ Hz} - 104.4 \text{ Hz} = 5.64 \text{ Hz} \quad \Diamond$$

---

53. A student holds a tuning fork oscillating at 256 Hz. He walks towards a wall at a constant speed of 1.33 m/s. (a) What beat frequency does he observe between the tuning fork and its echo? (b) How fast must he walk away from the wall to observe a beat frequency of 5.00 Hz?

Solution   For an echo, $f = f_0 \dfrac{v + v_s}{v - v_s}$ and the beat frequency $f_b = |f - f_0|$

Solving for f_b gives $f_b = f_0 \dfrac{2v_s}{v - v_s}$ when approaching the wall.

(a) $f_b = (256 \text{ Hz})\dfrac{2(1.33 \text{ m/s})}{343 \text{ m/s} - 1.33 \text{ m/s}} = 1.99 \text{ Hz} \quad \Diamond$

(b)   When moving away from wall, v_s changes sign. Solving for v_s gives

$$v_s = f_b \frac{v}{2f_0 - f_b} = (5.00 \text{ Hz})\frac{343 \text{ m/s}}{2(256 \text{ Hz}) - 5.00 \text{ Hz}} = 3.38 \text{ m/s} \quad \Diamond$$

---

59. If two adjacent natural frequencies of an organ pipe are determined to be 0.55 kHz and 0.65 kHz, calculate the fundamental frequency and length of this pipe. (Use $v = 340$ m/s.)

Solution   Because harmonic frequencies are given by $f_0 n$ for open pipes, and $f_0(2n-1)$ for closed pipes, the difference between all adjacent harmonics is constant. Therefore, we can find each harmonic below 650 Hz by subtracting

$$\Delta f_{\text{Harmonic}} = (650 \text{ Hz} - 550 \text{ Hz}) = 100 \text{ Hz}$$

from the previous value.

The harmonics are   {650 Hz, 550 Hz, 450 Hz, 350 Hz, 250 Hz, 150 Hz, and 50 Hz}

(a)   The fundamental frequency, then, is 50 Hz.   ◊

(b)   The wavelength can be calculated from the velocity: $\lambda = \dfrac{v}{f} = \dfrac{340 \text{ m/s}}{50 \text{ Hz}} = 6.8$ m

Because the step size Δf is twice the fundamental frequency, we know the pipe is closed, with an antinode at the open end, and a node at the closed end. The wavelength in this situation is four times the pipe length, so

$$L = \frac{\lambda}{4} = 1.7 \text{ m}   ◊$$

─────────────────────────────

65. Two train whistles have identical frequencies of 180 Hz. When one train is at rest in the station sounding its whistle, a beat frequency of 2 Hz is heard from a moving train. What two possible speeds and directions can the moving train have?

Solution   We know from our beat equation $f_b = |f_1 - f_2|$ that the moving train can have apparent frequencies of $f' = 182$ Hz or $f' = 178$ Hz.

If we assume the train is moving away from the station, the apparent frequency of the moving train is lower (178 Hz):

$$f' = f \frac{v}{v + v_s}$$

and the train is *moving away* at

$$v_s = v\left(\frac{f}{f'}-1\right)=(343 \text{ m/s})\left(\frac{180 \text{ Hz}}{178 \text{ Hz}}-1\right)=3.85 \text{ m/s} \quad \lozenge$$

If the train is pulling into the station, then the apparent frequency is 182 Hz.

Again from the Doppler shift, $\qquad f'=f\dfrac{v}{v-v_s}$

and the train is *approaching* at

$$v_s = v\left(1-\frac{f}{f'}\right)=(343 \text{ m/s})\left(\frac{180 \text{ Hz}}{182 \text{ Hz}}-1\right)=3.77 \text{ m/s} \quad \lozenge$$

67. Two wires are welded together. The wires are of the same material, but one is twice the diameter of the other one. They are subjected to a tension of 4.6 N. The thin wire has a length of 40 cm and a linear mass density of 2.0 g/m. The combination is fixed at both ends and vibrated in such a way that two antinodes are present with the central node being right at the weld. (a) What is the frequency of vibration? (b) How long is the thick wire?

Solution

(a) Since the first node is at the weld, the wavelength in the thin wire is 2L or 80 cm. The frequency and tension are the same in both sections, so

$$f=\frac{1}{2L}\sqrt{\frac{F}{\mu}}=\frac{1}{2(0.40 \text{ m})}\sqrt{\frac{4.6 \text{ N}}{0.0020 \text{ kg/m}}}=59.9 \text{ Hz} \quad \lozenge$$

(b) Since the thick wire is twice the diameter, it will have 4 times the cross-sectional area, and a linear density μ that is 4 times that of the thin wire.

$$\mu'=4(2.0 \text{ g/m})=0.0080 \text{ kg/m}$$

L' varies accordingly: $\qquad L'=\dfrac{1}{2f}\sqrt{\dfrac{F}{\mu'}}=\dfrac{1}{2(59.9 \text{ Hz})}\sqrt{\dfrac{4.6 \text{ N}}{0.0080 \text{ kg/m}}}=20.0 \text{ cm} \quad \lozenge$

Note that the thick wire is half the length of the thin wire.

69. A standing wave is set up in a string of variable length and tension by a vibrator of variable frequency. When the vibrator has a frequency f in a string of length L and tension F, there are n antinodes set up in the string. (a) If the length of the string is doubled, by what factor should the frequency be changed to get the same number of antinodes? (b) If the frequency and length are held constant, what tension will produce $n + 1$ antinodes? (c) If the frequency is tripled and the length halved, by what factor should the tension be changed to get twice as many antinodes?

Solution

(a) Combining the equations $v = f\lambda$ and $v = \sqrt{\dfrac{F}{\mu}}$ and noting that $\lambda_n = \dfrac{2L}{n}$

we find that

$$f_n = \frac{n}{2L}\sqrt{\frac{F}{\mu}} \qquad (1)$$

Keeping $n, f,$ and μ constant, we can create two equations:

$$f_n L = \frac{n}{2}\sqrt{\frac{F}{\mu}} \qquad \text{and} \qquad f_n' L' = \frac{n}{2}\sqrt{\frac{F}{\mu}}$$

Dividing the equations, we find $\qquad \dfrac{f_n}{f_n'} = \dfrac{L'}{L}$

If $L' = 2L$, then $f_n' = \frac{1}{2}f_n$

Therefore, in order to double the length but keep the same number of antinodes, the frequency should be halved. ◊

(b) From the same equation (1), we can hold L and f_n constant to get $\dfrac{n'}{n} = \sqrt{\dfrac{F}{F'}}$

From this relation, we see that the tension must be decreased to $F' = F\left[\dfrac{n}{n+1}\right]^2$ to produce $n + 1$ antinodes. ◊

(c) This time, we rearrange equation (1) to produce $\dfrac{2f_n L}{n} = \sqrt{\dfrac{F}{\mu}}$ and $\dfrac{2f_n' L'}{n'} = \sqrt{\dfrac{F'}{\mu}}$

Dividing, we get $\qquad \dfrac{F'}{F} = \left(\dfrac{f_n'}{f_n} \cdot \dfrac{n}{n'} \cdot \dfrac{L'}{L}\right)^2 = \left(\dfrac{3f}{f_n}\right)^2 \left(\dfrac{n}{2n}\right)^2 \left(\dfrac{L/2}{L}\right)^2 = \dfrac{9}{16}$ ◊

Chapter 19

Temperature

TEMPERATURE

INTRODUCTION

A quantitative description of thermal phenomena requires a careful definition of the concepts of temperature, heat, and internal energy. The laws of thermodynamics provide us with a relationship among heat flow, work, and the internal energy of a system.

The composition and structure of a body are important factors when dealing with thermal phenomena. For example, liquids and solids expand only slightly when heated, whereas gases expand appreciably when heated. If the gas is not free to expand, its pressure rises when heated. Certain substances may melt, boil, burn, or explode.

This chapter concludes with a study of ideal gases. We shall approach this study on two levels. The first will examine ideal gases on the macroscopic scale. Here we shall be concerned with the relationships among such quantities as pressure, volume, and temperature. On the second level, we shall examine gases on a microscopic scale, using a model that pictures the components of a gas as small particles. The latter approach, called the kinetic theory of gases, will help us to understand what happens on the atomic level to affect such macroscopic properties as pressure and temperature.

NOTES FROM SELECTED CHAPTER SECTIONS

19.1 Temperature and the Zeroth Law of Thermodynamics

Thermal physics is the study of the behavior of solids, liquids and gases, using the concepts of heat and temperature. Two approaches are commonly used in this area of science. The first is a *macroscopic* approach, called *thermodynamics*, in which one explains the bulk thermal properties of matter. The second is a *microscopic* approach, called *statistical mechanics*, in which properties of matter are explained on an atomic scale. Both approaches require that you understand some basic concepts, such as the concepts of temperature and heat. As we will see, *all* thermal phenomena are manifestations of the laws of mechanics as we have learned them. For example, thermal energy (or heat energy) is actually a consequence of the random motions of a large number of particles making up the system.

The concept of the *temperature* of a system can be understood in connection with a measurement, such as the reading of a thermometer. Temperature, a scalar quantity, is a property which can only be defined when the system is in thermal equilibrium with another system. Thermal equilibrium implies that two (or more) systems are at the same temperature.

The *zeroth law of thermodynamics* states that if two systems are in thermal equilibrium with a third system, they must be in thermal equilibrium with each other. The third system is usually a calibrated thermometer whose reading determines whether or not the systems are in thermal equilibrium. There are several types of thermometers which can be used.

19.2 Thermometers and Temperature Scales

Thermometers are devices used to define and measure the temperature of a system. All thermometers make use of a change in some physical property with temperature. Some of the physical properties used are (1) the change in volume of a liquid, (2) the change in length of a solid, (3) the change in pressure of a gas held at constant volume, (4) the change in volume of a gas held at constant pressure, (5) the change in electric resistance of a conductor, and (6) the change in color of a very hot object. For a given substance, a temperature scale can be established based on any one of these physical quantities.

The *gas thermometer* is a standard device for defining temperature. In the constant-volume gas thermometer, a low density gas is placed in a flask, and its volume is kept constant while it is heated. The pressure is measured as the gas is heated or cooled. Experimentally, one finds that the temperature is proportional to the absolute pressure.

The *thermodynamic temperature scale* is based on a scale for which the reference temperature is taken to be the *triple point of water*; that is, the temperature and pressure at which water, water vapor, and ice coexist in equilibrium. On this scale, the SI unit of temperature is the *kelvin*, defined as the fraction $\dfrac{1}{273.16}$ of the temperature of the triple point of water.

The *absolute temperature scale*, or Kelvin scale, is identical to the ideal-gas scale for temperatures above 1 K.

EQUATIONS AND CONCEPTS

Pressure is force per unit area and has units of N/m^2. The SI unit of pressure is the pascal (Pa).

$$1\,Pa \equiv 1\,N/m^2$$

One atmosphere of pressure is atmospheric pressure at sea level.

$$1\,atm = 1.013 \times 10^5\,Pa$$

The *Celsius temperature* T_C is related to the absolute temperature T (in kelvins) according to Equation 19.1, where 0°C corresponds to 273.15 K.

$$T_C = T - 273.15 \qquad (19.1)$$

The *Fahrenheit temperature* T_F can be converted to degrees Celsius using Equation 19.2. Note that 0°C = 32°F and 100°C = 212°F.

$$T_F = \frac{9}{5} T_C + 32°F \qquad (19.2)$$

If a body has a length L, the *change in its length* ΔL due to a change in temperature is proportional to the change in temperature and the length. The proportionality constant α is called the *average coefficient of linear expansion.*

$$\Delta L = \alpha L\, \Delta T \qquad (19.4)$$

or

$$\alpha = \frac{1}{L} \frac{\Delta L}{\Delta T}$$

If the temperature of a body of volume V changes by an amount ΔT at constant pressure, *the change in its volume* is proportional to ΔT and the original volume. The constant of proportionality β is the *average coefficient of volume expansion.* For an isotropic solid, $\beta \cong 3\alpha$.

$$\Delta V = \beta V \Delta T \qquad (19.6)$$

A sample of any substance contains a number of moles which equals the ratio of the mass of the sample to the molar mass characteristic of that particular substance.

$$n = \frac{m}{M} \qquad (19.9)$$

If n moles of a dilute gas occupy a volume V, the *equation of state* which relates the variables P, V, and T at equilibrium is that of an *ideal gas*,   Equation 19.10, where R is the *universal gas constant*.

$$PV = nRT \qquad (19.10)$$

$$R = 8.314 \text{ J/mol·K} \qquad (19.11)$$

or

$$R = 0.0821 \text{ L·atm/mol·K}$$

The *equation of state of an ideal gas* can also be expressed in the form of Equation 19.12, where N is the total number of gas molecules and k_B is *Boltzman's constant*.

$$PV = Nk_BT \qquad (19.12)$$

$$k_B = 1.38 \times 10^{-23} \text{ J/K} \qquad (19.13)$$

REVIEW CHECKLIST

▷   Understand the concepts of thermal equilibrium and thermal contact between two bodies, and state the zeroth law of thermodynamics.

▷   Discuss some physical properties of substances which change with temperature, and the manner in which these properties are used to construct thermometers.

▷   Describe the operation of the constant-volume gas thermometer and how it is used to define the ideal gas temperature scale.

▷   Convert between the various temperature scales, especially the conversion from degrees Celsius into kelvins, degrees Fahrenheit into kelvins, and degrees Celsius into degrees Fahrenheit.

▷   Provide a qualitative description of the origin of thermal expansion of solids and liquids; define the linear expansion coefficient and volume expansion coefficient for an isotropic solid, and learn how to deal with these coefficients in practical situations involving expansion or contraction.

SOLUTIONS TO SELECTED END-OF-CHAPTER PROBLEMS

1. A constant-volume gas thermometer is calibrated in dry ice (which is carbon dioxide in the solid state and has a temperature of $-80.0°C$) and in boiling ethyl alcohol ($78.0°C$). The two pressures are 0.900 atm and 1.635 atm. (a) What value of absolute zero does the calibration yield? What is the pressure at (b) the freezing point of water and (c) the boiling point of water?

Solution   Since we have a linear graph, the pressure is related to the temperature as $P = A + BT$, where A and B are constants. To find A and B, we use the given data:

$$0.900 \text{ atm} = A + (-80.0°C)B \qquad (1)$$

$$1.635 \text{ atm} = A + (78.0°C)B \qquad (2)$$

Solving (1) and (2) simultaneously, we find

$$A = 1.272 \text{ atm} \qquad \text{and} \qquad B = 4.652 \times 10^{-3} \text{ atm/°C}$$

Therefore, $P = 1.272 \text{ atm} + (4.652 \times 10^{-3} \text{ atm/°C})T$

(a)   At absolute zero, $P = 0 = 1.272 \text{ atm} + (4.652 \times 10^{-3} \text{ atm/°C})T$

which gives $T = -274 °C$ ◊

(b)   At the freezing point of water,

$$P = 1.272 \text{ atm} + 0 = 1.27 \text{ atm}   ◊$$

(c)   and at the boiling point,

$$P = 1.272 \text{ atm} + (4.652 \times 10^{-3} \text{ atm/°C})(100°C) = 1.74 \text{ atm}   ◊$$

---

7. Liquid nitrogen has a boiling point of $-195.81°C$ at atmospheric pressure. Express this temperature in (a) degrees Fahrenheit, and (b) kelvin.

Solution

(a) $T_F = \frac{9}{5}T_C + 32.000°F = \frac{9}{5}(-195.81) + 32.000 = -320°F$ ◊

(b) The ice point in kelvin is 273.15 K, so liquid N_2 boils at

$$273.15 \text{ K} - 195.81 \text{ K} = 77.3 \text{ K} \quad \Diamond$$

Note that a kelvin is the same size change as a degree Celsius.

---

13. A substance is heated from −12°F to 150°F. What is its change in temperature on (a) the Celsius scale and (b) the Kelvin scale?

Solution $\quad T_C = \frac{5}{9}(T_F - 32) \quad$ and $\quad T = T_C + 273.15$

(a) When $T_F = -12°F$,

$$T_C = \frac{5}{9}(-12 - 32)°C = -24.4°C \quad \Diamond$$

When $T_F = 150°F$,

$$T_C = \frac{5}{9}(150 - 32)°C = 65.6°C \quad \Diamond$$

(b) When $T_F = -12°F$,

$$T = (-24.4 + 273.15) \text{ K} = 248 \text{ K} \quad \Diamond$$

When $T_F = 150°F$,

$$T = (65.6 + 273.15) \text{ K} = 338 \text{ K} \quad \Diamond$$

---

17. A copper telephone wire is strung, with essentially no sag, between two poles that are 35.0 m apart. How much longer is the wire on a summer day with $T_C = 35.0°C$ than on a winter day with $T_C = -20.0°C$?

Solution The change in length between cold and hot conditions is

$$\Delta L = \alpha L_0 \Delta T = \left[17 \times 10^{-6} \, (°C)^{-1}\right](35.0 \text{ m})(35.0°C - (-20.0°C))$$

$$\Delta L = 3.27 \times 10^{-2} \text{ m} \quad \text{or} \quad \Delta L = 3.27 \text{ cm} \quad \Diamond$$

---

19. A structural steel I-beam is 15.0 m long when installed at 20.0°C. How much does its length change over the temperature extremes –30.0°C to 50.0°C?

Solution

$$\Delta L = \alpha L_0 \Delta T = \left(11 \times 10^{-6} \ (°C)^{-1}\right)(15.0 \ m)\left(50.0°C - (-30.0°C)\right) = 1.32 \times 10^{-2} \ m \quad \Diamond$$

27. At 20.0°C, an aluminum ring has an inner diameter of 5.000 cm, and a brass rod has a diameter of 5.050 cm. (a) To what temperature must the ring be heated so that it will just slip over the rod? (b) To what temperature must both be heated so that the ring just slips over the rod? Would this latter process work?

Solution

(a) $$L = L_0\left[1 + \alpha(T - T_0)\right]$$

or $$5.050 \ cm = (5.000 \ cm)\left[1 + (24 \times 10^{-6} \ (°C)^{-1})\Delta T\right]$$

From which $\Delta T = 417 \ C°$ and $T = 437°C$ $\Diamond$

(b) We must get $L_{Al} = L_{brass}$ for some ΔT, or

$$L_{0(Al)}(1 + \alpha_{Al} \ \Delta T) = L_{0(brass)}(1 + \alpha_{brass} \ \Delta T)$$

$$(5.000 \ cm)\left[1 + (24 \times 10^{-6}(°C)^{-1})\Delta T\right] = (5.050 \ cm)\left[1 + (19 \times 10^{-6}(°C)^{-1})\Delta T\right]$$

Solving for ΔT gives $\Delta T = 2079 \ C°$, so $T = 2.10 \times 10^3 \ °C$ $\Diamond$

This will not work because aluminum melts at 660°C $\Diamond$

29. A hollow aluminum cylinder 20.0 cm deep has an internal capacity of 2.000 L at 20.0°C. It is completely filled with turpentine, and then warmed to 80.0°C. (a) How much turpentine overflows? (b) If it is then cooled back to 20.0°C, how far below the surface of the cylinder's rim is the turpentine surface?

Solution

(a) When the temperature is increased from 20.0°C to 80.0°C, both the cylinder and the turpentine increase in volume by $\Delta V = \beta V \Delta T$.

The overflow, $V_{\text{over}} = \Delta V_{\text{Turp}} - \Delta V_{\text{Al}}$

$$V_{\text{over}} = (\beta V \, \Delta T)_{\text{Turp}} - (\beta V \, \Delta T)_{\text{Al}} = V \, \Delta T (\beta_{\text{Turp}} - 3\alpha_{\text{Al}})$$

$$V_{\text{over}} = (2.000 \text{ L})(60.0°C)(9.0 - 0.72) \times 10^{-4} (°C)^{-1}$$

$$V_{\text{over}} = 0.0994 \text{ L} \quad \lozenge$$

(b) The whole new volume of turpentine is

$$V' = 2000 \text{ cm}^3 + \left(9 \times 10^{-4}\right)\left(2000 \text{ cm}^3\right)(60 \text{ °C}) = 2108 \text{ cm}^3,$$

and the fraction lost is

$$\frac{99.4 \text{ cm}^3}{2108 \text{ cm}^3} = 4.71 \times 10^{-2}.$$

This also the fraction of the cylinder that will be empty after cooling:

$$\Delta h = \left(4.71 \times 10^{-2}\right)(20.0 \text{ cm}) = 0.943 \text{ cm} \quad \lozenge$$

33. The active element of a certain laser is made of a glass rod 30.0 cm long by 1.5 cm in diameter. If the temperature of the rod increases by 65°C, find the increase in (a) its length, (b) its diameter, and (c) its volume. (Take $\alpha = 9.0 \times 10^{-6}(°C)^{-1}$)

Solution

(a) $\Delta L = \alpha L_0 \, \Delta T = \left[9.0 \times 10^{-6}(°C)^{-1}\right](0.300 \text{ m})(65 \text{ C°}) = 1.76 \times 10^{-4} \text{ m}$ ◊

(b) The diameter is a linear dimension, so the same equation applies:

$$\Delta D = \alpha D_0 \, \Delta T = \left[9.0 \times 10^{-6}(°C)^{-1}\right](0.015 \text{ m})(65 \text{ C°}) = 8.78 \times 10^{-6} \text{ m} \quad ◊$$

(c) The original volume $V = \pi r^2 L = \dfrac{\pi}{4}(0.015 \text{ m})^2(0.300 \text{ m}) = 5.3 \times 10^{-5} \text{ m}^3$

Using the volumetric coefficient of expansion, β,

$$\Delta V = \beta V \, \Delta T \cong 3\alpha V \, \Delta T$$

$$\Delta V \cong 3\left(9.0 \times 10^{-6}(°C)^{-1}\right)\left(5.3 \times 10^{-5} \text{ m}^3\right)(65°C) = 93.0 \times 10^{-9} \text{ m}^3 \quad ◊$$

Related Calculation   The above calculation ignores ΔL^2 and ΔL^3 terms. Calculate the change in volume exactly, and compare your answer with the approximate solution above.

The volume will increase by a factor of

$$\frac{\Delta V}{V} = \left(1 + \frac{\Delta D}{D}\right)^2\left(1 + \frac{\Delta L}{L}\right) - 1 = \left(1 + \frac{8.78 \times 10^{-6} \text{ m}}{0.015 \text{ m}}\right)^2\left(1 + \frac{1.76 \times 10^{-4} \text{ m}}{0.300 \text{ m}}\right) - 1 = 1.76 \times 10^{-3}$$

So, $\Delta V = \dfrac{\Delta V}{V} V = \left(1.76 \times 10^{-3}\right)\left(5.30 \times 10^{-5} \text{ m}^3\right) = 93.0 \times 10^{-9} \text{ m}^3 = 93.0 \text{ mm}^3$ ◊

The answer is virtually identical, and the approximation $\beta \cong 3\alpha$ is a good one.

36. An auditorium has dimensions 10.0 m × 20.0 m × 30.0 m. How many molecules of air are needed to fill the auditorium at 20.0°C and 101 kPa pressure?

Solution

$PV = nRT$

$$n = \frac{PV}{RT} = \frac{(1.01 \times 10^5 \ \text{N} / \text{m}^2)[(10.0 \ \text{m})(20.0 \ \text{m})(30.0 \ \text{m})]}{(8.314 \ \text{J} / \text{mol} \cdot \text{K})(293 \ \text{K})}$$

$n = 2.49 \times 10^5$ mol

$$N = n(N_A) = (2.49 \times 10^5 \ \text{mol})\left(6.022 \times 10^{23} \ \frac{\text{molecules}}{\text{mol}}\right)$$

$N = 1.50 \times 10^{29}$ molecules   ◊

---

37. A full tank of oxygen (O_2) contains 12.0 kg of oxygen under a gauge pressure of 40.0 atm. Determine the mass of oxygen that has been withdrawn from the tank when the pressure reading is 25.0 atm. Assume the temperature of the tank remains constant.

Solution

$PV = nRT$ and at constant volume and temperature,

$$\frac{P}{n} = \text{constant} \quad \text{or} \quad \frac{P_1}{n_1} = \frac{P_2}{n_2}; \quad \text{and} \quad n_2 = \left(\frac{P_2}{P_1}\right)n_1$$

However, n is proportional to m, so

$$m_2 = \left(\frac{P_2}{P_1}\right)m_1 = \frac{26.0 \ \text{atm}}{41.0 \ \text{atm}}(12.0 \ \text{kg}) = 7.61 \ \text{kg}$$

The mass removed is

$$\Delta m = (12.0 - 7.61) \ \text{kg} = 4.39 \ \text{kg} \quad ◊$$

---

39. The mass of a hot-air balloon and its cargo (not including the air inside) is 200 kg. The air outside is at 10.0°C and 101 kPa. The volume of the balloon is 400 m³. To what temperature must the air in the balloon be heated before the balloon will lift off? (Air density at 10.0°C is 1.25 kg/m³.)

Solution   Consider 400 m³ of air at 10°C. We call this the air *displaced* by the balloon, with a mass m_d.

$$m_d = \rho V = \left(1.25 \ kg/m^3\right)\left(400 \ m^3\right) = 500 \ kg$$

The buoyancy force equals w_d, the weight of the displaced air:

$$w_d = m_d g = \left(500 \ kg\right)\left(9.8 \ m/s^2\right) = 4900 \ N$$

The weight of the balloon and cargo is

$$w_b = m_b g = \left(200 \ kg\right)\left(9.80 \ m/s^2\right) = 1960 \ N$$

At liftoff, the buoyancy force must equal the weight of the balloon and the air inside the balloon, w_a:

$$w_d = w_b + m_a g \qquad \text{or} \qquad m_a = \frac{w_d - w_b}{g} = \frac{4900 \ N - 1960 \ N}{9.8 \ m/s^2} = 300 \ kg$$

Dry air is approximately 20% O_2, and 80% N_2. So its molar mass is

$$0.80 \ (28 \ g/mol) + 0.20 \ (32.0 \ g/mol) = 29 \ g/mol,$$

or

$$n = \frac{m}{M} = \left(300 \ kg\right)\left(\frac{10^3 \ g/kg}{29 \ g/mol}\right) = 1.03 \times 10^4 \ mol$$

The pressure of this air is the ambient pressure; from $PV = nRT$ we have

$$T = \frac{PV}{nR} = \frac{\left(1.013 \times 10^5 \ N/m^2\right)\left(400 \ m^3\right)}{\left(1.03 \times 10^4 \ mol\right)\left(8.314 \ J/(mol \cdot K)\right)} = 471 \ K \qquad \Diamond$$

47. An automobile tire is inflated using air originally at 10°C and normal atmospheric pressure. During the process, the air is compressed to 28% of its original volume and the temperature is increased to 40°C. (a) What is the tire pressure? (b) After the car is driven at high speed, the tire air temperature rises to 85°C and the interior volume of the tire increases by 2%. What is the new tire pressure in pascals (absolute)?

Solution

(a) Taking $PV = nRT$ in the initial (i) and final (f) states, and dividing, we have

$$P_iV_i = nRT_i \qquad \text{and} \qquad P_fV_f = nRT_f \qquad \text{yield} \qquad \frac{P_fV_f}{P_iV_i} = \frac{T_f}{T_i}$$

So $\quad P_f = P_i\dfrac{V_iT_f}{V_fT_i} = \left(1.013\times10^5 \text{ Pa}\right)\left(\dfrac{V_i}{0.28V_i}\right)\left(\dfrac{273 \text{ K}+40 \text{ K}}{273 \text{ K}+10 \text{ K}}\right) = 4.00\times10^5 \text{ Pa} \qquad \Diamond$

(b) Introducing the hot (h) state, $\qquad \dfrac{P_hV_h}{P_fV_f} = \dfrac{T_h}{T_f}$

So $\quad P_h = P_f\left(\dfrac{V_f}{V_h}\right)\left(\dfrac{T_h}{T_f}\right) = \left(4.0\times10^5 \text{ Pa}\right)\left(\dfrac{V_f(358 \text{ K})}{1.02V_f(313 \text{ K})}\right) = 4.49\times10^5 \text{ Pa} \qquad \Diamond$

---

53. The rectangular plate shown in Figure P19.53 has an area A equal to lw. If the temperature increases by ΔT, show that the increase in area is $\Delta A = 2\alpha A\Delta T$, where α is the average coefficient of linear expansion. What approximation does this expression assume? (*Hint:* Note that each dimension increases according to $\Delta L = \alpha L\Delta T$.)

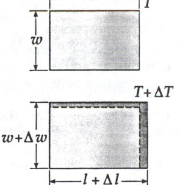

Figure P19.53

Solution   From the diagram in Figure 19.53, we see that the *change* in area is

$$\Delta A = l\,\Delta w + w\,\Delta l + \Delta w\,\Delta l$$

Since Δl and Δw are each small quantities, the product $\Delta w\Delta l$ will be very small.

Therefore, we assume $\qquad \Delta w\,\Delta l \approx 0.$

Since $\Delta w = w\alpha\,\Delta T$ and $\Delta l = l\alpha\,\Delta T$, we then have $\qquad \Delta A = lw\alpha\,\Delta T + wl\alpha\,\Delta T$

Finally, since $A = lw$, we find $\qquad \Delta A = 2\alpha A\,\Delta T \quad \lozenge$

---

55. A mercury thermometer is constructed as in Figure P19.55. The capillary tube has a diameter of 0.0040 cm, and the bulb has a diameter of 0.25 cm. Neglecting the expansion of the glass, find the change in height of the mercury column for a temperature change of 30°C.

Solution Neglecting the expansion of the glass, the volume of liquid in the capillary will be $\Delta V = A(\Delta h)$ where A is the cross-sectional area of the capillary and V is the volume of the bulb.

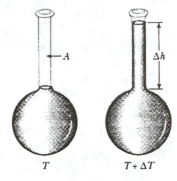

$$\Delta V = V\beta\,\Delta T$$

Figure P19.55

$$\Delta h = \left(\frac{V}{A}\right)\beta\,\Delta T = \left[\frac{\frac{4}{3}\pi R_{bulb}^3}{\pi R_{cap}^2}\right]\beta\,\Delta T$$

$$\Delta h = \frac{4}{3}\frac{(0.125\ \text{cm})^3}{(0.0020\ \text{cm})^2}\left(1.82\times10^{-4}\ (\text{°C})^{-1}\right)(30\ \text{°C})$$

$$\Delta h = 3.55\ \text{cm} \quad \lozenge$$

55A. A mercury thermometer is constructed as in Figure P19.55. The capillary tube has a diameter d_1, and the bulb has a diameter d_2. Neglecting the expansion of the glass, find the change in height of the mercury column for a temperature change ΔT.

Solution Neglecting the expansion of the glass, the volume of the liquid in the capillary will be equal to the change in volume of the liquid in the bulb.

$\Delta V = A(\Delta h)$ where A is the cross-sectional area of the capillary. Also, $\Delta V = V\beta(\Delta T)$, so

$$\Delta h = \frac{\Delta V}{A} = \left(\frac{V}{A}\right)\beta\,\Delta T = \frac{\frac{4}{3}\pi\left(\frac{d_2}{2}\right)^3\beta\,\Delta T}{\pi(d_1/2)^2} = \frac{2d_2^3\beta\,\Delta T}{3d_1} \quad \lozenge$$

---

56. A liquid has a density ρ. (a) Show that the fractional change in density for a change in temperature ΔT is $\Delta \rho / \rho = -\beta \Delta T$. What does the negative sign signify? (b) Fresh water has a maximum density of 1.000 g/cm³ at 4.0°C. At 10.0°C, its density is 0.9997 g/cm³. What is β for water over this temperature interval?

Solution

(a) $\quad \rho = \dfrac{m}{V} \quad$ and $\quad d\rho = -\dfrac{m}{V^2} dV$

For very small changes in V and ρ, this can be expressed as

$$\Delta \rho = -\frac{m}{V} \frac{\Delta V}{V} = -\rho \beta \, \Delta T \quad \text{where} \quad \rho = \frac{m}{V} \quad \text{and} \quad \frac{\Delta V}{V} = \beta \, \Delta T$$

The negative sign means that any increase in temperature causes the density to decrease and vice versa. ◊

(b) For water we have

$$\beta = \left| \frac{\Delta \rho}{\rho \, \Delta T} \right| = \left| \frac{(1.0000 \text{ g / cm}^3 - 0.9997 \text{ g / cm}^3)}{(1.0000 \text{ g / cm}^3)(10.0 - 4.0) \text{ C}^\circ} \right| = 5.00 \times 10^{-5} \ (^\circ\text{C})^{-1} \quad ◊$$

---

61. Starting with Equation 19.12, show that the total pressure P in a container filled with a mixture of several ideal gases is $P = P_1 + P_2 + P_3 + \dots$, where $P_1, P_2, \dots$, are the pressures that each gas would exert if it alone filled the container (these individual pressures are called the *partial pressures* of the respective gases). This is known as *Dalton's law of partial pressures.*

Solution   For each gas alone, $\quad P_1 = \dfrac{N_1 k_B T}{V_1}, \quad P_2 = \dfrac{N_2 k_B T}{V_2}, \quad P_3 = \dfrac{N_3 k_B T}{V_3}, \quad$ etc.

For the gases combined, $\quad N_1 + N_2 + N_3 + \dots = N = \dfrac{PV}{k_B T}$

Therefore, $\quad \dfrac{P_1 V_1}{k_B T} + \dfrac{P_2 V_2}{k_B T} + \dfrac{P_3 V_3}{k_B T} + \dots = \dfrac{PV}{k_B T}$

But $\quad V_1 = V_2 = V_3 = \dots = V, \quad$ so $\quad P_1 + P_2 + P_3 + \dots = P \quad ◊$

---

66. A vertical cylinder of cross-sectional area A is fitted with a tight-fitting, frictionless piston of mass m (Fig. P19.66). (a) If there are n mol of an ideal gas in the cylinder at a temperature T, determine the height h at which the piston is in equilibrium under its own weight. (b) What is the value for h if $n = 0.20$ mol, $T = 400$ K, $A = 0.0080$ m^2 and $m = 20.0$ kg?

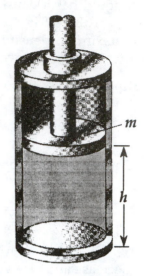

Solution

(a) We suppose that air at pressure P_{atm} is above the piston. For the piston's equilibrium, $\Sigma F_y = ma_y$ yields

$$-P_{atm}A - mg + PA = 0$$

where P is the pressure exerted by the gas contained.

Figure P19.66

Noting that $V = Ah$, and that n, T, m, g, A, and P_{atm} are given,

$PV = nRT$ becomes $\qquad P = \dfrac{nRT}{Ah} \qquad$ and $\qquad P_{atm}A - mg + \dfrac{nRT}{Ah}A = 0$

$$h = \frac{nRT}{P_{atm}A + mg} \qquad \lozenge$$

(b) $\quad h = \dfrac{(0.20 \text{ mol})(8.314 \text{ J} / \text{mol} \cdot \text{K})(400 \text{ K})}{\left(1.013 \times 10^5 \text{ N} / \text{m}^2\right)\left(0.0080 \text{ m}^2\right) + (20 \text{ kg})\left(9.8 \text{ m} / \text{s}^2\right)}$

$h = \dfrac{665 \text{ N} \cdot \text{m}}{810 \text{ N} + 196 \text{ N}} = 0.661 \text{ m} \qquad \lozenge$

---

67. An air bubble originating from a deep sea diver has a radius of 5.0 mm at some depth h. When the bubble reaches the surface of the water, it has a radius of 7.0 mm. Assuming the temperature of the air in the bubble remains constant, determine (a) the depth h of the diver and (b) the absolute pressure at this depth.

Solution

(a) The pressure at any depth is $P = P_o + \rho gh$

where $P_o = 1.013 \times 10^5 \text{ N/m}^2$ and $\rho = 10^3 \text{ kg/m}^3$

Also, for an ideal gas that remains at constant temperature, $PV = P_o V_a$

This gives $P = \left(\dfrac{V_o}{V}\right)P_o = \left(\dfrac{r_o}{r}\right)^3 P_o = \left(\dfrac{7}{5}\right)^3 P_o = 2.74 P_o$

Substituting this expression for P into the first equation and using the given numerical values, we have

$$h = \frac{P - P_o}{\rho g} = \frac{P_o}{\rho g}[2.74 - 1] = 18.0 \text{ m} \quad \lozenge$$

(b) $P = P_o + \rho gh = 1.013 \times 10^5 \text{ N/m}^2 + \left(10^3 \text{ kg/m}^3\right)\left(9.80 \text{ m/s}^2\right)(18 \text{ m})$

$P = 2.78 \times 10^5 \text{ Pa} = 278 \text{ kPa} \quad \lozenge$

---

73. A steel guitar string with a diameter of 1.00 mm is stretched between supports 80.0 cm apart. The temperature is 0.0°C. (a) Find the mass per unit length of this string. (Use the value $7.86 \times 10^3 \text{ kg/m}^3$ for the density.) (b) The fundamental frequency of transverse oscillations of the string is 200 Hz. What is the tension in the string? (c) If the temperature is raised to 30.0°C, find the resulting values of the tension and the fundamental frequency. [Assume that both the Young's modulus (Table 12.1) and the average coefficient of expansion (Table 19.2) have constant values between 0.0°C and 30.0°C.]

Solution

(a) $\mu = \rho\left(\dfrac{\pi d^2}{4}\right) = \tfrac{1}{4}\pi\left(1.00 \times 10^{-3} \text{ m}\right)^2\left(7.86 \times 10^{-3} \text{ kg/m}^3\right) = 6.17 \times 10^{-3} \text{ kg/m} \quad \lozenge$

(b) Since $f_1 = \dfrac{v}{2L}$, $\qquad v = \sqrt{\dfrac{F}{\mu}}$ $\qquad$ and $\qquad f_1 = \dfrac{1}{2L}\sqrt{\dfrac{F}{\mu}}$

$$F = \mu(2Lf_1)^2 = \left(6.173\times10^{-3}\text{ kg/m}\right)\left(2\times0.800\text{ m}\times200\text{ s}^{-1}\right)^2 = 632\text{ N} \quad \Diamond$$

(c) $L_{0°C} = L_{natural}\left(1 + \dfrac{F}{AY}\right)$

$$A = \left(\dfrac{\pi}{4}\right)\left(1\times10^{-3}\text{ m}\right)^2 = 7.854\times10^{-7}\text{ m}^2 \qquad \text{and} \qquad Y = 20\times10^{10}\text{ N/m}^2$$

Therefore, $\qquad \dfrac{F}{AY} = \dfrac{632\text{ N}}{\left(7.854\times10^{-7}\text{ m}^2\right)\left(20\times10^{10}\text{ N/m}^2\right)} = 4.024\times10^{-3}$

$$L_{0°C} = \dfrac{0.800\text{ m}}{1 + 4.024\times10^{-3}} = 0.7968\text{ m}$$

The unstressed length at $\quad 30.0°C = (0.7968\text{ m})[1 + 30.0°C(11\times10^{-6}\ (°C)^{-1})] = 0.7971\text{ m}$

Since $\quad 0.800\text{ m} = (0.7971\text{ m})\left[1 + \dfrac{F'}{A'Y}\right]$,

$$\dfrac{F'}{A'Y} = \dfrac{0.800}{0.7971} - 1 = 3.689\times10^{-3}$$

and

$$F' = A'Y\left(3.689\times10^{-3}\right) = \left(7.854\times10^{-7}\text{ m}^2\right)\left(20\times10^{10}\text{ N/m}^2\right)\left(3.689\times10^{-3}\right)(1 + \alpha\,\Delta T)^2$$

$$F' = (579.5\text{ N})\left(1 + 3.3\times10^{-4}\right)^2 = 580\text{ N} \quad \Diamond$$

Also, $\qquad \dfrac{f_1'}{f_1} = \sqrt{\dfrac{F'}{F}} \qquad$ so $\qquad f_1' = (200\text{ Hz})\sqrt{\dfrac{580}{632}} = 192\text{ Hz} \quad \Diamond$

Chapter 20

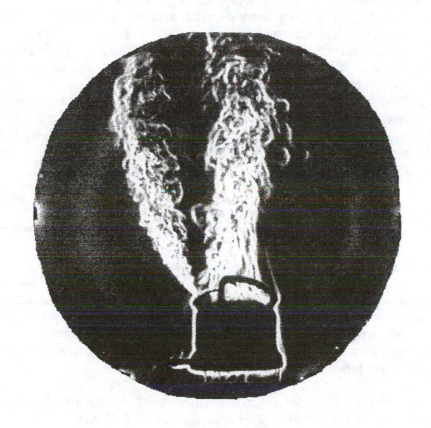

Heat and the First Law
of Thermodynamics

HEAT AND THE FIRST LAW OF THERMODYNAMICS

INTRODUCTION

This chapter focuses on the concept of heat, the first law of thermodynamics, processes by which thermal energy is transferred, and some important applications. The first law of thermodynamics is merely the law of conservation of energy. It tells us only that an increase in one form of energy must be accompanied by a decrease in some other form of energy. The first law places no restrictions on the types of energy conversions that can occur. Furthermore, it makes no distinction between heat and work. According to the first law, a system's internal energy can be increased either by transfer of thermal energy to the system or by work done on the system. An important difference between thermal energy and mechanical energy is not evident from the first law; it is possible to convert work completely to thermal energy but impossible to convert thermal energy completely to mechanical energy.

NOTES FROM SELECTED CHAPTER SECTIONS

20.1 Heat and Thermal Energy

When two systems at different temperatures are in contact with each other, energy will transfer between them until they reach the same temperature (that is, when they are in thermal equilibrium with each other). This energy is called heat, or thermal energy, and the term "heat flow" refers to an energy transfer as a consequence of a temperature difference.

The unit of heat is the *calorie* (cal), defined as the amount of heat necessary to increase the temperature of 1 g of water from 14.5°C to 15.5°C. The *mechanical equivalent of heat*, first measured by Joule, is given by 1 cal = 4.186 J.

20.2 Heat Capacity and Specific Heat

The *specific heat*, c, of any substance is defined as the amount of heat required to increase the temperature of 1 kg of that substance by one Celsius degree. Its units are J/kg·°C.

20.3 Latent Heat

The *heat of fusion* is a parameter used to characterize a solid-to-liquid phase change; the *heat of vaporization* characterizes the liquid-to-gas phase change.

20.4 Work and Heat in Thermodynamic Processes

The work done in the expansion from the initial state to the final state is the area under the curve in a *PV* diagram. See Figure 20.1.

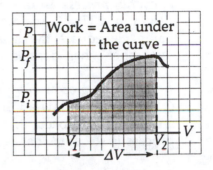

If the gas is compressed, $V_f < V_i$, and the work is negative. That is, work is done *on* the gas. If the gas expands, $V_f > V_i$, the work is positive, and the gas does work on the piston. If the gas expands at *constant pressure*, called an *isobaric process*, then $W = P(V_f - V_i)$.

The work done by a system depends on the process by which the system goes from the initial to the final state. In other words, the work done depends on the initial, final, and intermediate states of the system.

20.5 The First Law of Thermodynamics

In the first law of thermodynamics, $\Delta U = Q - W$, Q is the heat added to the system and W is the work done by the system. Note that by convention, Q is *positive* when heat enters the system and *negative* when heat is removed from the system. Likewise, W can be positive or negative as mentioned earlier. The initial and final states must be *equilibrium* states; however, the intermediate states are, in general, not equilibrium states since the thermodynamic coordinates undergo finite changes during the thermodynamic process. For an *infinitesimal change of the system*, we can express the *first law of thermodynamics* in the form $dU = dQ - dW$. It is important to note that dQ and dW *are not exact differentials*, since both Q and W are not functions of the system's coordinates. That is, both Q and W depend on the *path* taken between the initial and final equilibrium states, during which time the system interacts with its environment. On the other hand, dU is an *exact differential* and the internal energy U is a *state variable*. The function U is analogous to the potential energy function used in mechanics when dealing with conservative forces.

20.6 Some Applications of the First Law of Thermodynamics

An *isolated system* is one which does not interact with its surroundings. In such a system, $Q = W = 0$, so it follows from the first law that $\Delta U = 0$. That is, the internal energy of an isolated system cannot change.

A *cyclic process* is one that originates and ends up at the same state. In this situation, $\Delta U = 0$, so from the first law we see that $Q = W$. That is, the work done per cycle equals the heat added to the system per cycle. This is important to remember when dealing with heat engines in the next chapter.

An *adiabatic process* is a process in which no heat enters or leaves the system; that is, $Q = 0$. The first law applied to this process gives $\Delta U = -W$. A system may undergo an adiabatic process if it is thermally insulated from its surroundings, or if the process is so rapid that negligible heat has time to flow.

An *isobaric process* is a process which occurs at constant pressure. For such a process, the heat transferred and the work done are nonzero.

An *isovolumetric process* is one which occurs at constant volume. By definition, $W = 0$ for such a process (since $dV = 0$), so from the first law it follows that $\Delta U = Q$. That is, all of the heat added to the system kept at constant volume goes into increasing the internal energy of the system.

A process that occurs at constant temperature is called an *isothermal process*, and a plot of P versus V at constant temperature for an ideal gas yields a hyperbolic curve called an *isotherm*. The internal energy of an ideal gas is a function of temperature only. Hence, in an isothermal process of an ideal gas, $\Delta U = 0$.

20.7 Heat Transfer

There are three basic processes of heat transfer. These are (1) conduction, (2) convection, and (3) radiation.

Conduction is a heat transfer process which occurs when there is a *temperature gradient* across the body. That is, conduction of heat occurs only when the body's temperature is *not* uniform. For example, if you heat a metal rod at one end with a flame, heat will flow from the hot end to the colder end. If the heat flow is along x (that is, along the rod), and we define the *temperature gradient* as dT/dx, then a quantity of heat dQ will flow in a time dt along the rod. The rate of flow of heat along the rod, sometimes called the *heat current*, is proportional to the cross-sectional area of the rod, the temperature gradient, and k, the thermal conductivity of the material of which the rod is made.

When heat transfer occurs as the result of the motion of material, such as the mixing of hot and cold fluids, the process is referred to as *convection*. Convection heating is used in conventional hot-air and hot-water heating systems. Convection currents produce changes in weather conditions when warm and cold air masses mix in the atmosphere.

Heat transfer by *radiation* is the result of the continuous emission of electromagnetic radiation by all bodies.

EQUATIONS AND CONCEPTS

The *specific heat c* of a substance is a measure of the quantity of heat energy required to change the temperature of the substance by 1°C.

$$c \equiv \frac{Q}{m \, \Delta T} \tag{20.3}$$

The *heat energy Q* transferred between a system of mass m and its surroundings for a temperature change ΔT varies with the substance.

$$Q = mc \, \Delta T \tag{20.4}$$

A substance may undergo a phase change when heat is transferred between the substance and its surroundings. The heat Q required to change the phase of a mass m is called the *latent heat, L*. The phase change process occurs at constant temperature. The value of L depends on the nature of the phase change and the properties of the substance.

$$Q = mL \tag{20.6}$$

The *work done by a gas* which undergoes an expansion or compression from initial volume V_i to final volume V_f depends on the path taken between the initial and final states. The pressure is generally not constant, so you must exercise care in evaluating W from this equation. In general, the work done equals the area under the PV curve bounded by V_i and V_f, and the function P, as in Figure 20.1.

$$W = \int_{V_i}^{V_f} P\,dV \qquad (20.8)$$

The *first law of thermodynamics* is a generalization of the law of conservation of energy that includes possible changes in internal energy. It states that the *change* in internal energy of a system, ΔU, equals the quantity $Q - W$, where Q is the heat added to the system. The quantity $(Q - W)$ is independent of the path taken between the initial and final states.

$$\Delta U = Q - W \qquad (20.9)$$

In an *adiabatic process*, $Q = 0$, and the change in internal energy equals the negative of the work done *by* the gas.

$$\Delta U = -W \qquad (20.10)$$

In a constant volume (isovolumetric) process, the work done is zero and all heat added to the system goes into increasing the internal energy.

$$\Delta U = Q \qquad (20.11)$$

An *isothermal process* is one which occurs at constant temperature. During such a process, the change in internal energy results from both heat transfer and work done.

$$W = nRT \ln\left(\frac{V_f}{V_i}\right) \qquad (20.12)$$

SUGGESTIONS, SKILLS, AND STRATEGIES

Many applications of the first law of thermodynamics deal with the work done by (or on) a system which undergoes a change in state. For example, as a gas is taken from a state whose initial pressure and volume are P_i, V_i, and whose final pressure and volume are P_f, V_f, the work can be calculated if the process can be drawn on a PV diagram as in Figure 20.2. The work done during the expansion is given by the integral expression

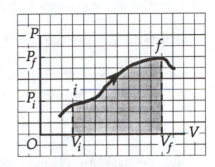

Figure 20.2

$$W = \int_{V_i}^{V_f} P\, dV$$

which numerically represents the *area* under the PV curve (the shaded region) shown in Figure 20.2. It is important to recognize that the work *depends on the path taken as the gas goes from i to f.* That is, W depends on the specific manner in which the pressure P changes during the process.

Problem-Solving Strategy: Calorimetry Problems

If you are having difficulty with calorimetry problems, consider the following factors:

- Be sure your units are consistent throughout. For instance, if you are using specific heats in cal/g·°C, be sure that masses are in grams and temperatures are Celsius throughout.

- Heat loss and gains are found by using $Q = mc\,\Delta T$ only for those intervals in which no phase changes are occurring. The equations $Q = mL_f$ and $Q = mL_v$ are to be used only when phase changes *are* taking place.

- Often sign errors occur in heat loss = heat gain equations. One way to check your equation is to examine the signs of all ΔT's that appear in it.

REVIEW CHECKLIST

▷   Understand the concepts of heat, internal energy, and thermodynamic processes.

▷   Define and discuss the calorie, heat capacity, specific heat, and latent heat.

▷   Understand how work is defined when a system undergoes a change in state, and the fact that work (like heat) depends on the path taken by the system. You should also know how to sketch processes on a PV diagram, and calculate work using these diagrams.

▷   State the first law of thermodynamics ($\Delta U = Q - W$), and explain the meaning of the three forms of energy contained in this statement.

▷   Discuss the implications of the first law of thermodynamics as applied to (a) an isolated system, (b) a cyclic process, (c) an adiabatic process, and (d) an isothermal process.

SOLUTIONS TO SELECTED END-OF-CHAPTER PROBLEMS

3.   Water at the top of Niagara Falls has a temperature of 10°C. If it falls through a distance of 50 m and all of its potential energy goes into heating the water, calculate the temperature of the water at the bottom of the falls.

Solution

Since potential energy is converted into thermal energy,

$$\Delta U = \Delta Q = mc\,\Delta T \qquad \text{or} \qquad mgy = mc\,\Delta T$$

$$\Delta T = \frac{gy}{c} = \frac{(9.80\ \text{m}/\text{s}^2)(50\ \text{m})}{4.186 \times 10^3\ \text{J}/\text{kg·°C}} = 0.117\ \text{°C}$$

$$T_g = T_i + \Delta T = 10\ \text{°C} + 0.117\ \text{°C} = 10.12\ \text{°C} \quad \Diamond$$

5. The temperature of a silver bar rises by 10.0°C when it absorbs 1.23 kJ of heat. The mass of the bar is 525 g. Determine the specific heat of silver.

Solution
$$\Delta Q = mc_{silver}\Delta T \qquad \text{or} \qquad c_{silver} = \frac{Q}{m\,\Delta T}$$

$$c_{silver} = \frac{1.23\times10^3 \text{ J}}{(0.525 \text{ kg})(10.0°C)} = 234 \text{ J/kg}\cdot°C \quad \lozenge$$

7. What is the final equilibrium temperature when 10 g of milk at 10°C is added to 160 g of coffee at 90°C? (Assume the heat capacities of the two liquids are the same as that of water, and neglect the heat capacity of the container.)

Solution
$$\Delta Q_{milk} = -\Delta Q_{coffee}$$

$$(mc\,\Delta T)_{milk} = -mc(\Delta T)_{coffee}$$

$$(10 \text{ g})(1 \text{ cal/g}\cdot°C)(10\,°C - T_f) = -(160 \text{ g})(1 \text{ cal/g}\cdot°C)(90\,°C - T_f)$$

$$T_f = \frac{1.45\times10^4 \text{ cal}}{170 \text{ cal/°C}} = 85.3°C \quad \lozenge$$

9. A 1.5-kg iron horseshoe initially at 600°C is dropped into a bucket containing 20 kg of water at 25°C. What is the final temperature? (Neglect the heat capacity of the container.)

Solution The iron loses as much heat as the water gains. Therefore,

$$\Delta Q_{iron} = -\Delta Q_{water} \qquad \text{or} \qquad (mc\,\Delta T)_{iron} = -(mc\,\Delta T)_{water}$$

$$(1.5 \text{ kg})(448 \text{ J/kg}\cdot°C)(T-600\,°C) = -(20 \text{ kg})(4186 \text{ J/kg}\cdot°C)(T-25\,°C)$$

$$T = 29.6\,°C \quad \lozenge$$

17. How much heat is required to vaporize a 1.0-g ice cube initially at 0°C? The latent heat of fusion of ice is 80 cal/g and the latent heat of vaporization of water is 540 cal/g.

Solution
$$\Delta Q = mL_f + mc\,\Delta T + mL_{vap} = m(L_f + c\,\Delta T + L_{vap})$$

$$\Delta Q = (1.0 \text{ g})\left[80\frac{\text{cal}}{\text{g}} + 100(1 \text{ cal/g} \cdot {}^\circ\text{C})(100{}^\circ\text{C}) + 540\frac{\text{cal}}{\text{g}}\right]$$

$$\Delta Q = 720 \text{ cal} = 0.720 \text{ kcal} \quad \lozenge$$

---

21. In an insulated vessel, 250 g of ice at 0°C is added to 600 g of water at 18°C. (a) What is the final temperature of the system? (b) How much ice remains when the system reaches equilibrium?

Solution

(a) When 250 g of ice is melted,

$$\Delta Q_f = mL_f = (0.250 \text{ kg})(3.33 \times 10^5 \text{ J/kg}) = 83.3 \text{ kJ}$$

The heat energy released when 600 g of water cools from 18°C to 0°C is

$$\Delta Q - mc\,\Delta T = (0.600 \text{ kg})(4186 \text{ J/kg} \cdot {}^\circ\text{C})(18 \text{ }{}^\circ\text{C}) = 45.2 \text{ kJ}$$

Since the heat required to melt 250 g of ice at 0°C *exceeds* the heat released by cooling 600 g of water from 18°C to 0°C, the final temperature of the system (water + ice) must be 0°C. $\lozenge$

(b) The thermal energy released by the water (45.2 kJ) will melt m grams of ice, where

$$Q = mL_f \qquad \text{or} \qquad m = \frac{Q}{L_f}$$

Solving,
$$m = \frac{45.2 \times 10^3 \text{ J}}{3.33 \times 10^5 \text{ J/kg}} = 0.136 \text{ kg}$$

Therefore, the ice remaining is equal to $m' = 0.250 \text{ kg} - 0.136 \text{ kg} = 0.114 \text{ kg}$ $\lozenge$

---

25. A 3.0-g copper penny at 25°C drops 50 m to the ground. (a) If 60% of its initial potential energy goes into increasing the internal energy, determine its final temperature. (b) Does the result depend on the mass of the penny? Explain.

Solution

(a) Let f equal the fraction of potential energy converted into internal energy.

$$f(\Delta U) = \Delta Q \quad \text{so} \quad f(mgy) = mc\,\Delta T$$

Therefore,
$$\Delta T = \frac{fgh}{c} = \frac{(0.60)(9.80 \text{ m/s}^2)(50 \text{ m})}{387 \text{ J/kg} \cdot \text{°C}} = 0.760\text{°C}$$

$$T_f = 25\text{°C} + 0.76\text{°C} = 25.8\text{°C} \quad \lozenge$$

(b) _No_: Both the change in potential energy and the heat absorbed are proportional to the mass; hence, the mass cancels in the energy relation. $\lozenge$

---

27. A 3.0-g lead bullet is traveling at 240 m/s when it embeds in a block of ice at 0°C. If all the heat generated goes into melting ice, what quantity of ice is melted? (The latent heat of fusion for ice is 80 kcal/kg, and the specific heat of lead is 0.030 kcal/kg·°C.)

Solution We suppose that the bullet is originally at 0 °C. Not all the ice melts, so the final temperature must also be 0 °C.

$$\Delta K = \Delta Q \quad \text{or} \quad \tfrac{1}{2}mv^2 = m_{ice}L_f$$

$$m_{ice} = \frac{mv^2}{2L_f} = \frac{(0.0030 \text{ kg})(240 \text{ m}/\text{s})^2}{3.33 \times 10^5 \text{ J}/\text{kg}}$$

$$m_{ice} = 2.59 \times 10^{-4} \text{ kg} = 0.259 \text{ g} \quad \lozenge$$

---

31. An ideal gas is enclosed in a cylinder that has a movable piston on top. The piston has a mass of 8000 g and an area of 5.0 cm² and is free to slide up and down, keeping the pressure of the gas constant. How much work is done as the temperature of 0.20 mol of the gas is raised from 20°C to 300°C?

Solution

$$W\big|_{\text{constant } P} = P\,\Delta V = P\left(\frac{nR}{P}\right)(T_h - T_c)$$

Therefore, $\qquad W = nR\,\Delta T = (0.20 \text{ mol})(8.314 \text{ J}/\text{mol}\cdot\text{K})(280 \text{ K}) = 466 \text{ J} \quad \Diamond$

---

33. A sample of ideal gas is expanded to twice its original volume of 1.0 m³ in a quasi-static process for which $P = \alpha V^2$, with $\alpha = 5.0$ atm/m⁶, as shown in Figure P20.33. How much work was done by the expanding gas?

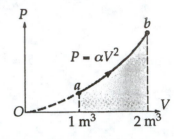

Solution $\qquad W_{ab} = \int_a^b P\,dV$

Figure P20.33

The work done by the gas is the area under the curve $P = \alpha V^2$, from V_a to V_b.

$$W_{ab} = \int_a^b \alpha V^2\,dV = \tfrac{1}{3}\alpha\left(V_b^3 - V_a^3\right)$$

$$V_b = 2V_a = 2(1.0 \text{ m}^3) = 2.0 \text{ m}^3$$

$$W_{ab} = \tfrac{1}{3}\left(5.0\,\frac{\text{atm}}{\text{m}^6}\times 1.013\times 10^5\,\frac{\text{Pa}}{\text{atm}}\right)\left[(2.0 \text{ m}^3)^3 - (1.0 \text{ m}^3)^3\right] = 1.18\times 10^6 \text{ J} \quad \Diamond$$

---

38. A thermodynamic system undergoes a process in which its internal energy decreases by 500 J. If at the same time, 220 J of work is done on the system, find the thermal energy transferred to or from it.

Solution $\quad \Delta U = Q - W$, where W is negative, because work is done *on* the system:

$$Q = \Delta U + W = -500 \text{ J} - 200 \text{ J} = -720 \text{ J}$$

+ 720 J of heat is transferred *from* the system. $\quad \Diamond$

---

41. Five moles of an ideal gas expand isothermally at 127°C to four times its initial volume. Find (a) the work done by the gas and (b) the thermal energy transferred to the system, both in joules.

Solution

(a) $W = nRT \ln\left(\dfrac{V_f}{V_i}\right) = (5.0 \text{ mol})(8.314 \text{ J/mol} \cdot \text{K})(400 \text{ K})\ln(4)$

so $W = 23.1 \text{ kJ}$ ◊

(b)   Since T = constant, $\Delta U = 0$

so $Q = W = 23.1 \text{ kJ}$ ◊

---

42. How much work is done by the steam when 1.0 mol of water at 100°C boils and becomes 1.0 mol of steam at 100°C at 1.0 atm pressure? Determine the change in internal energy of the steam as it vaporizes. Consider the steam to be an ideal gas.

Solution $W = P \, \Delta V = P(V_s - V_w)$

Substituting, $V_s = \dfrac{nRT}{P}$ and $V_w = n\left(\dfrac{M}{\rho}\right)$

We have $W = n\left[RT - \dfrac{PM}{\rho} \right]$

$$W = (1.0 \text{ mol})\left[(8.314 \text{ J/K} \cdot \text{mol})(373 \text{ K}) - \left(1.013 \times 10^5 \text{ N/m}^2\right)\left(\frac{18 \text{ g/mol}}{1.0 \times 10^6 \text{ g/m}^3}\right) \right]$$

$W = 3.10 \text{ kJ}$ ◊

$Q = mL_V = (18.0 \text{ g})\left(2.26 \times 10^6 \text{ J/kg}\right) = 40.7 \text{ kJ}$

$\Delta U = Q - W = 37.6 \text{ kJ}$ ◊

---

45. An ideal gas initially at 300 K undergoes an isobaric expansion at 2.50 kPa. If the volume increases from 1.00 m³ to 3.00 m³ and 12.5 kJ of thermal energy is transferred to the gas, find (a) the change in its internal energy and (b) its final temperature.

Solution

(a) $\Delta U = Q - W$ where $W = P\Delta V$ so that

$$\Delta U = Q - P\Delta V = 1.25\times10^4 \text{ J} - \left(2.50\times10^3 \text{ N/m}^2\right)\left(3.00 \text{ m}^3 - 1.00 \text{ m}^3\right) = 7500 \text{ J} \quad \Diamond$$

(b) Since $\dfrac{V_1}{T_1} = \dfrac{V_2}{T_2}$, $T_2 = \left(\dfrac{V_2}{V_1}\right)T_1 = \left(\dfrac{3.00 \text{ m}^3}{1.00 \text{ m}^3}\right)(300 \text{ K}) = 900 \text{ K} \quad \Diamond$

---

46. Two moles of helium gas initially at 300 K and 0.40 atm are compressed isothermally to 1.2 atm. Find (a) the final volume of the gas, (b) the work done by the gas, and (c) the thermal energy transferred. Consider the helium to behave as an ideal gas.

Solution

(a) $PV = nRT$, so $V_i = \dfrac{nRT}{P_i} = \dfrac{(2.00 \text{ mol})(8.31 \text{ J/mol}\cdot\text{K})(300 \text{ K})}{(0.40 \text{ atm})\left(1.01\times10^5 \text{ N/m}^2/\text{atm}\right)} = 0.123 \text{ m}^3$

For isothermal compression (or expansion), $PV = $ constant, so $P_iV_i = P_fV_f$.

$$V_f = V_i\left(\frac{P_i}{P_f}\right) = V_i\left(\frac{0.40 \text{ atm}}{1.2 \text{ atm}}\right) = \frac{0.123 \text{ m}^3}{3} = 0.0410 \text{ m}^3 \quad \Diamond$$

(b) $W = \int P\,dV = \int\dfrac{nRT}{V}\,dV = nRT\ln\left(\dfrac{V_f}{V_i}\right) = (4986 \text{ J})\ln\left(\dfrac{1}{3}\right) = -5.48 \text{ kJ} \quad \Diamond$

(c) $\Delta U = 0 = Q - W$ and $Q = -5.48 \text{ kJ} \quad \Diamond$

---

49. A 1.0-kg block of aluminum is heated at atmospheric pressure such that its temperature increases from 22°C to 40°C. Find (a) the work done by the aluminum, (b) the thermal energy added to it, and (c) the change in its internal energy.

Solution

(a) $W = P\,\Delta V \cong P(3\alpha V\,\Delta T)$

$$W = \left(1.013 \times 10^5 \text{ N/m}^2\right)(3)\left(24 \times 10^{-6}\right)(°C)^{-1}\left(\frac{1.0 \text{ kg}}{2.7 \times 10^3 \text{ kg/m}^3}\right)(18°C) = 0.0486 \text{ J} \quad \lozenge$$

(b) $Q = cm\,\Delta T = (900 \text{ J/kg} \cdot °C)(1.0 \text{ kg})(18°C) = 16.2 \text{ kJ} \quad \lozenge$

(c) $\Delta U = Q - W = 16.2 \text{ kJ} - 0.0486 \text{ kJ} = 16.2 \text{ kJ} \quad \lozenge$

---

55. A bar of gold is in thermal contact with a bar of silver of the same length and area (Fig. P20.55). One end of the compound bar is maintained at 80.0°C while the opposite end is at 30.0°C. When the heat flow reaches steady state, find the temperature at the junction.

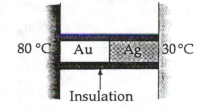

Figure P20.55

Solution   Call the gold bar Object 1 and the silver bar Object 2. When heat flow reaches a steady state, the flow rate through each will be the same: $H_1 = H_2$ or

$$\frac{k_1 A_1\,\Delta T_1}{L_1} = \frac{k_2 A_2\,\Delta T_2}{L_2}$$

In this case, $L_1 = L_2$ and $A_1 = A_2$, so $\qquad k_1\,\Delta T_1 = k_2\,\Delta T_2$

Let T_3 = temperature at the junction; then $\qquad k_1(80.0°C - T_3) = k_2(T_3 - 30.0°C)$

Solving, $\qquad T_3 = \dfrac{80 k_1 \cdot °C + 30 k_2 \cdot °C}{k_1 + k_2}$

$$T_3 = \frac{(80°C)\left(314\ \dfrac{W}{m \cdot °C}\right) + (30°C)\left(427\ \dfrac{W}{E \cdot °C}\right)}{(314 + 427)\ \dfrac{W}{m \cdot °C}} = 51.2\ °C \quad \lozenge$$

---

65. Around a crater formed by an iron meteorite, 75.0 kg of rock has melted under the impact of the meteorite. The rock has a specific heat of 0.800 kcal/kg·°C, a melting point of 500°C, and a latent heat of fusion of 48.0 kcal/kg. The original temperature of the ground was 0.0°C. If the meteorite hits the ground while moving at 600 m/s, what is the minimum mass of the meteorite? Assume no heat loss to the surrounding unmelted rock or the atmosphere during the impact. Disregard the heat capacity of the meteorite.

Solution   Assume that the total kinetic energy of the meteorite is converted into thermal energy of the rock.

$$K = Q_1 + Q_2$$

where $Q_1 = m_{rock} c_{rock} \Delta T_{rock}$ (to increase temperature to melting point)

and $Q_2 = m_{rock} L_f$ (to melt the rock)

$$\tfrac{1}{2} m v^2_{meteorite} = m_{rock}(c_{rock}\Delta T_{rock} + L_f)$$

$$m_{meteorite} = \frac{2 m_{rock}}{v^2}\left(c_{rock}\Delta T_{rock} + L_f\right)$$

Use $c_{rock} = 0.800 \dfrac{kcal}{kg\cdot°C} = 3.349 \dfrac{kJ}{kg\cdot°C}$ and $L_f = 48.0 \dfrac{kcal}{kg} = 201 \dfrac{kJ}{kg}$

$$m_{meteorite} = \frac{(2)(75.0 \text{ kg})}{(600 \text{ m/s})^2}\left[\left(3.349\times10^3 \ \frac{J}{kg\cdot°C}\right)(500°C) + 2.01\times10^5 \ \frac{J}{kg}\right]$$

$$m_{meteorite} = 781 \text{ kg} \quad \Diamond$$

---

69. An aluminum rod, 0.50 m in length and of cross-sectional area 2.5 cm², is inserted into a thermally insulated vessel containing liquid helium at 4.2 K. The rod is initially at 300 K. (a) If one half of the rod is inserted into the helium, how many liters of helium boil off by the time the inserted half cools to 4.2 K? (Assume the upper half does not cool.) (b) If the upper portion of the rod is maintained at 300 K, what is the approximate boil-off rate of liquid helium after the lower half has reached 4.2 K? (Note that aluminum has a thermal conductivity of 31 J/s·cm·K at 4.2 K, a specific heat of 0.21 cal/g·°C, and a density of 2.7 g/cm³. See Example 20.5 for data on helium.)

Chapter 20

Solution As you solve this problem, be careful not to confuse L (the *conduction length* of the rod) with L_v (the *heat of vaporization* of the helium).

(a) Before heat conduction has time to become important, we suppose the heat energy lost by half the rod equals the heat energy gained by the helium. Therefore,

$$(mL_v)_{He} = (mc\,\Delta T)_{Al} \quad \text{or} \quad (\rho V L_v)_{He} = (\rho V c\,\Delta T)_{Al}$$

so that
$$V_{He} = \frac{(\rho V c\,\Delta T)_{Al}}{(\rho L_v)_{He}} = \frac{(2.7\ g/cm^3)(62.5\ cm^3)(0.21\ cal/g\cdot{}^\circ C)(295.8{}^\circ C)}{(0.125\ g/cm^3)(4.99\ cal/g)}$$

and
$$V_{He} = 1.68\times10^4\ cm^3 = 16.8\ \text{liters} \quad \Diamond$$

(b) Heat energy will be conducted along the rod at a rate of $\dfrac{dQ}{dt} = H = \dfrac{kA\,\Delta T}{L}$.

During any time interval, this will boil a mass of helium according to

$$Q = mL_v \quad \text{or} \quad \frac{dQ}{dt} = \left(\frac{dm}{dt}\right)L_v.$$

Combining these two equations gives us the "boil-off" rate: $\quad \dfrac{dm}{dt} = \dfrac{kA\,\Delta T}{L\cdot L_v}$

Set the conduction length $L = 25$ cm, and use $k = 31\ J/s\cdot cm\cdot K = 7.41\ cal/s\cdot cm\cdot K$:

$$\frac{dm}{dt} = \frac{(7.41\ cal/s\cdot cm\cdot K)(2.5\ cm^2)(295.8\ K)}{(25\ cm)(4.99\ cal/g)} = 43.9\ g/s$$

or
$$\frac{dm}{dt} = \frac{43.9\ g/s}{0.125\ g/cm^3} = 351\ cm^3/s = 0.351\ \text{liters}/s \quad \Diamond$$

---

77. A vessel in the shape of a spherical shell has an inner radius a and outer radius b. The wall has a thermal conductivity k. If the inside is maintained at a temperature T_1 and the outside is at a temperature T_2, show that the rate of heat flow between the surfaces is

$$\frac{dQ}{dt} = \left(\frac{4\pi k\,ab}{b-a}\right)(T_1 - T_2)$$

405

Solution   For a spherical shell of radius r and thickness dr, the surface area of the shell will be $4\pi r^2$.

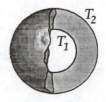

So $\dfrac{dQ}{dt} = -kA\dfrac{dT}{dx}$ becomes

$$\frac{dQ}{dt} = -k\left(4\pi r^2\right)\frac{dT}{dr}$$

Section view of
spherical shell

Since the rate of heat transfer, $\dfrac{dQ}{dt}$, is constant, we can integrate over r to find the difference in temperature across the wall of the shell:

$$-4\pi k(T_2 - T_1) = \left(\frac{dQ}{dt}\right)\int_a^b \frac{dr}{r^2} = \left(\frac{dQ}{dt}\right)\left(\frac{1}{a} - \frac{1}{b}\right)$$

and

$$\frac{dQ}{dt} = \left(\frac{4\pi kab}{b-a}\right)(T_1 - T_2) \quad \Diamond$$

---

79. The passenger section of a jet airliner is in the shape of a cylindrical tube of length 35 m and inner radius 2.5 m. Its walls are lined with a 6.0 cm thickness of insulating material of thermal conductivity 4.0×10^{-5} cal/s·cm·°C. The inside is to be maintained at 25°C while the outside is at −35°C. What heating rate is required to maintain this temperature difference? (Use the result from Problem 78.)

Solution   From problem 78, the rate of heat flow through the wall is:

$$\frac{dQ}{dt} = \frac{2\pi kL(T_1 - T_2)}{\ln(b/a)} = \frac{2\pi(3500 \text{ cm})\left(4.0 \times 10^{-5} \text{ cal/s} \cdot \text{cm} \cdot °C\right)(60°C)}{\ln(2.56/2.50)} = 2.23 \times 10^3 \text{ cal/s}$$

and $\dfrac{dQ}{dt} = 9.32$ kW $\Diamond = 9.32$ kW $\Diamond$

This is the rate of heat loss from the plane, and consequently the rate at which energy must be supplied in order to maintain an equilibrium temperature.

---

81. A "solar cooker" consists of a curved reflecting mirror that focuses sunlight onto the object to be heated (Fig. P20.81). The solar power per unit area reaching the Earth at some location is 600 W/m^2, and the cooker has a diameter of 0.60 m. Assuming that 40% of the incident energy is converted into thermal energy, how long would it take to completely boil off 0.50 liters of water initially at 20°C? (Neglect the heat capacity of the container.)

Figure 20.81

Solution   The power incident on the solar collector is

$$P_i = IA = (600 \text{ W/m}^2)\pi(0.30 \text{ m})^2 = 169.6 \text{ W}$$

For a 40% reflector, the collected power is

$$P_c = 67.9 \text{ J/s}$$

The total energy required to increase the temperature of the water to the boiling point and to evaporate it is

$$Q = mc\,\Delta T + mL_v = (0.500 \text{ kg})(4186 \text{ J/kg·°C})(80 \text{ °C}) + (0.500 \text{ g})(2.26 \times 10^6 \text{ J/kg}) = 1.30 \times 10^6 \text{ J}$$

The time required is

$$\Delta t = \frac{Q}{P_c} = \frac{1.30 \times 10^6 \text{ J}}{67.9 \text{ J/s}} = 1.91 \times 10^4 \text{ s} = 5.31 \text{ h} \quad \lozenge$$

Chapter 21

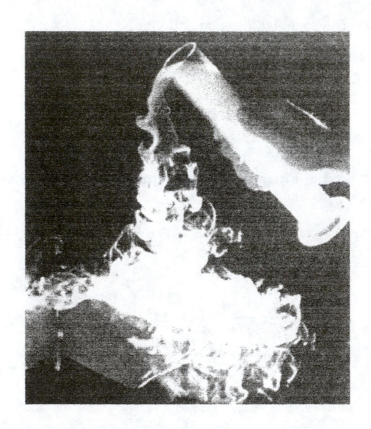

The Kinetic Theory of Gases

Chapter 21

THE KINETIC THEORY OF GASES

INTRODUCTION

In Chapter 19, we discussed the properties of an ideal gas using such macroscopic variables as pressure, volume, and temperature. We shall now show that such large-scale properties can be described on a microscopic scale, where matter is treated as a collection of molecules. Newton's laws of motion applied in a statistical manner to a collection of particles provide a reasonable description of thermodynamic processes. In order to keep the mathematics relatively simple, we shall consider only the molecular behavior of gases, where the interactions between molecules are much weaker than in liquids or solids. In the current view of gas behavior, called the *kinetic theory*, gas molecules move about in a random fashion, colliding with the walls of their container and with each other. Perhaps the most important consequence of this theory is that it shows the equivalence between the kinetic energy of molecular motion and the internal energy of the system. Furthermore, the kinetic theory provides us with a physical basis upon which the concept of temperature can be understood.

NOTES FROM SELECTED CHAPTER SECTIONS

21.1 Molecular Model of an Ideal Gas

A microscopic *model of an ideal gas* is based on the following assumptions:

- *The number of molecules is large, and the average separation between them is large* compared with their dimensions. Therefore, the molecules occupy a negligible volume compared with the volume of the container.

- *The molecules obey Newton's laws of motion, but the individual molecules move in a random fashion.* By random fashion, we mean that the molecules move in all directions with equal probability and with various speeds. This distribution of velocities does not change in time, despite the collisions between molecules.

- *The molecules undergo elastic collisions with each other.* Thus, the molecules are considered to be structureless (that is, point masses), and in the collisions both kinetic energy and momentum are conserved.

- *The forces between molecules are negligible except during a collision.* The forces between molecules are short-range, so that the only time the molecules interact with each other is during a collision.

- *The gas under consideration is a pure gas.* That is, all molecules are identical.

- *The gas is in thermal equilibrium with the walls of the container.* The collisions of molecules with the walls are perfectly elastic.

21.3 Adiabatic Processes for an Ideal Gas

An *adiabatic process* is one in which there is no heat transfer between a system and its surroundings.

A *quasi-static* adiabatic process is one which is slow enough to allow the system to be always near equilibrium, but fast compared with the time required for the system to exchange heat energy with its surroundings.

21.4 The Equipartition of Energy

The *theorem of equipartition of energy* states that the energy of a system in thermal equilibrium is equally divided among all degrees of freedom.

The heat capacities of gases containing complex molecules can sometimes be explained if one includes other degrees of freedom, namely, those associated with vibrational and rotational motions. The contribution from vibrations is especially important at high temperatures, while rotational effects are significant above about 50 K. The classical model does not provide an adequate description of molecular systems in all situations.

The heat capacities of solids generally increase nonlinearly with increasing temperature, and approach a value of about $3R$ at sufficiently high temperatures. This result is known as the *DuLong-Petit law*. The heat capacities approach zero at $T \rightarrow 0$, but the complete variation with temperature cannot be explained using classical concepts.

21.6 Distribution of Molecular Speeds

The speed distribution of gas molecules in thermal equilibrium is described by a quantity, N_v, called the Maxwell-Boltzman distribution function. The distribution of speeds in a gas depends on both the mass of the gas molecules and the temperature. If N is the total number of molecules, then the number with speeds between v and $v + dv$ is $(N_v dv)/N$. The fraction $(N_v dv)/N$ is the probability that a molecule has a speed in the range between v and $v + dv$.

EQUATIONS AND CONCEPTS

In the kinetic theory of an ideal gas, one finds that the pressure of the gas is proportional to the number of molecules per unit volume and the average translational kinetic energy per molecule.

$$P = \frac{2}{3}\left(\frac{N}{V}\right)\left(\frac{1}{2}m\overline{v^2}\right) \tag{21.2}$$

From Equation 21.2, and the equation of state for an ideal gas, $PV = NkT$, we find that *the absolute temperature of an ideal gas is a direct measure of the average molecular kinetic energy.*

$$T = \frac{2}{3k_B}\left(\frac{1}{2}m\overline{v^2}\right) \tag{21.3}$$

The *total internal energy U* of N molecules (or n mols) of a monatomic ideal gas is proportional to the absolute temperature.

$$U = \frac{3}{2}Nk_BT = \frac{3}{2}nRT \tag{21.10}$$

or

$$U = n\,C_VT$$

If we apply the first law of thermodynamics to a monatomic *ideal gas* in which heat is transferred at *constant volume*, we find that the *specific heat at constant volume* is equal to

$$C_V = \frac{3}{2}R \tag{21.12}$$

$$\tfrac{3}{2}R = 2.99 \text{ cal/mol} \cdot \text{K}$$

If heat is transferred to an ideal gas at *constant pressure*, the first law of thermodynamics shows that the *specific heat at constant pressure* is greater than the specific heat at constant volume by an amount R.

$$C_P - C_V = R \tag{21.16}$$

or

$$C_P = \frac{5}{2}R$$

The *ratio of specific heats* is a dimensionless quantity γ which, for a *monatomic ideal gas*, is equal to 1.67.

$$\gamma = \frac{C_P}{C_V} = 1.67 \tag{21.17}$$

If an *ideal gas* undergoes a *quasi-static, adiabatic expansion*, and we assume $PV = nRT$ is valid at any time, then the pressure and volume of the gas obey Equation 21.18.

$$PV^\gamma = \text{constant} \qquad (21.18)$$

The law of atmospheres predicts that the density of the atmosphere (molecules per unit volume, n) decreases exponentially with altitude, y. Atmospheric pressure also decreases exponentially.

$$n(y) = n_0 e^{-mgy/k_BT} \qquad (21.23)$$

$$n_0 = 2.69 \times 10^{25} \text{ molecules/m}^3$$

$$P = P_0 e^{-mgy/k_BT} \qquad (21.24)$$

The Boltzman distribution law (which is valid for a system of a large number of particles) states that the probability of finding the particles in a particular arrangement (distribution of energy values) varies exponentially as the negative of the reciprocal of the absolute temperature.

$$n(E) = n_0 e^{-E/k_BT} \qquad (21.25)$$

The Maxwell speed distribution function describes the most probable distribution of speeds of N gas molecules at temperature T (where k_B is the Boltzman constant).

$$N_v = 4\pi N \left(\frac{m}{2\pi k_B T} \right)^{3/2} v^2 e^{-mv^2/2k_BT} \qquad (21.26)$$

Specific expressions can be derived for the rms, average, and most probable speeds.

$$v_{\text{rms}} = 1.73\sqrt{k_B T / m} \qquad (21.27)$$

$$\bar{v} = 1.60\sqrt{k_B T / m} \qquad (21.28)$$

$$v_{\text{mp}} = 1.41\sqrt{k_B T / m} \qquad (21.29)$$

The average distance l between collisions is called the *mean free path* (where d is the molecular "diameter" and n_V is the density of molecules).

$$l = \frac{1}{\sqrt{2}\ \pi d^2 n_V}$$

(21.30)

The Van der Waals' equation of state is a modification of the ideal gas equation of state to take into account the volume of the gas molecules and the intermolecular forces when the molecules are close together. (The constants a and b are empirical and are chosen to provide best agreement on a particular gas.)

$$\left(P + \frac{a}{V^2}\right)(V - b) = RT$$

(21.32)

REVIEW CHECKLIST

▷ State and understand the assumptions made in developing the molecular model of an ideal gas.

▷ Recognize that the temperature of an ideal gas is proportional to the average molecular kinetic energy. State the theorem of equipartition of energy, noting that each degree of freedom of a molecule contributes an equal amount of energy, of magnitude $\frac{1}{2}k_BT$.

▷ Recognize that the internal energy of an ideal gas is proportional to the absolute temperature, and be able to derive the specific heat of an ideal gas at constant volume from the first law of thermodynamics.

▷ Define an adiabatic process, and be able to derive the expression $PV^\gamma =$ constant, which applies to a quasi-static, adiabatic process.

▷ Understand the meaning of the Maxwell speed distribution function, and recognize the differences among rms speed, average speed, and most probable speed.

SOLUTIONS TO SELECTED END-OF-CHAPTER PROBLEMS

5. A spherical balloon of volume 4000 cm³ contains helium at an (inside) pressure of 1.2×10^5 Pa. How many moles of helium are in the balloon, if each helium atom has an average kinetic energy of 3.6×10^{-22} J?

Solution

The gas temperature must be that implied by $\frac{1}{2}m\overline{v^2} = \frac{3}{2}k_BT$

$$T = \frac{2}{3}\left(\frac{3.6 \times 10^{-22} \text{ J}}{1.38 \times 10^{-23} \text{ J/K}}\right) = 17.4 \text{ K}$$

Now $PV = nRT$

$$n = \frac{PV}{RT} = \frac{(1.2 \times 10^5 \text{ N/m}^2)(4 \times 10^{-3} \text{ m}^3)}{(8.314 \text{ J/mol} \cdot \text{K})(17.4 \text{ K})}$$

$$n = 3.32 \text{ mol} \quad \Diamond$$

7. A cylinder contains a mixture of helium and argon gas in equilibrium at 150°C. What is the average kinetic energy of each gas molecule?

Solution

Both kinds of molecules have the same average kinetic energy, sharing it in collisions:

$$\frac{1}{2}m\overline{v^2} = \frac{3}{2}k_BT = \frac{3}{2}(1.38 \times 10^{-23} \text{ J/K})(273 + 150)\text{K}$$

$$\overline{K} = 8.76 \times 10^{-21} \text{ J} \quad \Diamond$$

14. (a) How many atoms of helium gas are required to fill a balloon to diameter 30.0 cm at 20.0°C and 1.00 atm? (b) What is the average kinetic energy of each helium atom? (c) What is the average speed of each helium atom?

Solution

(a) The volume is $V = \frac{4}{3}\pi r^3 = \frac{4}{3}\pi(0.150 \text{ m})^3 = 1.41\times10^{-2} \text{ m}^3$

  Now $PV = nRT$

$$n = \frac{PV}{RT} = \frac{(1.013\times10^5 \text{ N/m}^2)(1.41\times10^{-2} \text{ m}^3)}{(8.314 \text{ N}\cdot\text{m/mol}\cdot\text{K})(293 \text{ K})}$$

$$n = 0.588 \text{ mol}$$

$$N = nN_A = (0.588 \text{ mol})(6.02\times10^{23} \text{ molecule/mol})$$

$$N = 3.54\times10^{23} \text{ helium atoms} \quad \lozenge$$

(b) $\frac{1}{2}m\overline{v^2} = \frac{3}{2}k_BT = 1.5(1.38\times10^{-23} \text{ J/K})(293 \text{ K})$

$\overline{K} = 6.07\times10^{-21} \text{ J} \quad \lozenge$

(c) The mass of an atom is

$$m = \frac{M}{N_A} = \frac{4.0026 \text{ g/mol}}{6.02\times10^{23} \text{ molecule/mol}}$$

$$m = 6.65\times10^{-24} \text{ g} = 6.65\times10^{-27} \text{ kg}$$

So the kinetic energy is

$$\frac{1}{2}(6.65\times10^{-27} \text{ kg})\overline{v^2} = 6.07\times10^{-21} \text{ J}$$

and $v_{rms} = \sqrt{\overline{v^2}} = 1.35 \text{ km/s} \quad \lozenge$

19. One mole of hydrogen gas is heated at constant pressure from 300 K to 420 K. Calculate (a) the heat transferred to the gas, (b) the increase in its internal energy, and (c) the work done by the gas.

Solution

(a)   Since this is a constant-pressure process,

$$Q = nC_P \Delta T = (1.00 \text{ mol})(28.8 \text{ J/mol·K})(120 \text{ K}) = 3.46 \text{ kJ} \quad \lozenge$$

(b)   For any system, $\Delta U = nC_V \Delta T$,

so

$$\Delta U = (1.00 \text{ mol})(20.4 \text{ J/mol·K})(120 \text{ K}) = 2.45 \text{ kJ} \quad \lozenge$$

(c) $\Delta U = Q - W$;

Therefore,

$$W = Q - \Delta U = 3.46 \text{ kJ} - 2.45 \text{ kJ} = 1.01 \text{ kJ} \quad \lozenge$$

---

24. How much thermal energy is in the air in a 20.0 m³ room at (a) 0.0°C and (b) 20.0°C? Assume that the pressure remains at 1.00 atm.

Solution

Take air as diatomic molecules with $C_V = \frac{5}{2}R$.   Then

$$U = nC_V T = n\left(\tfrac{5}{2}\right)RT = \tfrac{5}{2}PV$$

The furnace raises the temperature, but expanding air escapes so that the room contains the same internal energy at all temperatures. $\lozenge$

$$U = \tfrac{5}{2}(1.013 \times 10^5 \text{ N/m}^2)(20.0 \text{ m}^3) = 5.06 \text{ MJ} \quad \lozenge$$

---

25. Two moles of an ideal gas ($\gamma = 1.40$) expand slowly and adiabatically from a pressure of 5.00 atm and a volume of 12.0 liters to a final volume of 30.0 liters. (a) What is the final pressure of the gas? (b) What are the initial and final temperatures?

Solution

(a) $P_1 V_1{}^{\gamma} = P_2 V_2{}^{\gamma}$

$$P_2 = \left(\frac{V_1}{V_2}\right)^{\gamma} P_1 = \left(\frac{12.0}{30.0}\right)^{1.40}(5.00 \text{ atm}) = 1.39 \text{ atm} \quad \Diamond$$

(b) $T_1 = \dfrac{P_1 V_1}{nR} = \dfrac{(5.00)(1.01 \times 10^5)(12.0 \times 10^{-3})}{(2)(8.31)} = 366 \text{ K} \quad \Diamond$

$T_2 = \dfrac{P_2 V_2}{nR} = 254 \text{ K} \quad \Diamond$

---

31. Air in a thundercloud expands as it rises. If its initial temperature was 300 K, and no heat is lost on expansion, what is its temperature when the initial volume is doubled?

Solution   It expands adiabatically, losing no heat but dropping in temperature as it does work on the air around it.

Combine $\qquad\qquad P_1 V_1{}^{\gamma} = P_2 V_2{}^{\gamma}$

and $\qquad\qquad P_1 = \dfrac{nRT_1}{V_1}$

with $\qquad\qquad P_2 = \dfrac{nRT_2}{V_2}$

to find $\qquad\qquad T_1 V_1{}^{\gamma-1} = T_2 V_2{}^{\gamma-1}$

$$T_2 = T_1 \left(\frac{V_1}{V_2}\right)^{\gamma-1} = 300 \text{ K}\left(\frac{1}{2}\right)^{(1.40-1)} = 227 \text{ K} \quad \Diamond$$

---

38. Consider 2 mol of an ideal diatomic gas. Find the total heat capacity at constant volume and at constant pressure if (a) the molecules rotate but do not vibrate and (b) the molecules rotate and vibrate.

Solution

(a) Count degrees of freedom. A diatomic molecule oriented along the y axis can possess energy by moving in x, y, and z directions and by rotating around the x and z axes. Outside forces can get no lever arms to make it spin around the y axis. The molecule will have average energy $\frac{1}{2}k_B T$ for each of these five degrees of freedom.

The gas will have internal energy $U = N\left(\frac{5}{2}\right)k_B T$

Therefore,
$$U = nN_A\left(\frac{5}{2}\right)\left(\frac{R}{N_A}\right)T = \left(\frac{5}{2}\right)nRT,$$

so the constant-volume heat capacity of the whole sample is

$$\frac{\Delta U}{\Delta T} = \left(\frac{5}{2}\right)nR = \left(\frac{5}{2}\right)(2 \text{ mol})(8.314 \text{ J/mol}\cdot\text{K}) = 41.6 \text{ J/K} \quad \Diamond$$

At constant pressure, $C_P = C_V + R$ for one mole; $nC_P = nC_V + nR$ for the sample:

$$nC_P = 41.6 \text{ J/K} + (2 \text{ mol})(8.314 \text{ J/mol·K}) = 58.2 \text{ J/K} \quad \Diamond$$

(b) Vibration adds a degree of freedom for kinetic energy and a degree of freedom for elastic energy.

Now the molecule's average energy is $\left(\frac{7}{2}\right)k_B T$,

The sample's internal energy is $N\left(\frac{7}{2}\right)k_B T = \left(\frac{7}{2}\right)nRT$,

and the sample's constant-volume heat capacity is

$$\left(\frac{7}{2}\right)nR = \left(\frac{7}{2}\right)(2 \text{ mol})(8.314 \text{ J/mol}\cdot\text{K}) = 58.2 \text{ J/K} \quad \Diamond$$

At constant pressure, its heat capacity is

$$nC_V + nR = 58.2 \text{ J/K} + (2 \text{ mol})(8.314 \text{ J/mol}\cdot\text{K}) = 74.8 \text{ J/K} \quad \Diamond$$

40. Fifteen identical particles have the following speeds: one has speed 2.0 m/s; two have speed 3.0 m/s; three have speed 5.0 m/s; four have speed 7.0 m/s; three have speed 9.0 m/s; two have speed 12.0 m/s. Find (a) the average speed, (b) the rms speed, and (c) the most probable speed of these particles.

Solution

(a) $\bar{v} = \dfrac{\Sigma n_i v_i}{\Sigma n_i}$

$\bar{v} = \dfrac{1\times 2 + 2\times 3 + 3\times 5 + 4\times 7 + 3\times 9 + 2\times 12}{1+2+3+4+3+2}$ m/s

$\bar{v} = 6.80$ m/s   ◊

(b) $\overline{v^2} = \dfrac{\Sigma n_i v_i^{\,2}}{\Sigma n_i}$

$\overline{v^2} = \dfrac{1\times 2^2 + 2\times 3^2 + 3\times 5^2 + 4\times 7^2 + 3\times 9^2 + 2\times 12^2}{15}$ m^2/s^2

$\overline{v^2} = 54.9$ m^2/s^2

$v_{rms} = \sqrt{\overline{v^2}}$

$v_{rms} = \sqrt{54.9 \text{ m}^2/\text{s}^2} = 7.41$ m/s   ◊

(c) More particles have $v_{mp} = 7.00$ m/s than any other speed.   ◊

44. Show that the most probable speed of a gas molecule is given by Equation 21.29. Note that the most probable speed corresponds to the point where the slope of the speed distribution curve, dN_v/dv, is zero.

Solution From the Maxwell speed distribution function,

$$N_v = 4\pi N\left(\frac{m}{2\pi k_B T}\right)^{3/2} v^2 \exp\left(-\frac{mv^2}{2k_B T}\right),$$

we locate the peak in the graph of N_v versus v by evaluating dN_v/dv and setting it equal to zero, to solve for the most probable speed:

$$\frac{dN_v}{dv} = 4\pi N\left(\frac{m}{2\pi k_B T}\right)^{3/2} v^2 \exp\left(-\frac{mv^2}{2k_B T}\right)\left(-\frac{m2v}{2k_B T}\right)$$

$$+4\pi N\left(\frac{m}{2\pi k_B T}\right)^{3/2} 2v \exp\left(-\frac{mv^2}{2k_B T}\right) = 0$$

This equation is solved by $v = 0$ and $v \rightarrow \infty$, but those correspond to minimum-probability speeds, so we divide by v and by the exponential function.

$$v_{mp}\left(-\frac{m2v_{mp}}{2k_B T}\right) + 2 = 0$$

$$v_{mp} = \left(\frac{2k_B T}{m}\right)^{1/2} \quad \text{as in Equation 21.29.} \quad \lozenge$$

47. In an ultrahigh vacuum system, the pressure is measured to be 1.00×10^{-10} torr (where 1 torr = 133 Pa). If the gas molecules have a molecular diameter of 3.00×10^{-10} m and the temperature is 300 K, find (a) the number of molecules in a volume of 1.00 m^3. (b) the mean free path of the molecules, and (c) the collision frequency, assuming an average speed of 500 m/s.

Solution   (a) $PV = \left(\dfrac{N}{N_A}\right)RT$ and $N = \dfrac{PVN_A}{RT}$,   so that

$$N = \frac{(1.00 \times 10^{-10})(133)(6.02 \times 10^{23})}{(8.31)(300)} = 3.21 \times 10^{12} \text{ molecules} \quad \Diamond$$

(b) $L = \dfrac{1}{n\sigma\sqrt{2}} = \dfrac{V}{N\sigma\sqrt{2}} = \dfrac{1.00 \text{ m}^3}{(3.21 \times 10^{12} \text{ molecules})\pi(3.00 \times 10^{-10} \text{ m})^2\sqrt{2}}$

$L = 7.78 \times 10^5 \text{ m} = 778 \text{ km} \quad \Diamond$

(c) $f = \dfrac{v}{L} = \dfrac{500 \text{ m/s}}{7.78 \times 10^5 \text{ m}} = 6.42 \times 10^{-4} \text{ s}^{-1} \quad \Diamond$

---

51. The constant b that appears in van der Waals' equation of state for oxygen is measured to be 31.8 cm^3/mol. Assuming a spherical shape, estimate the diameter of the molecule.

Solution   The constant b represents the volume occupied by the molecules. For one molecule we estimate the volume as

$$\frac{b}{N} = \frac{31.8 \text{ cm}^3 / \text{mol}}{6.02 \times 10^{+23} \text{ molecule} / \text{mol}} = 5.28 \times 10^{-23} \text{ cm}^3 / \text{molecule} = 5.28 \times 10^{-29} \text{ m}^3 / \text{molecule}$$

We then approximate the molecule as a sphere, and calculate the radius: $\dfrac{b}{N} = \dfrac{4}{3}\pi r^3$

$$r = \left(\frac{3 \times 5.28 \times 10^{-29} \text{ m}^3}{4\pi}\right)^{1/3} = 2.33 \times 10^{-10} \text{ m}$$

so the diameter = $2r = 4.66 \times 10^{-10}$ m $\Diamond$

54. A cylinder containing n mol of an ideal gas undergoes a reversible adiabatic process. (a) Starting with the expression $W = \int P\, dV$ and using $PV^\gamma =$ constant, show that the work done is

$$W = \left(\frac{1}{\gamma - 1}\right)(P_iV_i - P_fV_f)$$

(b) Starting with the first-law equation in differential form, prove that the work done is also equal to $nC_V(T_i - T_f)$. Show that this result is consistent with the equation in part (a).

Solution

(a) $PV^\gamma = k$ so $\qquad W = \int_i^f P\, dV = k\int_{V_i}^{V_f}\frac{dV}{V^\gamma} = \left.\frac{kV^{1-\gamma}}{1-\gamma}\right|_{V_i}^{V_f}$

$$W = \frac{P_fV_f^\gamma V_f^{1-\gamma} - P_iV_i^\gamma V_i^{1-\gamma}}{1-\gamma} = \frac{P_iV_i - P_fV_f}{\gamma - 1} \qquad \lozenge$$

(b) $dU = dQ - dW$ and $dQ = 0$ for an adiabatic process. Therefore,

$$W = -\Delta U = -nC_V\Delta T = nC_V(T_i - T_f) \qquad \lozenge$$

To show consistency between these two equations, consider that

$$\gamma = \frac{C_P}{C_V} \qquad \text{and} \qquad C_P - C_V = R.$$

Therefore, $\qquad \dfrac{1}{\gamma - 1} = \dfrac{C_V}{R}$

Using this, the result in part (a) becomes $\qquad W = (P_iV_i - P_fV_f)\dfrac{C_V}{R}$

Also, for an ideal gas $\qquad \dfrac{PV}{R} = nT$ so that $\qquad W = nC_V(T_i - T_f) \qquad \lozenge$

55. Twenty particles, each of mass m and confined to a volume V, have the following speeds: two have speed v; three have speed $2v$; five have speed $3v$; four have speed $4v$; three have speed $5v$; two have speed $6v$; one has speed $7v$. Find (a) the average speed, (b) the rms speed, (c) the most probable speed, (d) the pressure they exert on the walls of the vessel, and (e) the average kinetic energy per particle.

Solution

(a)   The average speed v_{av} is just the weighted average of all the speeds:

$$\bar{v} = \frac{\Sigma n_i v_i}{\Sigma n_i}$$

$$\bar{v} = \frac{2(v) + 3(2v) + 5(3v) + 4(4v) + 3(5v) + 2(6v) + 1(7v)}{2+3+5+4+3+2+1} = 3.65v \quad \Diamond$$

(b)   First find the average of the square of the speeds:

$$\overline{v^2} = \frac{\Sigma n_i v_i^2}{\Sigma n_i}$$

$$\overline{v^2} = \frac{2(v)^2 + 3(2v)^2 + 5(3v)^2 + 4(4v)^2 + 3(5v)^2 + 2(6v)^2 + 1(7v)^2}{2+3+5+4+3+2+1} = 15.95v^2$$

The root-mean square speed is then

$$v_{rms} = \sqrt{\overline{v^2}} = 3.99v \quad \Diamond$$

(c)   The most probable speed is the one that most of the particles have. That is, five particles have speed $3.00v$. $\Diamond$

(d) $PV = \frac{1}{3} Nm\overline{v^2}$.

Therefore, $P = \left(\frac{20}{3}\right)\dfrac{m(15.95)v^2}{V} = 106\,\dfrac{mv^2}{V} \quad \Diamond$

(e)   The average kinetic energy for each particle is

$$K = \frac{1}{2}m\overline{v^2} = \frac{1}{2}m(15.95v^2) = 7.98mv^2 \quad \Diamond$$

60. The compressibility, κ, of a substance is defined as the fractional change in volume of that substance for a given change in pressure:

$$\kappa = -\frac{1}{V}\frac{dV}{dP}$$

(a) Explain why the negative sign in this expression ensures that κ is always positive. (b) Show that if an ideal gas is compressed isothermally, its compressibility is given by $\kappa_1 = 1/P$. (c) Show that if an ideal gas is compressed adiabatically, its compressibility is given by $\kappa_2 = 1/\gamma P$. (d) Determine values for κ_1 and κ_2 for a monatomic ideal gas at a pressure of 2.00 atm.

Solution   Since pressure increases as volume decreases (and vice versa),

(a) $\dfrac{dV}{dP} < 0$ and $-\left(\dfrac{1}{V}\right)\dfrac{dV}{dP} > 0$ ◊

(b)   For an ideal gas, $V = \dfrac{nRT}{P}$ and $\kappa_1 = -\dfrac{1}{V}\dfrac{d}{dP}\left(\dfrac{nRT}{P}\right)$

   If the compression is isothermal, T is constant and $\kappa_1 = -\dfrac{nRT}{V}\left(\dfrac{-1}{P^2}\right) = \dfrac{1}{P}$ ◊

(c)   For an adiabatic compression, $PV^\gamma = C$ (where C is a constant) and

$$\kappa_2 = -\left(\frac{1}{V}\right)\frac{d}{dP}\left(\frac{C}{P}\right)^{1/\gamma} = \left(\frac{1}{V\gamma}\right)\frac{C^{1/\gamma}}{P\left(P^{1/\gamma+1}\right)} = \frac{V}{V\gamma P} = \frac{1}{\gamma P}$$ ◊

(d) $\kappa_1 = \dfrac{1}{P} = \dfrac{1}{2.00 \text{ atm}} = 0.500 \text{ atm}^{-1}$ ◊

   $\gamma = \dfrac{C_P}{C_V}$ and for a monatomic ideal gas, $\gamma = \dfrac{5}{3}$,   so that

   $\kappa_2 = \dfrac{1}{\gamma P} = \dfrac{1}{\left(\dfrac{5}{3}\right)(2.00 \text{ atm})} = 0.300 \text{ atm}^{-1}$ ◊

Chapter 21

61. One mole of a gas obeying van der Waals' equation of state is compressed isothermally. At some critical temperature, T_c, the isotherm has a point of zero slope, as in Figure 21.17. That is, at $T = T_c$,

$$\frac{\partial P}{\partial V} = 0 \quad \text{and} \quad \frac{\partial^2 P}{\partial V^2} = 0$$

Using Equation 21.32 and these conditions, show that at the critical point,

$$P_c = \frac{a}{27b^2}, \quad V_c = 3b, \quad \text{and} \quad T_c = \frac{8a}{27Rb}.$$

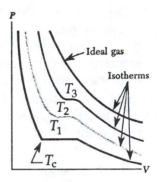

Figure 21.17

Solution   Van der Waals' equation is $\quad \left(P + \frac{a}{V^2}\right)(V - b) = RT \quad$ (1)

Holding T constant, take the partial derivative $\frac{\partial}{\partial V}$ of each side to get

$$\left(\frac{\partial P}{\partial V} - \frac{2a}{V^3}\right)(V - b) + \left(P + \frac{a}{V^2}\right) = 0$$

Setting $\frac{\partial P}{\partial V} = 0$ gives $\quad P + \frac{a}{V^2} = \frac{2a}{V^3}(V - b) \quad$ (2)

Taking the second derivative $\frac{\partial^2}{\partial V^2}$ of (1) gives

$$\left(\frac{\partial^2 P}{\partial V^2} + \frac{6a}{V^4}\right)(V - b) + 2\left(\frac{\partial P}{\partial V} - \frac{2a}{V^3}\right) = 0$$

Setting $\frac{\partial^2 P}{\partial V^2} = \frac{\partial P}{\partial V} = 0$ gives $\frac{2a}{V^3} - \frac{6ab}{V^4} = 0$,and $\quad \frac{3b}{V} = 1 \quad$ (3)

Solving equations (1), (2), and (3), gives

$$P_c = \frac{a}{27b^2}, \quad V_c = 3b, \quad \text{and} \quad T_c = \frac{8a}{27Rb} \quad \Diamond$$

67. There are roughly 10^{59} neutrons and protons in an average star and about 10^{11} stars in a typical galaxy. Galaxies tend to form in clusters of (on the average) about 10^3 galaxies, and there are about 10^9 clusters in the known part of the Universe. (a) Approximately how many neutrons and protons are there in the known Universe? (b) Suppose all this matter were compressed into a sphere of nuclear matter such that each nuclear particle occupied a volume of 10^{-45} m^3 (about the "volume" of a neutron or proton). What would be the radius of this sphere of nuclear matter? (c) How many moles of nuclear particles are there in the observable Universe?

Solution

(a)   1 Universe $\cong$ 1 Universe (10^9 clusters/Universe)

$\qquad\qquad \cong 10^9$ clusters (10^3 galaxies/cluster)

$\qquad\qquad \cong 10^{12}$ galaxies (10^{11} stars/galaxy)

$\qquad\qquad \cong 10^{23}$ stars (10^{59} nucleons/star)

$\qquad\qquad \cong 10^{82}$ nucleons   ◊

(b)   volume $\cong (10^{82}$ nucleons)(10^{-45} m^3/nucleon) $= 10^{37}$ m$^3 = \frac{4}{3}\pi r^3$

$$r \cong \left(\frac{3\times 10^{37}\ \text{m}^3}{4\pi}\right)^{1/3} = 1.34\times 10^{12}\ \text{m} \cong 10^{12}\ \text{m}\quad ◊$$

This would fit inside the orbit of Saturn.

(c)   Universe $\cong 10^{82}$ nucleons $\left(\dfrac{1\ \text{mol}}{6.02\times 10^{23}\ \text{particles}}\right)$

$\qquad\qquad \cong 10^{58}$ mole   ◊

70. For a Maxwellian gas, use a computer or programmable calculator to find the numerical value of the ratio $\{N_v(v)/N_v(v_{mp})\}$ for the following values of v: $v = (v_{mp}/50)$, $(v_{mp}/10)$, $(v_{mp}/2)$, $2v_{mp}$, $20v_{mp}$, $50\,v_{mp}$. Give your result to three significant figures.

Solution Into $N_v(v) = 4\pi N\left(\dfrac{m}{2\pi k_B T}\right)^{3/2} v^2 \exp\left(-\dfrac{mv^2}{2k_B T}\right)$

substitute $v_{mp} = \left(\dfrac{2k_B T}{m}\right)^{1/2}$ to evaluate

$$N_v(v_{mp}) = 4\pi N\left(\dfrac{m}{2\pi k_B T}\right)^{3/2} \dfrac{2k_B T}{m} \exp(-1)$$

$$= 4\pi^{-1/2}\, e^{-1}\, N\left(\dfrac{m}{2k_B T}\right)^{1/2} = 4\pi^{-1/2}\, e^{-1}\, N v_{mp}^{-1}$$

Then

$$\frac{N_v(v)}{N_v(v_{mp})} = \frac{4\pi^{-1/2}\, N v_{mp}^{-3}\, v^2 \exp\left(-\dfrac{v^2}{v_{mp}^2}\right)}{4\pi^{-1/2}\, N v_{mp}^{-1}\, e^{-1}}$$

$$= \left(\frac{v^2}{v_{mp}^2}\right)\exp\left(1 - \frac{v^2}{v_{mp}^2}\right)$$

For $v = \dfrac{v_{mp}}{50}$, $\quad \dfrac{N_v(v)}{N_v(v_{mp})} = \left(\dfrac{1}{50}\right)^2 \exp\left(1 - \left(\dfrac{1}{50}\right)^2\right) = 1.09 \times 10^{-3}$ ◊

For $v = \dfrac{v_{mp}}{10}$, $\quad \dfrac{N_v(v)}{N_v(v_{mp})} = \left(\dfrac{1}{10}\right)^2 \exp\left(1 - \dfrac{1}{100}\right) = 2.69 \times 10^{-2}$ ◊

For $v = \dfrac{v_{mp}}{2}$, $\quad \dfrac{N_v(v)}{N_v(v_{mp})} = \left(\dfrac{1}{2}\right)^2 \exp\left(1 - \dfrac{1}{4}\right) = 0.529$ ◊

For $v = v_{mp}$, $\dfrac{N_v(v)}{N_v(v_{mp})} = (1)^2 \exp(1-1) = 1.00$ ◊

The other values are computed similarly:

$\dfrac{v}{v_{mp}}$	$\dfrac{N_v(v)}{N_v(v_{mp})}$
$\dfrac{1}{50}$	1.09×10^{-3}
$\dfrac{1}{10}$	2.69×10^{-2}
$\dfrac{1}{2}$	0.529
1	1.00
2	0.199
10	1.01×10^{-41}
50	1.25×10^{-1082}

To find the last, we go

$$50^2 \, e^{1-2500} = 50^2 \, e^{-2499}$$

$$= 10^{\log 2500} \, e^{\ln 10(-2499/\ln 10)}$$

$$= 10^{(\log 2500 - 2499/\ln 10)}$$

$$= 10^{-1080} \quad ◊$$

Chapter 22

Heat Engines, Entropy, and the Second Law of Thermodynamics

HEAT ENGINES, ENTROPY,
AND THE SECOND LAW OF THERMODYNAMICS

INTRODUCTION

The first law of thermodynamics, studied in Chapter 20, is a statement of conservation of energy, generalized to include heat as a form of energy transfer. This law tells us only that an increase in one form of energy must be accompanied by a decrease in some other form of energy. It places no restrictions on the types of energy conversions that can occur. Furthermore, it makes no distinction between heat and work. However, such a distinction exists. One manifestation of this difference is the fact that it is impossible to convert thermal energy to mechanical energy entirely and continuously. Contrary to the first law, only certain types of energy conversions can take place. The second law of thermodynamics establishes which processes in nature can and which cannot occur.

From an engineering viewpoint, perhaps the most important application of the second law of thermodynamics is the limited efficiency of heat engines. The second law says that a machine capable of continuously converting thermal energy completely to other forms of energy in a cyclic process cannot be constructed.

NOTES FROM SELECTED CHAPTER SECTIONS

22.1 Heat Engines and the Second Law of Thermodynamics

A heat engine is a device that converts thermal energy to other useful forms, such as electrical and mechanical energy.

A heat engine carries some working substance through a cyclic process during which (1) heat is absorbed from a source at a high temperature, (2) work is done by the engine, and (3) heat is expelled by the engine to a source at a lower temperature.

The engine absorbs a quantity of heat, Q_h, from a hot reservoir, does work W, and then gives up heat Q_c to a cold reservoir. Because the working substance goes through a cycle, its initial and final internal energies are equal, so $\Delta U = 0$. Hence, from the first equation we see that *the net work, W, done by a heat engine equals the net heat flowing into it.*

If the working substance is a gas, *the net work done for a cyclic process is the area enclosed by the curve representing the process on a PV diagram.*

The **thermal efficiency**, *e*, of a heat engine is the ratio of the net work done to the heat absorbed at the higher temperature during one cycle.

The *second law of thermodynamics* can be stated in several ways:

Clausius Statement: No thermodynamic process can occur whose only result is to transfer heat from a colder to a hotter body. Such a process is only possible if work is done on the system.

Kelvin-Planck Statement: It is impossible for a thermodynamic process to occur whose only final result is the complete conversion of heat extracted from a hot reservoir into work. That is, *it is impossible to construct a heat engine that, operating in a cycle, produces no other effect than the absorption of heat from a reservoir and the performance of an equal amount of work.*

The Kelvin-Planck statement is equivalent to stating that *it is impossible to construct a perpetual motion machine of the second kind* (that is, a machine which violates the second law). Perpetual motion machines of the *first kind* are those which violate the first law of thermodynamics, which requires that energy be conserved.

22.2 Reversible and Irreversible Processes

A process is *reversible* if the system passes from the initial to the final state through a succession of equilibrium states. Then, the process can be made to run in the opposite direction by an infinitesimally small change in conditions. Otherwise the processes is *irreversible*.

22.3 The Carnot Engine

The most efficient cyclic process is called the *Carnot cycle*, described in the *P-V* diagram of Figure 22.1. The Carnot cycle consists of two adiabatic and two isothermal processes, all being reversible.

- The process $A \rightarrow B$ is an isotherm (constant T), during which time the gas expands at constant temperature T_h, and absorbs heat Q_h from the hot reservoir.

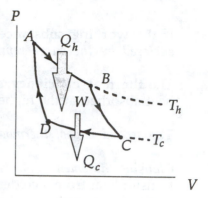

- The process $B \rightarrow C$ is an adiabatic expansion ($Q = 0$), during which time the gas expands and cools to a temperature T_c.

- The process $C \rightarrow D$ is a second isotherm, during which time the gas is compressed at constant temperature T_c, and gives up heat Q_c to the cold reservoir.

Figure 22.1

- The final process $D \rightarrow A$ is an adiabatic compression in which the gas temperature increases to a final temperature of T_h.

In practice, no working engine is 100% efficient, even when losses such as friction are neglected. One can obtain some theoretical limits on the efficiency of a real engine by comparison with the ideal Carnot engine. A *reversible engine* is one which will operate with the same efficiency in the forward and reverse directions. The Carnot engine is one example of a reversible engine.

Carnot's theorems, which are consistent with the first and second laws of thermodynamics, can be stated as follows:

> *Theorem I.* No real (irreversible) engine can have an efficiency greater than that of a reversible engine operating between the same two temperatures.

> *Theorem II.* All reversible engines operating between T_h and T_c have the *same* efficiency given by Equation 22.2.

A schematic diagram of a heat engine is shown in Figure 22.2a, where Q_h is the heat extracted from the hot reservoir at temperature T_h, Q_c is the heat rejected to the cold reservoir at temperature T_c, and W is the work done by the engine.

22.6 Heat Pumps and Refrigerators

A refrigerator is a heat engine operating in reverse, as described in Figure 22.2b. During one cycle of operation, the refrigerator absorbs heat Q_c from the cold reservoir, expels heat Q_h to the hot reservoir, and the work done on the system is $W = Q_h - Q_c$.

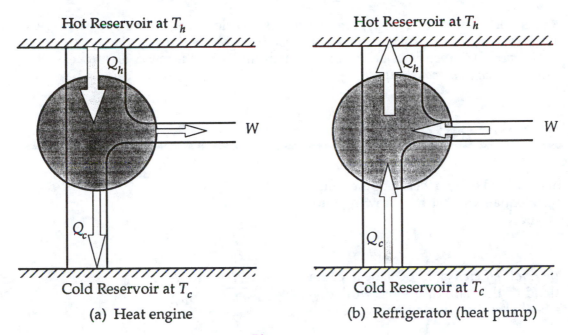

Hot Reservoir at T_h

Q_h

W

Q_c

Cold Reservoir at T_c

(a) Heat engine

Hot Reservoir at T_h

Q_h

W

Q_c

Cold Reservoir at T_c

(b) Refrigerator (heat pump)

Figure 22.2

22.7 Entropy
22.8 Entropy Changes in Irreversible Processes

Entropy is a quantity used to measure the degree of *disorder* in a system. For example, the molecules of a gas in a container at a high temperature are in a more disordered state (higher entropy) than the same molecules at a lower temperature.

When heat is added to a system, dQ_r is *positive* and the entropy *increases*. When heat is removed, dQ_r is *negative* and the entropy *decreases*. Note that only *changes* in entropy are defined by Equation 22.8; therefore, the concept of entropy is most useful when a system undergoes a *change in its state*.

When using Equation 22.9 to calculate entropy changes, note that ΔS may be obtained even if the process is irreversible, since ΔS depends only on the initial and final equilibrium states, not on the path. In order to calculate ΔS for an irreversible process, you must devise a reversible process (or sequence of reversible processes) between the initial and final states, and compute dQ_r/T for the reversible process. The entropy change for the irreversible process is the *same* as that of the reversible process between the same initial and final equilibrium states.

The second law of thermodynamics can be stated in terms of entropy as follows: *The total entropy of an isolated system always increases in time if the system undergoes an irreversible process.* If an isolated system undergoes a *reversible* process, the total entropy *remains constant.*

EQUATIONS AND CONCEPTS

The net *work done* by a heat engine during one cycle equals the net heat flowing into the engine.

$$W = Q_h - Q_c \qquad (22.1)$$

The *thermal efficiency, e,* of a heat engine is defined as the ratio of the net work done to the heat absorbed during one cycle of the process.

$$e = \frac{W}{Q_h} = 1 - \frac{Q_c}{Q_h} \qquad (22.2)$$

The *coefficient of performance* of a refrigerator or a heat pump used in the cooling cycle is defined by the ratio of the heat absorbed, Q_c, to the work done. A good refrigerator has a high coefficient of performance.

$$\text{COP (refrigerator)} = \frac{Q_c}{W} \qquad (22.7)$$

No real engine operating between the temperatures T_c and T_h can be more efficient than an engine operating reversibly in a Carnot cycle between the same two temperatures.

$$e_c = 1 - \frac{T_c}{T_h} \qquad (22.4)$$

The efficiency of the Otto cycle (gasoline engine) depends on the compression ratio (V_1/V_2) and the ratio of molar specific heats.

$$e = 1 - \frac{1}{(V_1/V_2)^{\gamma-1}} \qquad (22.5)$$

The effectiveness of a heat pump is described in terms of the coefficient of performance, COP.

$$\text{COP (heat pump)} \equiv$$

$$\frac{\text{heat transferred}}{\text{work done by pump}} = \frac{Q_h}{W} \qquad (22.6)$$

If a system changes from one equilibrium state to another, under a reversible, quasi-static process, and a quantity of heat dQ_r is added (or removed) at the absolute temperature T, the *change in entropy* is defined by the ratio dQ_r/T.

$$dS = \frac{dQ_r}{T} \qquad (22.8)$$

The *change in entropy* of a system which undergoes a reversible process between the states i and f depends only on the properties of the initial and final equilibrium states.

$$\Delta S = \int_i^f \frac{dQ_r}{T} \qquad (22.9)$$

The change in entropy of a system for any arbitrary *reversible cycle* is identically *zero*.

$$\oint \frac{dQ_r}{T} = 0 \qquad (22.10)$$

The change in entropy of an ideal gas which undergoes a quasi-static and reversible process depends only on the initial and final states.

$$\Delta S = nC_v \ln \frac{T_f}{T_i} + nR \ln \frac{V_f}{V_i} \qquad (22.12)$$

During a *free expansion* (irreversible, adiabatic process), the entropy of a gas *increases*.

$$\Delta S = nR \ln \frac{V_f}{V_i} \qquad (22.13)$$

During the process of irreversible heat transfer between two masses, the entropy of the system will increase.

$$\Delta S = m_1 c_1 \ln \frac{T_f}{T_1} + m_2 c_2 \ln \frac{T_f}{T_2} \qquad (22.15)$$

All isolated systems tend toward disorder and the increase in entropy is proportional to the probability, W, of occurrence of the event. Entropy is a measure of microscopic disorder.

$$S \equiv N k_B \ln W \qquad (22.18)$$

REVIEW CHECKLIST

▷ Understand the basic principle of the operation of a heat engine, and be able to define and discuss the *thermal efficiency* of a heat engine.

▷ State the second law of thermodynamics, and discuss the difference between reversible and irreversible processes. Discuss the importance of the first and second laws of thermodynamics as they apply to various forms of energy conversion and thermal pollution.

▷ Describe the processes which take place in an ideal heat engine taken through a *Carnot cycle*.

▷ Calculate the efficiency of a Carnot engine, and note that the efficiency of real heat engines is always less than the Carnot efficiency.

▷ Calculate entropy changes for reversible processes (such as one involving an ideal gas).

▷ Calculate entropy changes for irreversible processes, recognizing that the entropy change for an irreversible process is equivalent to that of a reversible process between the same two equilibrium states.

SOLUTIONS TO SELECTED END-OF-CHAPTER PROBLEMS

6. A particular engine has a power output of 5.0 kW and an efficiency of 25%. If the engine expels 8000 J of thermal energy in each cycle, find (a) the heat absorbed in each cycle and (b) the time for each cycle.

Solution

(a) We have $\quad e = \dfrac{W}{Q_h} = \dfrac{Q_h - Q_c}{Q_h} = 1 - \dfrac{Q_c}{Q_h} = 0.25 \quad$ and $\quad Q_h = \dfrac{Q_c}{1 - e}, \quad$ with $\quad Q_c = 8000$ J

We have $\qquad\qquad\qquad Q_h = \dfrac{8000 \text{ J}}{1 - 0.25} = 10.7$ kJ $\quad \lozenge$

(b) $\quad W = Q_h - Q_c = 2667$ J $\qquad$ and from $\qquad P = \dfrac{W}{t}, \qquad$ we have

$$t = \frac{W}{P} = \frac{2667 \text{ J}}{5000 \text{ J/s}} = 0.533 \text{ s} \quad \lozenge$$

13. One of the most efficient engines ever built (42%) operates between 430°C and 1870°C. (a) What is its maximum theoretical efficiency? (b) How much power does the engine deliver if it absorbs 1.4×10^5 J of thermal energy each second?

Solution $\qquad T_c = 430°\text{C} = 703 \text{ K} \qquad$ and $\qquad T_h = 1870°\text{C} = 2143 \text{ K}$

(a) $\quad e_c = \dfrac{\Delta T}{T_h} = \dfrac{1440 \text{ K}}{2143 \text{ K}} = 0.672 \qquad$ or $\qquad 67.2\% \quad \lozenge$

(b) $\quad Q_h = 1.40 \times 10^5$ J $\qquad W = 0.42 Q_h = 5.88 \times 10^4$ J

$$P = \frac{W}{t} = \frac{5.88 \times 10^4 \text{ J}}{1 \text{ s}} = 58.8 \text{ kW} \quad \lozenge$$

17. A steam engine is operated in a cold climate where the exhaust temperature is 0°C. (a) Calculate the theoretical maximum efficiency of the engine using an intake steam temperature of 100°C. (b) If, instead, superheated steam at 200°C is used, find the maximum possible efficiency.

Solution

(a) $\quad e_c = \dfrac{\Delta T}{T_h} = \dfrac{100}{373} = 0.268 = 26.8\%$ ◊

(b) $\quad e_c = \dfrac{\Delta T}{T_h} = \dfrac{200}{473} = 0.423 = 42.3\%$ ◊

It is important to remember that the temperatures to which Carnot's equation refers are absolute temperatures.

———————————————————

22. In a cylinder of an automobile engine, just after combustion, the gas is confined to a volume of 50 cm³, and has an initial pressure of 3.0×10^6 Pa. The piston moves outward to a final volume of 300 cm³ and the gas expands without heat loss. If $\gamma = 1.40$ for the gas, what is the final pressure?

Solution   Adiabatic expansion is the power stroke of our industrial civilization.

For adiabatic expansion,

$$P_A V_A{}^\gamma = P_B V_B{}^\gamma$$

Therefore, $\qquad P_B = P_A \left(\dfrac{V_A}{V_B}\right)^\gamma = 3.0 \times 10^6 \text{ Pa} \left(\dfrac{50 \text{ cm}^3}{300 \text{ cm}^3}\right)^{1.40}$

and $\qquad P_B = 2.44 \times 10^5 \text{ Pa}$ ◊

———————————————————

23. How much work is done by the gas in Problem 22 in expanding from $V_1 = 50$ cm³ to $V_2 = 300$ cm³?

Solution   Since $Q = 0$, $\qquad\qquad\qquad\qquad \Delta U = Q - W = -W$

and $\qquad\qquad\qquad\qquad\qquad\qquad W = -\Delta U = -nC_V\, \Delta T = -nC_V(T_B - T_A)$

From $\gamma = \dfrac{C_P}{C_V} = \dfrac{C_V + R}{C_V}$, we get $\quad (\gamma - 1)C_V = R$

and $\qquad\qquad\qquad\qquad\qquad\qquad C_V = \dfrac{R}{0.4} = 2.5R$

Therefore, $\qquad\qquad\qquad\qquad\quad W = n(2.5R)T_A - 2.5nRT_B = 2.5P_A V_A - 2.5P_B V_B$

From Problem 22, $\quad P_B = 2.44 \times 10^5\,\text{Pa}$, so

$$W = 2.5(3 \times 10^6\,\text{Pa})(50 \times 10^{-6}\,\text{m}^3) - 2.5(2.44 \times 10^5\,\text{Pa})(300 \times 10^{-6}\,\text{m}^3) = 192\,\text{J} \quad \Diamond$$

---

24.   An ideal refrigerator or ideal heat pump is equivalent to a Carnot engine running in reverse. That is, heat Q_c is absorbed from a cold reservoir and heat Q_h is rejected to a hot reservoir. (a) Show that the work that must be supplied to run the refrigerator or pump is

$$W = \frac{T_h - T_c}{T_c} Q_c$$

(b) Show that the coefficient of performance of the ideal refrigerator is $\quad \text{COP} = \dfrac{T_c}{T_h - T_c}$

Solution   (a)   For a complete cycle $\Delta U = 0$, and $\quad W = Q_h - Q_c = Q_c\!\left(\dfrac{Q_h}{Q_c} - 1\right)$

We have already shown for a Carnot cycle (and only for a Carnot cycle) that

$$\frac{Q_h}{Q_c} = \frac{T_h}{T_c}$$

Therefore, $\qquad\qquad\qquad\qquad W = \dfrac{T_h - T_c}{T_c} Q_c \quad \Diamond$

(b)   From Equation 22.7, $\text{COP} = \dfrac{Q_c}{W}$.

Using the result from part (a), this becomes $\quad \text{COP} = \dfrac{T_c}{T_h - T_c} \quad \Diamond$

27. How much work is required, using an ideal Carnot refrigerator, to remove 1.0 J of thermal energy from helium gas at 4.0 K and reject this thermal energy to a room-temperature (293 K) environment?

Solution

$$(COP)_{\text{He refrig}} = \frac{T_c}{\Delta T} = \frac{4.0}{289} = 0.0138$$

$$W = \frac{Q_c}{COP} = \frac{1.0 \text{ J}}{0.0138} = 72.3 \text{ J} \quad \lozenge$$

---

31. Calculate the change in entropy of 250 g of water heated slowly from 20°C to 80°C. (*Hint:* Note that $dQ = mc\, dT$.)

Solution To do the heating reversibly, put the water pot successively into contact with reservoirs at temperatures 20°C + δ, 20°C + 2δ, ... 80°C, where δ is some small increment. Then

$$\Delta S = \int_{\text{reversible path}} \frac{dQ}{T} = \int_{T_i}^{T_f} mc\frac{dT}{T}$$

Here T means the absolute temperature. We would ordinarily think of dT as the change in the celsius temperature, but one celsius degree of temperature change is the same size as one kelvin of change, so dT is also the change in absolute T. Then

$$\Delta S = mc \ln T \Big|_{T_i}^{T_f} = mc \ln\left(\frac{T_f}{T_i}\right)$$

$$\Delta S = (0.250 \text{ kg})(4186 \text{ J/kg} \cdot \text{K}) \ln\left(\frac{353 \text{ K}}{293 \text{ K}}\right) = 195 \text{ J/K} \quad \lozenge$$

---

37. One mole of H_2 gas is contained in the left-hand side of the container shown in Figure P22.37, which has equal volumes left and right. The right-hand side is evacuated. When the valve is opened, the gas streams into the right side. What is the final entropy change? Does the temperature of the gas change?

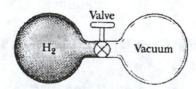

Figure P22.37

Solution

(a) This is an example of free expansion; from Equation 22.13 we have

$$\Delta S = nR \ln\left(\frac{V_f}{V_i}\right) \qquad \Delta S = (1.00 \text{ mole})\left(8.31\frac{\text{J}}{\text{mole} \cdot \text{K}}\right)\ln\left(\frac{2}{1}\right)$$

$$\Delta S = 5.76 \text{ J/K} \qquad \text{or} \qquad \Delta S = 1.38 \text{ cal/K} \quad \lozenge$$

(b) The gas is expanding into an evacuated region. Therefore, $W = 0$; and it expands so fast that heat has no time to flow; $Q = 0$. But $\Delta U = Q - W$, so in this case $\Delta U = 0$. For an ideal gas, the internal energy is a function of the temperature, so if $\Delta U = 0$, the temperature remains constant. $\quad \lozenge$

---

43. A house loses thermal energy through the exterior walls and roof at a rate of 5000 J/s = 5.00 kW when the interior temperature is 22°C and the outside temperature is –5°C. Calculate the electric power required to maintain the interior temperature at 22°C for the following two cases: (a) The electric power is used in electric resistance heaters (which convert all of the electricity supplied to thermal energy). (b) The electric power is used to operate the compressor of a heat pump (which has a coefficient of performance equal to 60% of the Carnot cycle value).

Solution

(a) $P_{\text{electric}} = \dfrac{\Delta E}{\Delta t}$ and if all of the electricity is converted into thermal energy, $\Delta E = \Delta Q$.

Therefore, $\qquad P_{\text{electric}} = \dfrac{\Delta Q}{\Delta t} = 5000 \text{ W} \quad \lozenge$

(b) For a heat pump, $\quad (\text{COP})_{\text{Carnot}} = \dfrac{T_h}{\Delta T} = \dfrac{295}{27} = 10.92$

Actual COP $= (0.6)(10.92) = 6.55 = \dfrac{Q_h}{W} = \dfrac{Q_h/t}{W/t}$

Therefore, to bring 5000 W of heat into the house only requires input power

$$\frac{W}{t} = \frac{Q_h/t}{\text{COP}} = \frac{5000 \text{ W}}{6.56} = 763 \text{ W} \quad \lozenge$$

---

45. One mole of an ideal monatomic gas is taken through the cycle shown in Figure P22.45. The process AB is a reversible isothermal expansion. Calculate (a) the net work done by the gas, (b) the thermal energy added to the gas, (c) the thermal energy expelled by the gas, and (d) the efficiency of the cycle.

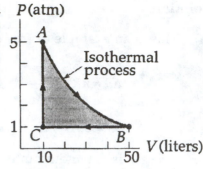

Figure P22.45

Solution

(a) For the isothermal process AB,

$$W_{AB} = P_A V_A \ln\left(\frac{V_B}{V_A}\right)$$

$$W_{AB} = (5\ \text{atm})\left(1.013\times10^5\ \frac{\text{Pa}}{\text{atm}}\right)(10\ \text{L})\left(10^{-3}\ \frac{\text{m}^3}{\text{L}}\right)\ln\left(\frac{50\ \text{L}}{10\ \text{L}}\right) = 8.15\ \text{kJ}$$

Likewise, $W_{BC} = P_B\,\Delta V = (1.01\times10^5\ \text{Pa})\big[(10-50)\times10^{-3}\big]\,\text{m}^3 = -4.05\ \text{kJ}$

$W_{CA} = 0$ because $\Delta V_{CA} = 0$, so

$$W = W_{AB} + W_{BC} = 4.11\ \text{kJ} \quad \Diamond$$

(b) AB is an isothermal process; since RT remains constant, the internal energy of the gas also remains constant, and the heat added equals the work done: $Q_{AB} = W_{AB}$.

$$Q_{AB} = W_{AB} = 8.15\ \text{kJ}$$

Process CA also adds heat to the gas. If P increases at constant V, RT must also increase. Since no work is done during this process, the additional internal energy of the gas must come from the addition of heat.

$$Q_{CA} = nC_v\,\Delta T = n\left(\frac{3R}{2}\right)\left(\frac{P_A V_A}{R} - \frac{P_C V_C}{R}\right) = n\frac{3V}{2}\Delta P$$

Solving, $Q_{CA} = (1.00)\dfrac{3\big(10\times10^{-3}\ \text{m}^3\big)}{2}(4\ \text{atm})\big(1.013\times10^5\ \text{Pa}/\text{atm}\big) = 6.08\ \text{kJ}$

and the thermal energy added is $Q_{in} = Q_{CA} + Q_{AB} = 8.15\ \text{kJ} + 6.08\ \text{kJ} = 14.2\ \text{kJ} \quad \Diamond$

(c)   Since processes AB and CA absorb heat, process BC must expell heat.

$Q_h = W + Q_c$, so $Q_c = Q_h - W = 14.2 \text{ kJ} - 4.11 \text{ kJ} = 10.1 \text{ kJ}$ ◊

Alternatively, we can apply the equations $C_P = \dfrac{5R}{2}$ and $Q_{BC} = nC_p \Delta T$.

$$Q_{BC} = nC_p \Delta T = n\left(\frac{5R}{2}\right)\left(\frac{P_C V_C}{R} - \frac{P_B V_B}{R}\right) = n\frac{5P}{2}\Delta V$$

Solving, $Q_{CA} = (1.00)\dfrac{5(1 \text{ atm})\left(1.013\times10^5 \text{ Pa / atm}\right)}{2}\left(40\times10^{-3} \text{ m}^3\right) = 10.1 \text{ kJ}$ ◊

(d) $e = \dfrac{W}{Q_h} = \dfrac{4.10 \text{ kJ}}{14.2 \text{ kJ}} = 0.289$ or   28.9%   ◊

---

47. Figure P22.47 represents n mol of an ideal monatomic gas being taken through a reversible cycle consisting of two isothermal processes at temperatures $3T_0$ and T_0 and two constant-volume processes. For each cycle, determine in terms of n, R, and T_0 (a) the net thermal energy transferred to the gas and (b) the efficiency of an engine operating in this cycle.

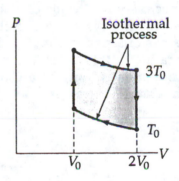

Figure P22.47

Solution

(a)   For an isothermal process,

$$Q = nRT \ln\left(\frac{V_2}{V_1}\right)$$

Therefore, $Q_1 = nR(3T_0) \ln 2$ and $Q_3 = nR(T_0) \ln\left(\tfrac{1}{2}\right)$

For the constant volume processes, we have

$$Q_2 = nC_V(T_0 - 3T_0)   \quad \text{and} \quad   Q_4 = nC_V(3T_0 - T_0)$$

The net heat transferred is then

$$Q = Q_1 + Q_2 + Q_3 + Q_4 \quad \text{or} \quad Q = 2nRT_0 \ln 2 \quad \Diamond$$

(b) Heat > 0 is the heat added to the system. Therefore,

$$Q_h = Q_1 + Q_4 = 3nRT_0(1 + \ln 2)$$

Since the change in temperature for the complete cycle is zero, $\Delta U = 0$ and $W = Q$.

Therefore, the efficiency is

$$e = \frac{W}{Q_h} = \frac{Q}{Q_h} = \frac{2 \ln 2}{3(1 + \ln 2)} = 0.273 \quad \text{or} \quad 27.3\% \quad \Diamond$$

---

49. One mole of a monatomic ideal gas is taken through the reversible cycle shown in Figure P22.49. At point A, the pressure, volume, and temperature are P_0, V_0, and T_0, respectively. In terms of R and T_0, find (a) the total heat entering the system per cycle, (b) the total heat leaving the system per cycle, (c) the efficiency of an engine operating in this reversible cycle, and (d) the efficiency of an engine operating in a Carnot cycle between the same temperature extremes.

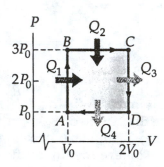

Figure P22.49

Solution

At point A, $P_0V_0 = nRT_0$, $n = 1$

At point B, $3P_0V_0 = RT_B$,   and $T_B = 3T_0$

At point C, $(3P_0)(2V_0) = RT_C$ and $T_C = 6T_0$

At point D, $P_0(2V_0) = RT_0$ and $T_D = 2T_0$

The heat transfer for each step in the cycle is found using $\quad C_V = \dfrac{3R}{2} \quad$ and $\quad C_P = \dfrac{5R}{2}$

$$Q_{AB} = C_V(3T_0 - T_0) = 3RT_0 \qquad Q_{BC} = C_P(6T_0 - 3T_0) = 7.5RT_0$$

$$Q_{CD} = C_V(2T_0 - 6T_0) = -6RT_0 \qquad Q_{DA} = C_P(T_0 - 2T_0) = -2.5RT_0$$

Therefore,

(a) $\quad Q_{in} = Q_h = Q_{AB} + Q_{DA} = 10.5RT_0 \quad \Diamond$

(b) $\quad Q_{out} = Q_c = |Q_{CD} + Q_{DA}| = 8.5RT_0 \quad \Diamond$

(c) $\quad e = \dfrac{Q_h - Q_c}{Q_h} = 0.190 \qquad$ or $\qquad 19\% \quad \Diamond$

(d) $\quad$ Carnot Eff, $\quad e_C = 1 - \dfrac{T_C}{T_h} = 1 - \dfrac{T_0}{6T_0} = 0.833 \qquad$ or $\qquad 83.3\% \quad \Diamond$

---

51. A system consisting of n mol of an ideal gas undergoes a reversible, *isobaric* process from a volume V_0 to a volume $3V_0$. Calculate the change in entropy of the gas. (*Hint:* Imagine that the system goes from the initial state to the final state first along an isotherm and then along an adiabatic path—there is no change in entropy along the adiabatic path.)

Solution   The isobaric process (AB) is shown along with an isotherm (AC) and an adiabat (CB) in the PV diagram.

Since the change in entropy is path independent,

$$\Delta S_{AB} = \Delta S_{AC} + \Delta S_{CB}$$

But because (CB) is adiabatic, $\Delta S_{CB} = 0$.

For isotherm (AC), $\qquad P_A V_A = P_C V_C$

For adiabat (CB), $\qquad P_C V_C{}^\gamma = P_B V_B{}^\gamma$.

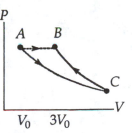

Combining these gives

$$V_C = \left(\frac{P_B V_B{}^\gamma}{P_A V_A}\right)^{1/(\gamma-1)} = \left[\left(\frac{P_0}{P_0}\right)\frac{(3V_0)^\gamma}{V_0}\right]^{1/(\gamma-1)} = 3^{\gamma/\gamma-1} V_0$$

Therefore,

$$\Delta S_{AC} = \left(\frac{P_A V_A}{T}\right)\ln\left(\frac{V_C}{V_A}\right) = nR\ln\left[3^{(\gamma/\gamma-1)}\right] = \frac{nR\gamma\ln 3}{\gamma-1} \quad \lozenge$$

---

55. An idealized Diesel engine operates in a cycle known as the *air-standard Diesel cycle*, shown in Figure P22.55. Fuel is sprayed into the cylinder at the point of maximum compression, B. Combustion occurs during the expansion $B \to C$, which is approximated as an isobaric process. The rest of the cycle is the same as in the gasoline engine, described in Figure 22.11. Show that the efficiency of an engine operating in this idealized Diesel cycle is

$$e = 1 - \frac{1}{\gamma}\left(\frac{T_D - T_A}{T_C - T_B}\right)$$

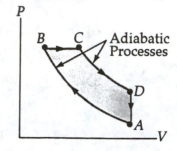

Figure P22.55

Solution   The heat transfer over the paths CD and BA are zero since they are adiabats.

Over path BC: $Q_{BC} = nC_P(T_C - T_B) > 0$

Over path DA: $Q_{DA} = nC_V(T_A - T_D) < 0$

Therefore, $Q_c = |Q_{DA}|$ and $Q_h = Q_{BC}$

Hence, the efficiency is

$$e = 1 - \frac{Q_c}{Q_h} = 1 - \left(\frac{T_D - T_A}{T_C - T_B}\right)\frac{C_V}{C_P}$$

$$e = 1 - \frac{1}{\gamma}\left(\frac{T_D - T_A}{T_C - T_B}\right) \quad \lozenge$$